For EU product safety concerns:
IN TECH d.o.o., Prolaz Marije Krucifikse Kozulić 3, 51000 Rijeka, Croatia,
info@intechopen.com or visit our website at intechopen.com.

Meet the editor

Dr Andrew Akwasi Oteng-Amoako is an emeritus chief research scientist of the Forestry Research Institute of Ghana. He was the former Deputy Director of the Institute and held position of Assistant Director Papua New Guinea Forest Research Institute.
His research interests include wood anatomy, wood quality, wood structure and preservative treatment, wood identification, bamboo and rattan science and technology. He has more than 120 publications, including five books. He is the co-editor of Plant Resources of Tropical Africa Volumes 1 and 2 and the editor of 100 Tropical African Timber Trees from Ghana.
Dr Oteng-Amoako holds undergraduate degree from the University of Hamburg in Germany, a Masters degree from Oregon State University at Corvallis, Oregon, and a Doctor of Philosophy degree from the College of Environmental Science and Forestry, Syracuse of the State University of New York. He is a fellow of Ghana Institution of Foresters and a member of the World Bamboo Council.

Contents

Preface

Forestry research is the application of scientific methods to investigate and understand management and practice of forestry and to effectively and sustainably use the resources derived from the forests. This book *New Advances and Contributions to Forestry Research* consists of 14 chapters divided into three sections and is authored by 48 researchers from 16 countries and all five continents.

In the section *Whither the Use of Forest Resources* five chapters authored by 16 researchers from seven countries, discuss negative and positive practices in forestry. A paper on phenology by four authors from Finland and Estonia describes seasonal growth in trees that leads to wood formation and understanding of biosphere – atmosphere interactions. Two papers authored by 10 researchers from Kenya and Ireland describe how the use of forest and its resources has seen extensive harvesting which has resulted in deforestation culminating in nutrient loss from the soil and pollution of water bodies. A paper authored by four Mexicans and a Cuban, evaluates socio-economic and environmental study of white tail deer in the Mixteca region of New Mexico. The forest should be handled with care and the author from Iran describes the need to design structures including industrial parks and plantations to conform with nature.

The forest is a complex habitat for man, animals, insects and other microorganisms and their activities may impact either negatively or positively on the forest. Four papers in the section *Forest and Organisms Interactions* are authored by six researchers from three countries. The first paper by an author from Tasmania, describes trapping systems developed in Tasmania, Australia to detect exotic forest insect pests in urban environment. Timber harvesting and wild fires in Australia, according to the four authors of a paper, have effected changes in relative density and distribution of wallabies and kangaroos, the most predominant and important mammals in the forests of Australia. What is the economic importance of the numerous microorganisms including bacteria, viruses and fungi that dominate the earth planet? This is extensively reviewed by the author from Egypt, while another researcher from Bolivia investigates decline of *Potylepsis tarapacana* forest, an important forest ecosystem in Bolivia, caused by a fungal disease, *Leptosphaeria polylepsis*.

Development of tree plantations has been man's response to forest degradation and deforestation caused by human, other organisms and natural disasters. In this last section *Amelioration of Dwindling Forest Resources through Plantation Development* five papers are authored by 20 researchers from five countries. Three authors in the paper from China give a comprehensive review on research, development and utilization of *Jatropha curcas,* the seed to seedling, plantation establishment, chemical properties, and more importantly its use as a biofuel. Beech is an important timber tree in Europe especially in Croatia, and the author discusses ways of optimizing the quantity and quality of beech wood in a plantation setting. Judicious management practices, according to the author, can reduce incidence of false heartwood in beech trees thereby increasing the quality of wood yield from beech plantation. The four researchers from Japan and China give a world-wide review on distribution, composition and cytogenetics of spruce in China which accounts for about 43.3% of the world's spruce forests. What is unique about the Eucalyptus tree with more than 800 species that dominate tree flora in Australia? The paper written by six authors from Brazil and Argentina discusses essential oils found in the leaves of Eucalyptus which have been demonstrated to have one of the most effective defenses against noxious organisms. The applications of these oils for use as insecticide, medicine, pest control, perfumes and pharmaceuticals are fully discussed. The use of a pilodyn to determine the basic density and strength properties in Eucalptus timber tree grown in China, is discussed in a comprehensive paper by seven authors from China.

New Advances and Contributions to Forestry Research will appeal to forest scientists, researchers and allied professionals. It will be of interest to all those who care about forest and subscribe to the adage that the last tree on our planet dies with the last man on the earth.

This publication was only possible with kind contributions made by 48 very hardworking international forest researchers and scientists from 16 countries and five continents. They deserve my congratulations for a job very well done! I am grateful to INTECH for giving me the opportunity to be the Editor of this innovative but challenging project. I am thankful to Mrs. Ana Skalamera, my Publishing Process Manager, for continuously prompting me on the project deadlines. I cannot help being grateful to the Director and my colleagues of the Forestry Research Institute of Ghana of the Council for Scientific and Industrial Research, for providing me conducive working environment for the project. Finally, my greatest appreciation to the Almighty God to whom be the Glory!

I recommend the book to you and entreat all to purchase or download a copy of the publication for your library, enjoy reading it and contribute to save our forests and save life.

Andrew Akwasi Oteng-Amoako, Ph. D.
Emeritus Chief Research Scientist,
Forest Research Institute of Ghana,
Council for Scientific and Industrial Research, Kumasi,
Ghana

Section 1

Whither the Use of Forest Resources

1

Effects of Deforestation on Water Resources: Integrating Science and Community Perspectives in the Sondu-Miriu River Basin, Kenya

Frank O. Masese[1*], Phillip O. Raburu[1],
Benjamin N. Mwasi[2] and Lazare Etiégni[3]
[1]*Department of Fisheries & Aquatic Science, Moi University, Eldoret*
[2]*Department of Environmental Health, Moi University, Eldoret*
[3]*Department of Forestry & Wood Science, Moi University, Eldoret*
Kenya

1. Introduction

Rivers play a major role as sources of water for both domestic and industrial use in many parts around the world. In developing countries, where infrastructure for water supply has not been fully developed, rivers provide a direct source of water for domestic use with minimal or no treatment at all. For water scarce countries, including Kenya (WRI, 2007), this means that water catchment areas should be managed properly so as to retain their capacity to supply good quality water all year round. Thus, understanding the possible consequences of land use and land cover changes on water resources is a requisite for better water resources management. However, this is not to be as many river catchments are undergoing rapid change mediated by human encroachment.

Africa boosts over 4 million first-order streams that were originally in forested catchments. However, loss of indigenous forests and their subsequent conversion to agricultural use in East Africa, for example, is one of the major threats to surface water quality (FAO, 2010). Major water catchment areas in Kenya have lost their forest cover over the years with the closed canopy forest cover currently standing at a paltry 2.0% (The World Bank, 2007). Most of these forests are montane forests and they constitute the nation's water towers. The Mau Forest Complex, the most important of them, is the source of many rivers draining the Kenyan side of the Lake Victoria basin, with other rivers draining into Lakes Nakuru, Baringo and Natron. The Mau Forest Complex has witnessed considerable land use and land cover changes. For instance, between 1973 and 2000, there was a 32% decrease in forest cover and a 203% increase in agricultural cover in the Mara River basin (Mati et al., 2008). Other river catchments on the Kenyan side of the Lake Victoria basin have also undergone similar changes. Increased intensity of agriculture and

* Corresponding Author

deforestation have been linked to increasing magnitude and frequency of runoff events and reduced baseflows, increased pesticide contamination, erosion and sedimentation of streams and rivers (Matie et al., 2008; Okungu and Opango, 2005; Osano et al., 2003; Raini, 2009). With the inevitable challenge of climate change amid a rapidly increasing human population, averaging 3% per annum, these problems are likely to be exacerbated jeopardizing environmental management efforts, biodiversity conservation and sustainable social and economic development.

To address the problems of deforestation and land use change in Kenya, a number of approaches have been used; including forceful eviction of settlers from protected forests and catchments and awareness creation among small scale farmers, who make the largest bulk of land owners, for the use of best management practices that include agroforestry and minimal tillage, to minimize the negative effects on water resources. Several studies have been conducted that focus on assessing the effects of land use change on water resources, including water quality (e.g., Kibichii et al., 2007) and water quantity (Mango et al., 2011; Mati et al., 2008). Other studies have also focused on the use of aquatic biota to develop protocols to help monitor changes in water quality in streams and rivers in the basin (e.g., Masese et al., 2009a; Raburu et al., 2009). However, studies that integrate or combine the effects of land use and land use change on water quality and resident aquatic biota are scarce, limiting their use as indicators of surface water quality. The practice of using aquatic biota as indicators of changes in water quality arising from land use practices is a well developed system which gives resource managers a scientific basis for effecting water management guidelines and practices. Since such system does not exist in Kenya and in the wider Eastern Africa region, there is a need to develop biological criteria using aquatic communities as indicators of water quality.

Indigenous knowledge and community perspectives on deforestation and land use change and their effects on water resources have also not been recognized in efforts to conserve and manage key water catchment areas in Kenya. The approach used by the Kenyan Government has been to forcefully evict people from forests and key water catchment areas around the country. However, questions have been raised on the success of this command-and-control approach to environmental conservation (Norgrove and Hulme, 2006; Okeyo-Owuor et al., 2011). This approach also negates the fact that communities that have lived with forests for ages and entirely depend on them for their daily livelihoods have by necessity developed a sense of ownership and systems that conserve the forest for posterity. In the Mau Forest Complex, the Ogiek community has a long history of sustainably living with the forest. As hunters and gatherers the Ogiek have a system of territoriality that prohibits members of one clan or family from invading another's territory for hunting, thus reducing overexploitation of the forest. However, because of immigration, other communities have over the years moved in to clear sections of the forest for farming and settlement. This has led to degradation of the forest and the recent calls for restoration and conservation. This paper discusses (i) the effects of deforestation on water quality and macroinvertebrate communities in streams and rivers draining into the Kenyan part of the Lake Victoria basin; (ii) the Ogiek community's perspectives on land use change and its effects on water resources from an indigenous knowledge point of view.

2. Materials and methods

2.1 Study area

The Sondu-Miriu River Basin is located at latitude 0°17'S and 0°22'S, longitude 34° 04' E and 34° 49' E (Fig. 1). It forms the fourth largest basin of Kenyan rivers that drain into Lake Victoria, covering an area of 3470 km^2. The main tributaries of the river are the Kapsonoi and Yunith. The river originates from the Mau Forest Complex, an expansive water tower in Kenya where several rivers that drain into Lakes Bogoria, Nakuru and Natron originate. However, forest excisions and the subsequent conversion to agricultural use have reduced its forest cover. The Sondu-Miriu River catchment is characterized by diverse land use types and developments including forestry, large-scale and small-scale agriculture, urban and sub-urban settlements, agro-based industries and hydroelectric power generation. Because of the combined effects of these human activities, and the increase in their scale and intensity over the years, they impose multiple threats to water quality, aquatic biodiversity and general ecology of the river. Evidently, the water quality status in the Sondu-Miriu River has recorded increasing rates of sedimentation over the years (Fig. 2).

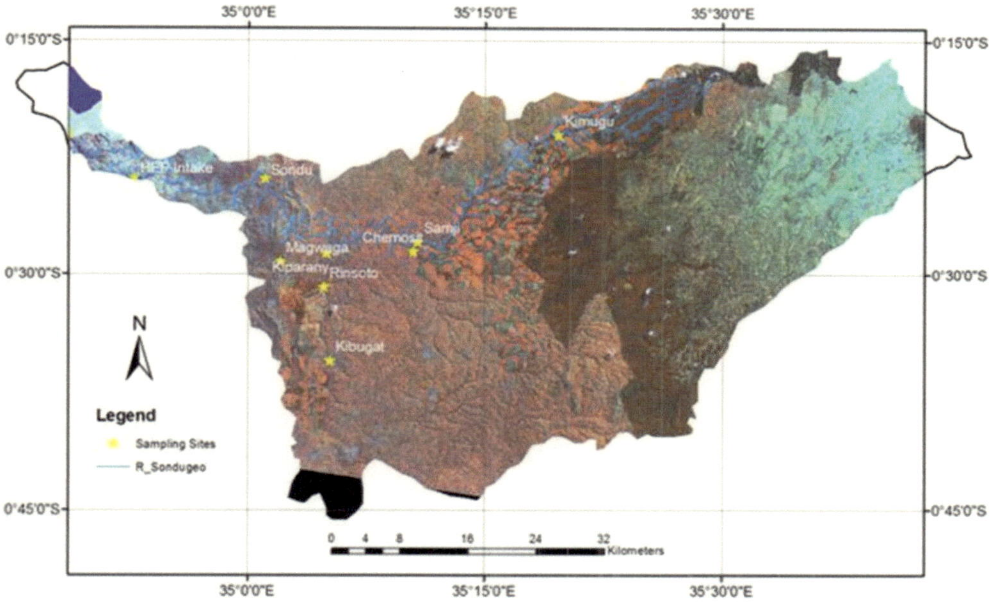

Fig. 1. The Sondu-Miriu River basin indicating sampling sites.

2.2 Study design

The Sondu-Miriu River Basin can be divided into three zones on the basis of altitude and climate. Altitude in the upper zone ranges from 1686 to 2003m above sea level (a.s.l) with humid climatic conditions. The middle zone falls within an altitude range of 1496 to1630 m a.s.l and sub-humid climatic conditions. The lower zone, whose altitude ranges from 1137 to 1394 m a.s.l and falls within the semi-humid climatic regime. The upper part, is mostly covered by forests and woodlands, while the remaining part is under tea, both plantations

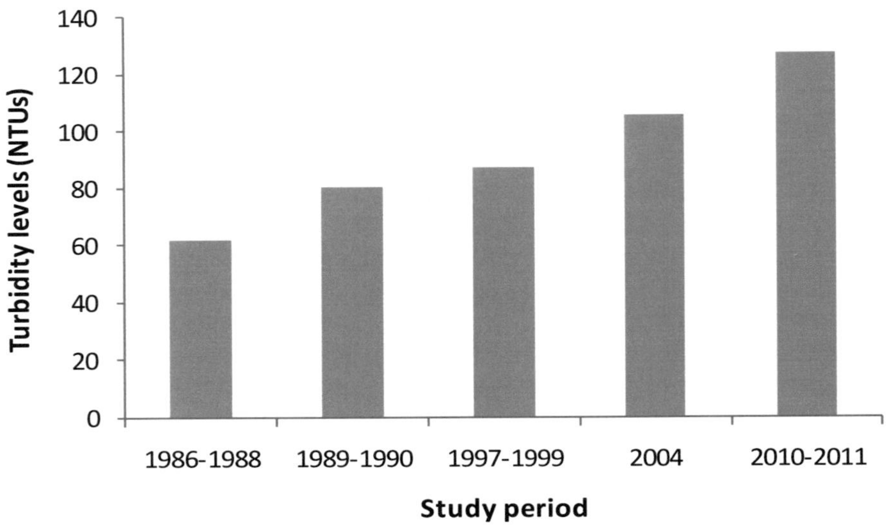

Fig. 2. Historical trends in mean turbidity (NTUs) levels in the Sondu-Miriu River. Sources of data: Ochumba and Manyala 1992; Mwashote and Shimbira 1994; Ojwang 2004 unpublished data, and this study.

and small-holder farms. The middle zone, which is mostly hilly, is covered by herbaceous vegetation. However, most of the natural vegetation has been replaced by exotic tree species, mainly *Eucalyptus* sp., inter-planted with crops. The lower zone is generally semi-arid, with bare soils covered by sparsely distributed shrubs dominated by acacias. This zone is settled by people practicing subsistence agriculture of both crop and livestock. For this study, only the upper and middle zones were considered. A total of 8 sampling sites were selected for the study (Table 1).

2.3 Land cover mapping

The main objective of the study is to determine the relationship between land cover characteristics on one hand and stream flow and aquatic organisms characteristics on the other hand. Land cover characteristics are represented by changes in land cover, mainly the loss of forests and other natural vegetation conditions to farming and settlements, including large scale tea plantation, subsistence farms, homesteads as well as urban and road infrastructure. Land cover changes were obtained by classifying satellite-based remotely sensed data from Landsat TM images acquired in January 1986 and January 2009. Stream flow characteristics analyzed were water quality and stream discharge parameters. Aquatic organisms are represented by diversity and abundance of macro-invertebrates found along the river.

A field survey identified ten main land cover classes including water, natural forests, plantation forests, woodlands, bushlands, bare surfaces, tea plantations, subsistence farms and built environments (homesteads, urban areas and roads). Due to inter-class similarity and intra-class variability in spectral characteristics, the study area was defined into three zones namely upper, middle and lower. However, because of obvious differences brought about by a waterfall in the lower sections of the river, only two zones, the upper and middle, were included in the water quality and macroinvertebrates data.

2.4 Sampling and sample analysis

Sampling was done from September 2009 to April, 2011 to capture both dry and rainy seasons. In each sampling station, electronic meters were used to measure conductivity, temperature, pH and DO *in situ*. Alkalinity and water hardness were determined colorimetrically on site immediately after sampling. Triplicate water samples were collected, fixed with sulphuric acid to below pH 2 and transported to the laboratory for nutrient analysis using standard procedures (APHA, 1998).

2.4.1 Macroinvertebrate assemblages

Triplicate samples of macroinvertebrates were collected from pools, riffles and runs using a dip net. They were placed in polyethene bags and immediately preserved using 75% ethanol and shipped to the laboratory where they were sorted, identified to lowest taxon level possible and counted. Assemblage attributes were determined for each site using diversity and richness measures and the relative abundance of various taxa. Potential macroinvertebrate metrics for IBI development were categorized by their relationship to community structure, taxonomic composition, individual condition and biological processes (Table 2) using groups previously used in riverine ecosystems in the ecoregion (Kobingi et al.; 2009, Masese et al.; 2009a; Raburu et al., 2009; Aura et al., 2010). A ''metric'' is an attribute that changes in some predictable way in response to increased human disturbance and that has been included as component of a multi-metric IBI (Karr and Chu, 1999). Testable hypotheses for these classes of attributes were proposed regarding the direction (increase, decrease, no change) of change to increasing levels of human disturbance (Table 2). Twenty-two metrics were selected *a priori* based on their demonstrated ability to evaluate environmental condition in rivers in the region (Aura et al., 2010; Kobingi et al., 2009; Raburu et al., 2009) and evaluated to identify key ones that responded to changes in macroinvertebrate condition in the Sondu-Miriu River Basin.

2.5 Indigenous knowledge about deforestation

In December 2008 a comprehensive study was undertaken to incorporate indigenous knowledge in the mapping of the critical areas within Mau Forest Complex. During this survey indigenous knowledge data were collected from the communities living within and around the forests using structured questionnaires, Focused Group Discussions (FGD) and Key Informants Interviews (KII).

2.6 Data analysis

To describe the variation in environmental variables, means of all measured environmental variables were calculated for all sites. One-way ANOVA was used to detect differences among different physico-chemical parameters, land uses variables and macro-invertebrate attributes. Metrics were evaluated for responsiveness to changes in water quality and land use by correlation analysis. Summary statistics on indigenous knowledge are presented in tables and charts.

Biophysical Characteristics	Sampling sites and thir attributes							
	Upper Zone				Middle zone			
	Kibugat	Chemosit	Jamji	Kimugu	Sondu Bridge	Magwagwa	Rinsoto	Kiparanye
Altitude	1686	1701	1709	2003	1496	1567	1591	1630
Topography	Steep slopes	Undulating hills	Undulati ng hills	Undulating hills	Flat surrounded by hills	Gentle undulating hills	Steep slopes	Rolling hills
Natural land cover	Woodland				Woodlands	Shrubland with planted woodlands	Woodland	Planted woodlands
Major land use	Tea and	Tea plantations	Tea plantatio ns	Tea plantations	Subsistence farming	Subsistence/smal l-holder tea	Tree and tea plantations	Subsistence/small holder tea
Level of degradation	Little evidence	Little evidence	Little evidence	Little evidence	No evidence	Reduced water levels	Little	Little evidence, clear water

Table 1. Site physical characteristics in the upper and middle zones of the Sondu-Miriu River Basin as characterized in this study.

Metric	Metric definition	Predicted response
Number Ephemeroptera taxa	Total number of mayfly taxa	Decrease
Number Plecoptera taxa	Total number of stonefly taxa	Decrease
Number Trichoptera taxa	Total number of caddisfly taxa	Decrease
Number Ephemeropter-Plecoptera-Trichoptera genera	Total number of taxa from mayfly, stonefly and caddisfly orders	Decrease
Total number of taxa	All different taxa at a site	Decrease
Percent EPT individuals	% individuals from mayfly, stonefly and caddisfly orders	Decrease
Percent non-insect individuals	% of individuals no belonging to the insect orders	Increase
Percentage Diptera individuals	% midge individuals	Increase
EPT: Diptera individuals ratio	Ratio of individuals belonging to mayfly, stonefly and caddisfly orders to that of midges	Decrease
Percent Coleoptera individuals	% of beetle individuals	Decrease
Shannon diversity index	Value of Shannon diversity index	Decrease
Number intolerant taxa	Total number of taxa belonging to pollution intolerant taxa	Decrease
Percent intolerant individuals	% of individuals in pollution sensitive taxa	Decrease
Percentage tolerant individuals	% of individuals in pollution tolerant taxa	Increase
Percentage filterer individuals	Filter fine organic material	Increase
Percentage scraper individuals	Feed on epiphytes	Decrease
Percentage predator individuals	Carnivores- scavangers, engulf or pierce prey	Decrease
Percent Shredder individuals	Feed on leaf litter	Decrease
Percentage gatherer individuals	Collect fine deposited organic material	Increase

Table 2. Metrics for macroinvertebrates that were considered for development of an index of biotic integrity for the Sondu-Miriu River Basin and the predicted responses to pollution.

3. Results

3.1 Land use/ cover characteristics

The ISODATA unsupervised classification algorithm was used to create 12 spectral classes for each zone for both 1986 and 2009. These classes were combined into 3-5 major land cover types using field data (Tables 3 and 4). Based on initial spectral analyses of representative signatures, only 3 of the 7 Landsat TM which had high inter-class separation were used. These are TM bands 2, 3 and 5. Class statistics for the two time periods were computed and compared for each zone.

Land use/ cover type	1986 (Ha)	2009 (Ha)	Change (1986-2000)	Change (%)
Forests	19486	15452	-4034	-20.7
Bushlands	11850	10627	-1223	-10.3
Farms and settlement	11141	16398	5256	32.1

Table 3. Land use/ cover areas in the upper Sondu-Miriu River basin- Kimugu Sampling Site.

Land use/ cover type	1986	2009	Change (1986-2000)	Change (%)
Water	8409	7006	-1403	-16. 7
Woodland/ Wetlands	24332	18458	-5874	-24.1
Sparse	8213	6992	-1221	-14.9
Bare	13452	11888	-1564	-11.6
Farms/ settlement	21439	31501	10062	46.9

Table 4. Land use/ cover areas for the lower Sondu-Miriu River Basin- River-Mouth Sampling Site.

3.2 Water quality

The changes in physico-chemical water quality parameters and nutrients downstream are given in Table 5. With the intensification of human activity downstream, corresponding with the increase in the land area under agriculture, changes in the water physico-chemical parameters were also observed. For instance, temperature, turbidity and TSS values were higher at the lower reaches.

Site	Turbidity (NTUs)	Conductivity (µS/cm)	DO (mg/L)	Temperature (°C)	TN mg/L	TP mg/L	TSS mg/L
Kimugu	64.3±32.6	63.4±25.2	7.0±0.3	16.5±0.7	2.6±0.25	0.4±0.18	31.4±20.6
Jamji	76.2±24.3	103.7±18.8	7.40.2	17.2±0.5	2.1±0.18	0.4±0.14	25.8±15.3
Chemosit	118.4±32.6	43.0±25.2	6.9±0.4	19.5±0.7	1.9±0.25	0.7±0.18	21.8±20.6
Kibugat	173.3±28.1	79.0±22.9	6.9±.03	21.5±0.6	2.5±0.23	0.3±0.17	32.6±18.8
Kipranye	90.9±23.1	95.6±17.8	7.1±0.2	19.5±0.5	2.0±0.17	0.4±0.13	30.2±14.5
Rinsoto	192.8±29.8	90.5±22.9	6.8±0.3	21.6±0.7	2.6±0.23	0.4±0.17	64.8±18.7
Magwagwa	163.1±29.8	58.8±23.0	6.6±0.3	21.7±0.7	2.2±0.22	0.5±0.17	43.0±18.8
Sondu Bridge	117.03±24.3	128.7±18.6	7.1±0.3	20.3±0.5	1.8±0.18	0.3±0.13	42.6±15.3

Table 5. Physico-chemical characteristics and nutrient levels at the various sites along the Sondu-Miriu River.

3.3 Macroinvertebrate assemblages

A total of 16 orders, 47 families and 49 genera were encountered during the study period. Whereas their distribution was varied with a few predominating upstream, mid-stream and downstream reaches, many of the macroinvertebrates displayed basin-wide distribution. Some of these included odonates belonging to Genera *Gomphus* sp. and *Agrion* sp., plecopterans *Nemoura* sp and *Neoperla* sp., hemipterans *Belostoma* sp. and *Gerris* sp., the pulmonate *Sphaeriun* sp., dipterans *Tipula* sp and *Chironomidae* and the ephemeropterans *Baetis* sp, *Afronurus* sp., *Caenis* sp. and *Adenophlebia* sp. and lastly the trichopteran *Hydropsyche* sp. Table 6 summarizes the species richness and diversity indices of the macroinvertebrates found along the river system. Taxon richness was low in the uppermost stations in both the Yurith and Kipsonoi sub-catchments, increasing significantly downstream with the mid stations registering a relatively high number of taxa at Kipranye, Magwagwa and Sondu Bridge Stations.

		Macroinvertebrate Diversity Measures			
Station	Number of taxa	Dominance	Shannon Index	Evenness	Simpson Richness (1/D)
Kimugu	25	0.26	2.51	0.32	8.89
Jamji	22	0.24	1.91	0.35	7.31
Kipranye	26	0.18	2.25	0.37	5.67
Sotik	19	0.25	1.66	0.34	4.1
Kibugat	18	0.11	2.41	0.39	9.09
Rinsoto	16	0.17	2.02	0.35	6.04
Magwagwa	22	0.12	2.45	0.45	6.71
Sondu Bridge	24	0.16	2.28	0.35	5.15

Table 6. The diversity measures of macroinvertebrate communities at the study stations along the Sondu-Miriu River.

3.3.1 Index of biotic integrity

Table 7 show metrics that qualified for the determination of the macroinvertebrate Index of Biotic Integrity (MIBI) and the scoring criteria derived from the data collected in Sondu-Miriu during the study period.

	Scoring criteria		
MIBI metrics	5	3	1
Number Ephemeroptera genera	≥ 6	3 - 5	≤ 2
Number Plecoptera genera	2	1	0
Number Trichoptera genera	≥ 5	3 - 4	≤ 2
Total number of genera	> 40	20 - 40	< 20
Number intolerant ganera	≥ 10	5 - 9	≤ 4
% EPT-BCH individuals	> 40	18 - 40	≤ 18
% Non-insect individuals	< 10	10 - 16	> 16
% Tolerant individuals	< 31	31 - 70	> 70
% Gatherer individuals	< 5	5 - 35	> 35
% Predator individuals	> 30	15 - 30	< 15

Table 7. Ten component metrics of the macroinvertebrate Index of Biotic Integrity (MIBI) and metric values corresponding with scores based on the 1, 3, 5 scoring system.

3.4 Community perspectives on deforestation

A total of 76 households were randomly sampled in two administrative Divisions namely Elburgon and Keringet. A typical household in the two divisions had 7 members with the majority having at least primary school level of education (Figure3). Most of the respondents were also elderly (Figure 3).

In Keringet Division, which was the main focus of the survey , about 88.2 % of the respondents were farmers with a minority (11.8%) engaged in small scale business as their primary source of livelihood. It was also reported that 62.5% and 37.5% of the respondents are engaged in business and agriculture as secondary source of income, respectively. Some of the respondents interviewed in Keringet Division have been resident in the area since 1918.

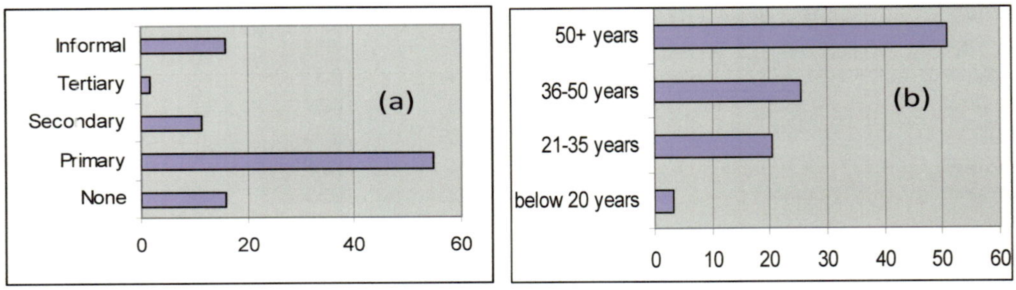

Fig. 3. Demographic characteristics of respondents in this survey: (a) level of education and (b) age distribution.

3.4.1 Value of Mau Forest in Keringet Division

The Mau forest is valuable to residents as acknowledged by a majority (94%) of the respondents during this survey. Control of soil erosion, rainfall, source of building materials and firewood are the most important uses (Figure 4). The Mau Forest has been a source of medicinal plants to the resident Ogiek community which they use to treat many ailments. In addition, communities graze their livestock and farm millet and pumpkin within the forest. Other uses include gathering of honey, fruits and hunting.

3.4.2 Water resources

The Mau Forest is an important source of streams and rivers in Keringet Division that are relied upon by the local community for water supply. Other streams and rivers mentioned by the community include Kiplapo, Cheptemet, Buchechet, Kiphoobo, Anguruwet, Oinetopilongotisiek, Oinetoptiepoison and Oinetoptieposere. Many of the streams are protected by the community by discouraging grazing of livestock within the forest, discouraging cutting of trees at their sources and discouraging cultivation at the water catchment areas. However, some streams, like Oinetopkongotisiek and Oinetopmogireri, have dried up. Some have also experienced changes because of deforestation.

3.4.3 Changes in climatic and weather parameters

The Keringet area has witnessed some changes in weather patterns in the past 20 years as reported by respondents in this survey. Before 2001 when large sections of the Mau Forest were cleared to create room for farming and settlement, the Keringet section of Mau Forest experienced moderate fluctuations in weather patterns as opposed to the current irregular state depicted in Figures 5 and 6.

During focus group discussions and key informant interviews, community members indicated that the status of the catchment area of most streams and rivers has changed. The elderly from among the Ogiek community narrated that the changes in the vegetation cover of the Mau Forest started in the 1970s when exotic commercially viable tree species were extensively introduced in the section of the forest at the expense of indigenous woody perennials. The peak of the changes occurred in 1996 when community settlement schemes began prior to the official government degazattement of sections of the forest in 2001.

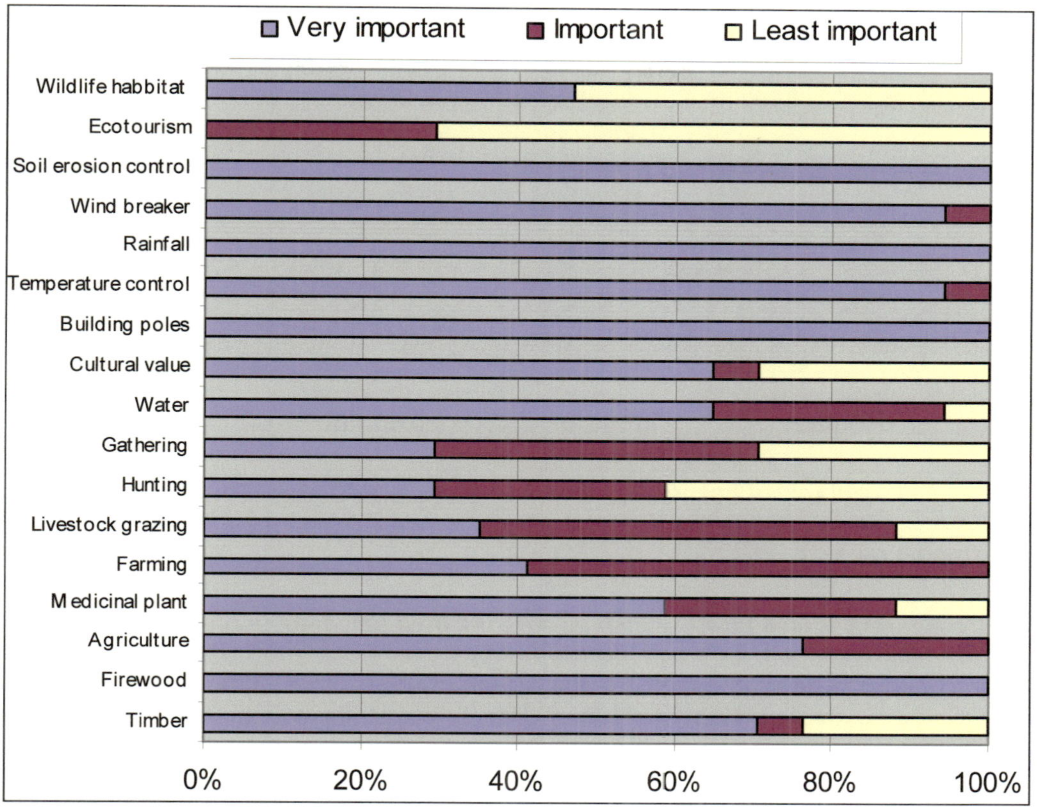

Fig. 4. Community perspectives on the importance of the Mau Forest.

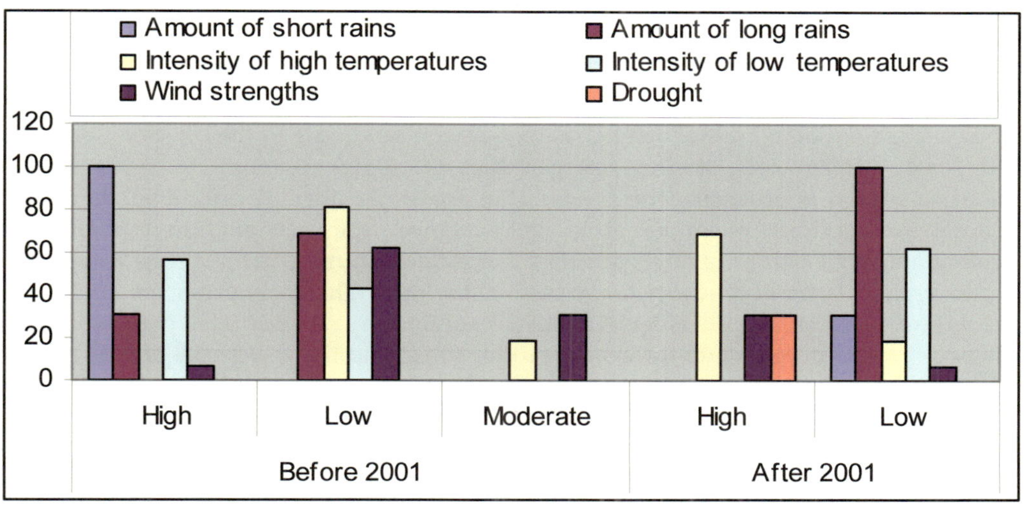

Fig. 5. Community perception on the changes in weather parameters in Keringet Division.

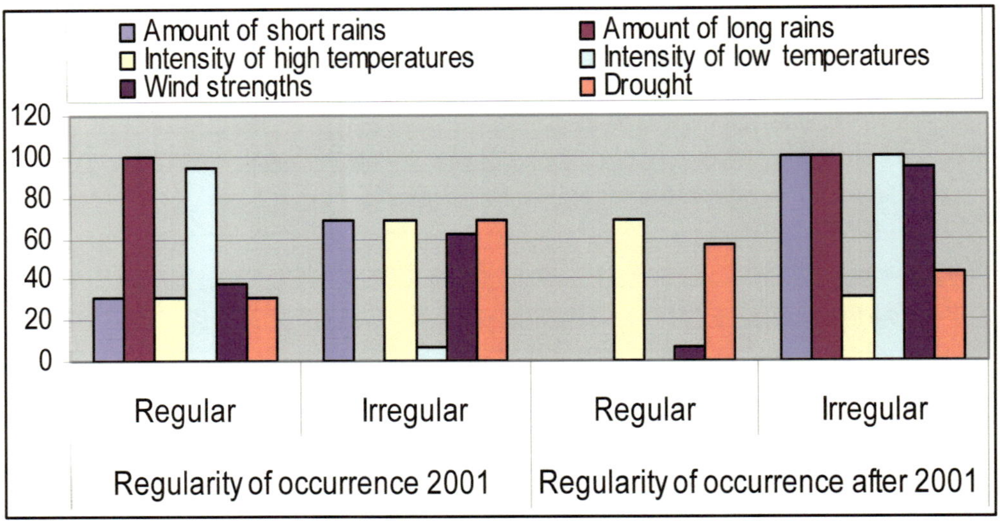

Fig. 6. Community perceptions on climatic parameters in Keringet Division.

The informants narrated how the changes have occurred gradually since the 1970's. Prior to this, the Mau Forest was rich in indigenous woody perennials with highly predictable and reliable weather patterns as opposed to the current situation. Stream flows were also regular and the water in streams and rivers clearer. Changes in quantity and quality of streams from Mau Forest became apparent in 1984 resulting in some sections of the streams gradually drying up. The land bordering the Mau Forest was initially very productive with farmers not using commercially supplied fertilizers. However, this phenomenon has changed and farmers have to apply fertilizers on their farms. Other changes include the rise in temperature levels, which has shortened the growing season of maize by about 4 months. The water table has also been going down with some springs and wells around the forest drying up.

3.4.4 Challenges to forest conservation

Lack of awareness amongst the members of the community on the environmental significance of the Mau Forest emerged as the greatest hindrance to the conservation of the Mau Forest. Nonetheless, it also emerged that the Ogiek community had a socially structured system of protecting the greater Mau Forest Complex. For example, they had a territorial management system that prohibited a member from a certain clan or family from invading another's clan or family's territory. For example during hunting, one was to seek permission from a territory leader to pursue his/her prey. Charcoal burning and cutting of trees were also prohibited. An elder of the Ogiek community who has lived in the area since 1929 gave greater insights into the changes in the forest and the way of life of the Ogiek. It was learnt that before the excision, the Ogiek community was living and zealously protected the forest along the Mau Narok, Buret and Nandi Forests, since they were hunters and gatherers harvesting only honey and wild meat. However, in 1976 their livelihood strategies started changing. They were taught how to plant maize, potatoes and other crops by the immigrants. The ensuing clearing of forests for agriculture led to degradation and mass migration of wildlife that the Ogiek depended upon.

4. Discussion

The longitudinal deterioration in water quality in the Sondu-Miriu River reflects the cumulative effects of human activities both on the riparian and in the catchment areas. This phenomenon has been reported in a number of studies conducted to investigate the influence of land use on water quality in rivers in the region (Kibichii et al., 2007; Masese et al., 2009b; McCartney, 2010). Turbidity increases downstream mainly originated from agricultural areas and erosion from unpaved roads. Previous studies in the river have also indicated that the water quality has been deteriorating (Figure 2) as a result of the intensification of agricultural activities and clearing of forests, as also corroborated by community members living in the upper reaches of the Sondu-Miriu River basin.

Macroinvertebrates assemblages encountered along the Sondu-Miriu River are typical of riverine communities in the region (Raburu et al., 2009). However, there were variations in composition and distribution and this is explained by tolerance to poor environmental conditions exhibited by the various taxa. The non-insect taxa gained more diversity and abundance as one moved downstream. This could be explained by their tolerance to pollution and higher turbidity levels. Other groups that were abundant, both in terms of taxon richness and abundance were soft bodied macroinvertebrates like oligochaetes, especially in sites receiving organic pollution. These groups are considered to be among the most tolerant to organic pollution in the Lake Victoria Basin (Kobing et al., 2009; Masese et al., 2009b). Other tolerant species include *Chironomus* sp. and *Lumbricus* sp. which are found in degraded sites because they possess high glycogen content and display reduced activity which allows them to withstand increased conductivity levels. In contrast, high abundance of Ephemeroptera, Plecoptera and Trichoptera dominated stations at the upper reaches where per cent land use under forestry is higher than in the middle and lower reaches. These sites are also less impacted by organic waste and general human disturbance. This confirms their utility as sensitive indicators of poor water quality (e.g., Masese et al., 2009a; Raburu et al., 2009).

The macroinvertebrate-based index of biotic integrity developed in this study had previously been used in the basin. Thus, the sensitivity of the metrics included in the final index has already been proven. Their use in this study was, however, to test their utility in detecting the effect of land use change of forestry to agriculture, which is a major problem in many of the river catchments in the region (e.g., Mango et al., 2011; Raini, 2009).

4.1 Integrating science and community perspectives

It was clear from the survey that the Ogiek Community has traditionally utilized the forest sustainably with well structured systems that guard against over-exploitation and conflicts. However, during interviews and discussions with community members it emerged that majority of the forest uses identified by the community as important are all consumptive in nature. This poses a great challenge to management given that their livelihoods are closely linked to the forest. However, by virtue of their long history of living with forests, the Ogiek Community had a good record of events and changes that have taken place in the forest over the years and this can be used as a basis during restoration and conservation efforts. This is more pertinent considering that not many studies have been previously conducted in the forest to assess the status of water quality and other forest resources.

There was congruence in the views held by the community and what has emerged in most studies in land use studies about the main reasons for some streams in the forest drying up. This is a good score on part of a community whose presence in the forest has been perceived as destructive. With awareness creation among these communities on the importance of forests, their participation in conservation efforts can be enhanced. Forceful evictions of communities from the forests where they have lived for generations has not been well received by residents living within and in areas adjoining protected areas (Norgrove and Hulme, 2006; Okeyo-Owuor et al., 2011). This has further entrenched, negative feelings further jeopardizing conservation efforts. There is a feeling among local communities that their interests should be given priority allowing them free access to forest resources. Their exclusion leads to the loss of ownership, making them adopt more destructive practices of forest exploitation practices in protest, instead of the traditional ones which are often more sustainable in nature (Norgrove and Hulme, 2006).

5. Conclusions

In the management of aquatic resources in Kenya, biological assessment has not been widely used to evaluate the level of degradation of streams and rivers. In the Lake Victoria basin, this is largely attributable to lack of long-term monitoring programs that can generate reference data sets for the initial development and subsequent evaluation and refinement of biological criteria and indices. However, this paper makes a significant first step towards developing a tool for monitoring human induced influences on river water quality at the catchment level. Following its effectiveness, there is the potential for developing similar indices for basin-wide and national monitoring of streams and rivers as a cost-effective means of maintaining the integrity and sustainability of our national water resources.

This study also indicates that indigenous knowledge by communities living in conservation areas can be used to identify critical areas for restoration. Their knowledge of local forest resources such as tree species and streams become useful during mapping. Their recording of events and changes in the structure and functioning of the forest can be used to benchmark restoration efforts and also to assess their success. The capacity of the communities also needs to be enhanced by offering adequate awareness creation on the significance of forests and the need for their conservation. Community representation in various groups and committees concerned with forest conservation and management should be enhanced to dispel feelings of exclusion. Meaningful restoration efforts must genuinely involve community members and their leaders.

6. Acknowledgements

We wish to acknowledge Kenya's National Council for Science and Technology for funding this study. Our sincere gratitude also goes to Chepkoiel University College and KMFRI (Kenya Marine and Fishery Research Institute) technicians who assisted us during sampling and sample analysis.

7. References

APHA (American Public Health Association) (1998). Standard methods for the examination of water and wastewater. Washington, DC: American Public Health Association, American Water Works Association, and Water Pollution Control Federation.

Aura CM, Raburu PO and Herman J (2010). A preliminary macroinvertebrate Index of Biotic Integrity for bioassessment of the Kipkaren and Sosiani Rivers, Nzoia River basin, Kenya. Lakes & Reservoirs: Research and Management 15:119–128.

FAO (2010). Food and Agriculture Organization of the United Nations: Global Forest Resources Assessment Main report, FAO Forestry Paper 163, Food and Agriculture Organization of the United Nations, Rome.

Karr JR and Chu EW (1999). *Restoring Life in Running Waters. Better Biological Monitoring*. Washington, D.C. Island Press. 206pp.

Kibichii S, Shivoga WA, Muchiri M and Miller SN (2007). Macroinvertebrate assemblages along a land-use gradient in the upper River Njoro watershed of Lake Nakuru drainage basin, Kenya. Lakes and Reservoirs: Research and Management 12: 107–117.

Kobingi N, Raburu PO, Masese FO and Gichuki J (2009). Assessment of pollution impacts on the ecological integrity of the Kisian and Kisat rivers in Lake Victoria drainage basin, Kenya. African Journal of Environmental Science and Technology 3: 097-107.

Mango LM, Melesse AM, McClain ME, Gann D and Setegn SG (2011). Land use and climate change impacts on the hydrology of the upper Mara River Basin, Kenya: results of a modelling study to support better resource management. Hydrology and Earth System Sciences 15: 2245–2258.

Masese FO, Raburu PO and Muchiri M (2009a). A preliminary benthic macroinvertebrate index of biotic integrity (B-IBI) for monitoring the Moiben River, Lake Victoria Basin, Kenya. African Journal of Aquatic Science 34: 1–14.

Masese FO, Muchiri M and Raburu PO (2009b). Macroinvertebrate assemblages as biological indicators of water quality in the Moiben River, Kenya. African Journal of Aquatic Science 34: 15–26.

Mati BM, Mutie S, Gadain H, Home P and Mtalo F (2008). Impacts of land-use/ cover changes on the hydrology of the transboundary Mara River, Kenya/Tanzania. Lakes and Reservoirs: Research and Management 2008 13: 169–177.

Mwashote BM and Shimbira W (1994). Some limnological characteristics of the lower Sondu-Miriu River, Kenya. In: *Okemwa, E.; Wakwabi, E.O.; Getabu, A. (Ed.) Proceedings of the Second EEC Regional Seminar on Recent Trends of Research on Lake Victoria Fisheries, Nairobi*: ICIPE Science Press, p. 15-27.

Norgrove L and Hulme D (2006). Confronting Conservation at Mount Elgon, Uganda. Development and Change 37: 1093-1116.

Ochumba PBO and Manyala JO (1992). Distribution of fishes along the Sondu-Miriu River of Lake Victoria, Kenya with special reference to upstream migration, biology and yield. Aquaculture and Fish Management 23:701–719.

Okeyo-Owuor JB, Masese FO, Mogaka H, Okwuosa E, Kairu G, Nantongo P, Agasha A and Biryahwaho B (2011). Status, Challenges and New Approaches for Management of the Trans-Boundary Mt. Elgon Ecosystem: A Review. In: *Towards Implementation of Payment for Environmental Services (PES): a collection of findings linked to the ASARECA funded research activities,* 60-82 pp. VDM Verlag Dr. Müller, Saarbrücken

Okungu J and Opango P (2005). Pollution loads into Lake Victoria from the Kenyan catchment. In: *Knowledge and Experiences gained from Managing the Lake Victoria Ecosystem,* Mallya GA, Katagira FF, Kang'oha G, Mbwana SB, Katunzi EF, Wambede JT, Azza N, Wakwabi E, Njoka SW, Kusewa M, Busulwa H (eds).

Regional Secretariat, Lake Victoria Environmental Management Project (LVEMP): Dar es Salaam; 90-108.

Osano O, Nzyuko D and Admiraal W (2003). The fate of chloroacetalinide herbicides and their degradation products in the Nzoia Basin, Kenya. Ambio: Journal of the Environment 32: 424–427.

Raburu PO, Masese FO and Mulanda CA (2009). Macroinvertebrate Index of Biotic Integrity (M-IBI) for monitoring rivers in the upper catchment of Lake Victoria Basin, Kenya. Aquatic Ecosystem Health and Management 12: 197–205.

Raini JA (2009). Impact of land use changes on water resources and biodiversity of Lake Nakuru catchment basin, Kenya. African Journal of Ecology 47 : 39–45.

The World Bank. (2007). Strategic Environmental Assessment of the Kenya Forests Act 2005

The International Bank for Reconstruction and Development / The World Bank 1818 H Street, NW Washington, DC 20433.

WRI (2007). World Resources Institute, Department of Resource Surveys and remote Sensing, Ministry of Environment and Natural resources, Kenya, Central Bureau of Statistics, Ministry of Planning and Development, Kenya; and International Livestock Research Institute: Nature's Benefits in Kenya: An Atlas of Ecosystems and Human Well-Being, World Resources Institute, Washington, DC, and Nairobi.

2

Systematic Approach to Design with Nature

Ali Sepahi
Isfahan University
Iran

1. Introduction

One of the main goals in the field of Environmental Design in the 21st Century should be to find objective and systematic design processes for the implementation of the concept of design with nature for large scale projects such as regional parks, residential villages, reforestation and green belts around industrial sites. The author also believes that to design with nature, one should design the way nature does. Nature designs in terms of natural laws rather than taste, personal preference or school of design. Some projects, however, involve man-made as well as natural laws and parameters. Fortunately with regards to site planning, surface modeling and site analysis software have been available by which the site can be analyzed and placements made based on topo-edaphic variables. For planting design, however, very little work has been done. In fact, to the author's knowledge only one system for urban forests has been developed (Kirnbauer et al, 2009) which uses climatic and soil data to select the plant palette. Plant placement, however, is not based on topographic variables.

2. The systematic approach

The approach mainly consists of choosing the suitable parameters and variables, based on which the design evolves more or less automatically. The choice of parameters and variables is the most important step and requires careful consideration and consultation. This is due to the fact that from here on, the design will evolves from the interaction of these parameters and variables and the designer's involvement will be minimal. In this chapter the systematic approach is first explained for site planning in general and then more specifically for planting design.

2.1 Site planning

The approach is demonstrated for site planning of a 38 ha villa complex on an undulating topography (Fig. 1) near Isfahan, Iran (by the author, in 1998) using the available surface modeling and site analysis modules of Landcadd. The procedure consists of three steps.

2.1.1 Step 1 – Determining the parameters and variables

The suitable range for each variable for the different design elements is determined. For the present project, three variables (based on the available data) were considered. Some examples are presented in Table 1.

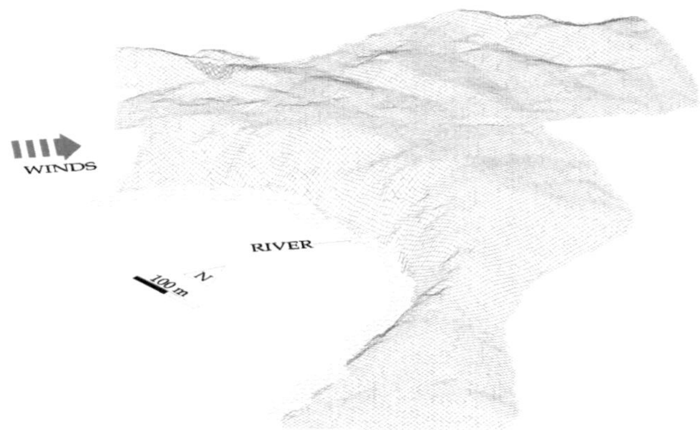

Fig. 1. Topography of the 38 ha project site for a villa complex.

Design element	Slope (%)	Aspect	Elevation (m)
	Variable		
Parking lots	0-10		
Sports courts	0-10	N, NE, E (less wind)	<110 (less wind)
Vilas	0-15		>65
Hotel and restaurant	0-15	S, SE, SW (good view)	>110 (good view)
Mass irrigated tree planting	15-30		
Mass non-irrigated shrubs	30-45		

Table 1. Design elements and the corresponding classes of variables.

The three parameters considered were: Minimum distance of 15m between villa buildings, 8% maximum slope for the roads and 30 km speed limit (for the radii of road curvature).

2.1.2 Step 2 – Site analysis

Surface modeling and site analysis software generate aspect, slope and elevation variables from survey data (x, y, z) from project sites. They do this in terms of grid cells, the size of which is determined by the designer, based on the topography and the intended use of the site. Other variables such as soil depth can be added, using the elevation analysis (Sepahi, 2005). Figure 2 presents the classes (AutoCAD layers) for slope and elevation generated for the site using Landcadd software. Regarding aspect, the software generates nine AutoCAD layers for N, NE, E, SE, S, SW, W, NW and FLAT. A short review of the application of Remote Sensing to site analysis regarding topography, soil and plant cover is presented by Sepahi (2009).

2.1.3 Step 3 – Placement of the design elements

AutoCAD layers bearing the hatch patterns representing the suitable classes of slope, aspect and elevation for each element were 'frozen', which led to the disappearance of their hatches from the computer monitor. This resulted in patches of land with blank (not hatched) grid cells in which the AutoCAD blocks representing the respective elements (such as villas) were inserted. After laying out the access roads to the patches, finer adjustments, such as alignment of elements along the roads were made. For the hotel and restaurant, for instance, the AutoCAD layers corresponding to the suitable attribute (Table 1) i.e. SLOPE-0-10, SLOPE-10-15, ASPEC-SOUT, ASPEC-SWST, ASPEC-SEST and ELEVE-110-MAX were frozen resulting in a few options, of which the most suitable were chosen. The blank patches would be more clearly visible if all the AutoCAD layers were assigned one color such as grey. It should be noted that Landcadd has a command by which, for a given point on the site, the scope of the observable terrain is indicated.

The layout of the roads was a function of maximum allowable slope, speed limit and the topography. Civil engineering software is available for such a task. For a preliminary road layout, however, a simple procedure can be used. A circle is drawn at the origin (O) of the road (Fig. 3-a) with radius R = CI/MS, in which CI is the contour interval and MS is the maximum allowable slope (e.g., 0.08 for 8%). The circle intersects the adjacent contour line at two points (A and B). The point which is to the direction of the destination - point A in this case- is chosen and an AutoCAD polyline is drawn from the origin to it. The circle is then moved to this point and the process is repeated. At locations where the slope of the land is less than the MS, the circle will not intersect the next contour and any line drawn will have a slope less than MS. At the end of the road layout, the FIT Command of AutoCAD is used to smooth the path of the road (Fig. 3-b). Finer modifications are then made regarding the radii of the curves based on the speed limit (indicated by an arrow in Figure 3-b). This approach to road layout does away with the common disagreement between the landscape architects and engineers, i.e. aesthetics vs. engineering principles. In fact the author doubts if there is such a thing as an aesthetic road that is not soundly engineered.

The systematic approach provides the bulk of the conceptual site planning. Final decisions and refinements will eventually be made on the site. The main advantages of the approach are: conserving the natural topography by avoiding massive land leveling; organic distribution of the elements (naturalistic aesthetics) and the elements fitting comfortably in the terrain. In Figure 4 the site planning for the villa complex is presented.

2.2 Planting design

To the author, design with nature with respect to planting design, implies achieving three objectives: *conserving nature, establishing a sustainable ecosystem* and achieving a *natural appearance*. Conserving nature involves many issues. Those related to this topic are: maintaining the site's topography and contour planting to reduce erosion.

Sustainability is a complex concept. A version sufficient for mass planting involves:

- selection of a plant palette suitable for the site's climate and soil
- placing individual plants at suitable locations within the site
- ensuring compatibility among the species
- avoiding extensive plant loss due to natural causes and attracting varied wildlife through a diverse plant palette

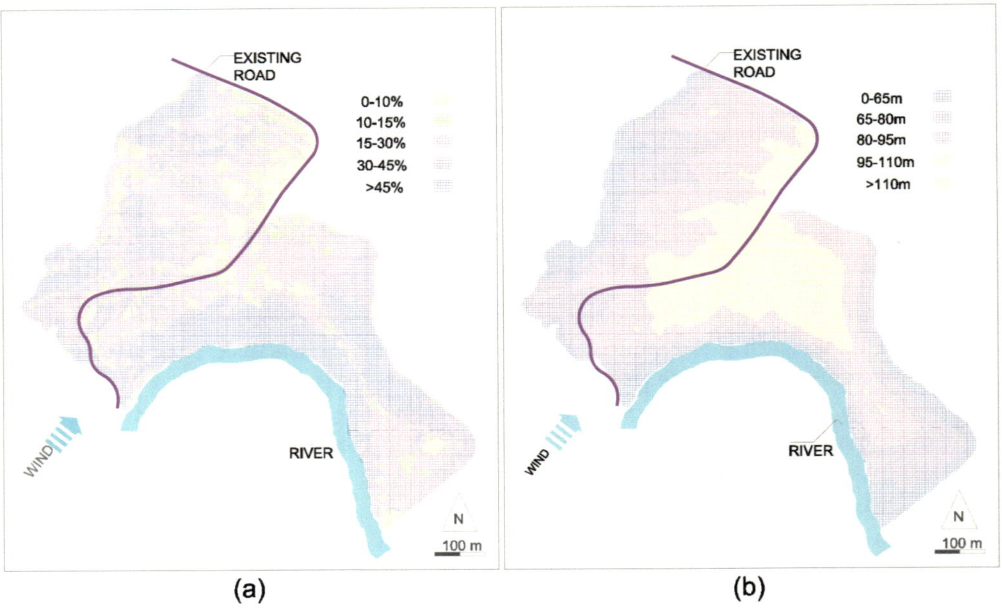

Fig. 2. Slope analysis (a) and elevation analysis (b) of the 38 hectare project site.

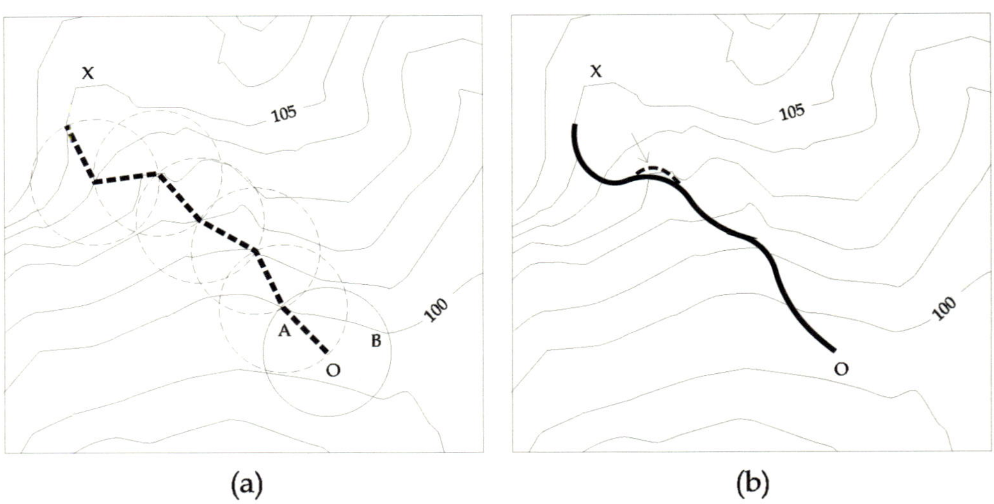

Fig. 3. Road layout from the origin O to the destination X.

Natural appearance is an issue related to aesthetics and involves:

- diversity of species
- visual association of species (seen together in nature)
- unity, brought about by one species being dominant
- organic distribution of the species, in contrast to geometric patterns
- blending of the site into the natural surroundings

Fig. 4. Site planning of the 38 ha villa complex near Isfahan, Iran.

The above objectives can be realized by selecting a plant palette from a native (model) plant community suitable for the climate and soil of the project site, and placing the individual plants based on the specific topographic and soil characteristics of the locations within the site. The author believes, as some other workers cited by Thompson (1998) that if landscape planning is undertaken along ecological lines, the aesthetic aspects will be taken care of automatically. Not only should there not be a "'tension between aesthetics and scientific foundations in Landscape Architecture" as Harding Hooper et al (2008) put it, application of scientific findings should be an integral part of the design process.

Three methods were presented (Sepahi, 2000, 2005, 2009, in print) to arrive at an objective and systematic design process for planting designs resembling native plant communities. The approach can be summarized in four main steps. The steps, however, do not correspond to those in the respective articles. The three methods differ only with respect to Step-3. The main purpose of the present chapter is to present an overall view of these methods. Hence, the same set of data with a few variables and species (Sepahi, in print) is considered to briefly explain the three methods. For discussions on the justification, literature review and mathematical details, the reader is referred to the original articles.

2.2.1 Step 1 – Site analysis of the project site

For demonstration, Landcadd's surface modeling and site analysis modules (Eagle Point, 2005a and 2005b) were used to generate data on aspect, slope and elevation along with soil depth, using survey data from a 14.4 ha project site (Fig. 5). Topo-edaphic data for the first row of grid cells (cell N° 1 at the top left of the figure) is presented in Table 2.

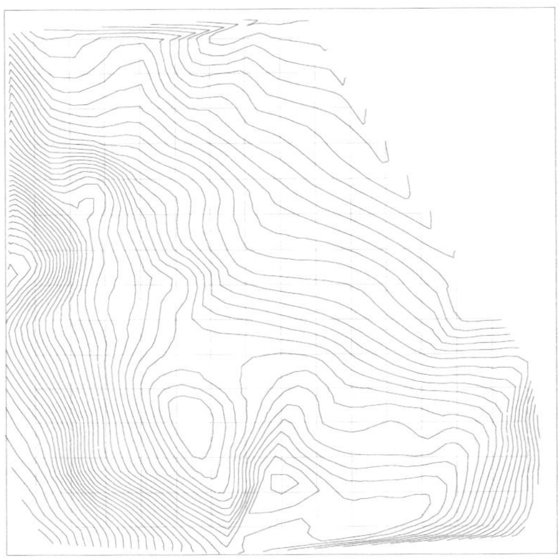

Fig. 5. Topographic map of the 14.4 ha project site with 30x30 m grid cells and 2m contour intervals.

Grid cell N°	Variable: Aspect (degree)	Slope (%)	Elev. (m)	Soil depth (cm)
1	45	25	1495	130
2	315	15	1495	150
3	0	15	1495	130
4	90	25	1495	110
5	45	25	1495	90
6	0	15	1485	50
7	0	15	1485	50
8	0	15	1485	50

Table 2. Topo-edaphic variables of the first row of grid cells in Fig. 5.

2.2.2 Step 2 – Selecting a plant palette

A plant palette from a native (model) plant community suitable for the site's climate and overall soil characteristics (with supplementary irrigation, if required) is chosen. For demonstration, a plant palette of seven species was chosen from an Englemann Spruce-Subalpine Fir biogeoclimatic zone in the Boston Bar area, B.C., Canada, as the model community. Data from 10m-radius sample plots, provided by the Resource Inventory Branch, Ministry of Forestry, Victoria, B.C. was used (Table 3).

2.2.3 Step 3 – Determining species composition of the grid cells

In this step the expected percent crown cover (abundance) for each species for every grid cell is determined. This is done differently in the three methods explained below.

The Regression Method - The method (Sepahi, 2005) can accommodate any number of environmental variables and species, thus different levels of biodiversity and natural representation can be attempted. For each of the seven species (Table 3) a multiple regression of the percent crown cover on the topo-edaphic variables is run. The format of the resulting seven equations would be:

$$Y = b_0 + b_1X_1 + \ldots + b_nX_n \quad (1)$$

in which Y represents the expected percent crown cover (EP), X the value for the topo- edaphic variable and b the corresponding partial regression coefficient. The values of the topo-edaphic variables for each grid cell (Table 2) are inserted in every one of the seven multiple regression equations and the expected percent crown covers for the seven species for the grid cell are calculated. These are later translated into the recommended number of trees in Step 4.

The Least Difference Method - With the imminent availability of a large volume of data from model communities through Remote Sensing technology, the Least Difference Method

	Topo-edaphic variables				Species[1] percent crown cover								
Sample Plot No	Aspect (degrees)	Slope (%)	Elev. (m)	Soil depth (cm)	ABIELAS	ALNUCRI	ALNUVIR	LONIINV	PICEENE	PICEENG	PINUCON	TS	TA
1	78	2	1354	141.0	17.9	0.0	2.3	0.0	0.0	26.7	0.3	47.2	131.9
2	165	12	1275	106.0	16.0	0.0	10.0	0.0	0.0	5.0	55.0	86.0	141.2
3	353	70	1330	102.0	12.3	0.0	0.0	0.0	0.0	3.0	0.0	15.3	62.5
4	350	4	1090	149.0	0.4	0.0	23.0	0.0	5.0	0.0	14.0	42.4	46.3
5	352	15	1200	55.0	38.0	0.0	1.0	3.0	0.0	38.0	0.0	80.0	90.2
.	.	.	.	.	.	.	.	.	.	.	.	.	.
.	.	.	.	.	.	.	.	.	.	.	.	.	.
36	34	23	1230	160.0	38.0	0.0	0.0	0.0	0.0	8.0	10.0	56.0	146.6

1: Abbreviations used by the Ministry of Forestry, BC, Canada
TS: Total percent crown cover of the selected seven species
TA: Total percent crown cover of all the species (originally recorded) in the sample plot

Table 3. Data from Englemann Spruce-Subalpine Fir biogeoclimatic zone in, B.C. Canada.

(Sepahi, 2009) was presented that emulates nature more closely and is not based on a mathematical model. The approach is based on a simple argument that if two patches of land are similar, plant composition suitable for one (a sample plot in the model community) is also suitable for the other (a grid cell on the project site). Topo-edaphic variables of each grid cell (Table 2) are compared with those of all the 36 sample plots in the model community (Table 3) one plot at a time. The sample plot, most similar to the grid cell is chosen and its percent crown covers are assigned (as EPs) to the grid cell. The method can accommodate any number of environmental variables and species.

The Variable Classification Method- This method (Sepahi, 2012) draws on the fact that plants respond to ranges, rather than to specific values of environmental variables. It is not based on a mathematical model and does not involve statistical analysis. Although it can be fully computerized, it can also be applied semi-automatically, using the available site analysis software. The values for the different variables in Table 2 and Table 3 are grouped into different classes. The method is suitable when few variables and classes (e.g. low, medium and high) are considered.

Regarding aspect, Landcadd generates nine aspects: N, NE, E, SE, S, SW, W, NW and FLAT. The ideal way to reduce this number would be to customize the software to generate five aspects of NE, SE, SW, NW and FLAT. This is based on the fact that in the Northern Hemisphere, the growth gradient is along the NE-SW axis rather than the N-S axis (Urban et al., 2000). A simpler and close enough alternative is to merge every two consecutive aspects (AutoCAD layers) to produce four aspects of: N-NE, E-SE, S-SW, W-NW plus one FLAT. This approach was followed for the present demonstration. In Table 4 the resulting classes for the variables in Table 2 are presented. Although a bit more involved (see Sepahi, 2012) in principle, the percent crown covers of the sample plots are allocated to the grid cells with matching topo-edaphic variables.

	Variable			
Grid cell N°	Aspect (degree)	Slope (%)	Elev. (m)	Soil depth (cm)
1	N-NE	L	L	H
2	W-NW	L	L	H
3	N-NE	L	L	H
4	E-SE	L	L	H
5	N-NE	L	L	H
6	N-NE	L	L	L
7	N-NE	L	L	L
8	N-NE	L	L	L

H and L: represent high and low classes respectively.

Table 4. Topo-edaphic variables of Table 2, classified into different classes.

2.2.4 Step 4 – Determining plant number and placement

The expected percent crown covers calculated in step 3 (by any of the three methods) are translated into the number of plants for the seven species using equation 2.

$$N = \frac{EP \times GA \times TA}{100 \times CA \times TS \times \cos\alpha} \quad (2)$$

Where, EP is the expected percent crown cover for the species, GA is the area of a grid cell, CA is the crown area (from the crown diameter assigned to the species), TA and TS are from the last two columns of Table 3 and α is the angle of the slope of the grid cell.

Once the numbers of the plants of the species for the grid cells are determined, different schemes can be used for their placement. For demonstration, one type of contour planting is explained here. The MEASURE command of AutoCAD is used to mark the contour lines at the desired intervals (Fig. 6). The icons (AutoCAD blocks) for the species are, then, placed at random on contour lines within the grid cells in the drawing. Figure 6 presents the placement of icons for cell N°5, using the Regression Method.

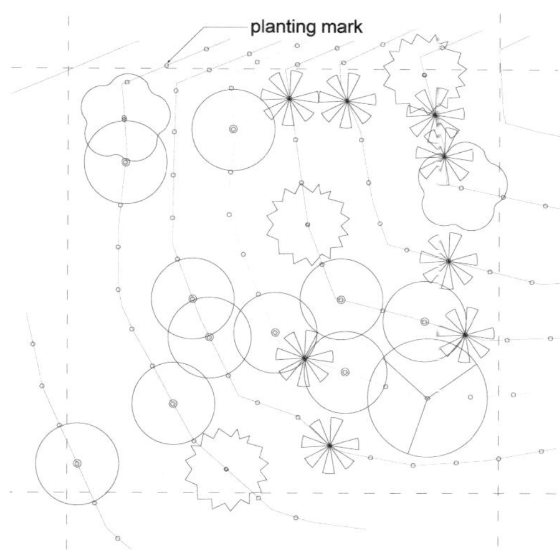

Fig. 6. Plant composition of the grid cell for cell N° 5 using the Regression Method.

The choice of the Regression Method, the Least Difference Method or the Variable Classification Method is based on the number of sample plots, number of variables and the mode of data processing (Table 5). Data processing could be automatic (using a computer program) or manual (using a spreadsheet).

The three Methods place individual plants based on topo-edaphic variables, thus realizing the ecological potential of the project site more fully than the present common practice. They draw on an already available wealth of data on plant communities in forestry departments. They provide an organic distribution of species for the bulk of the planting. Designers can make necessary changes in specific areas (e.g. around buildings, sitting areas) using a larger or different plant palette. Designers can also use their ingenuity for modifications or attaining different levels of natural representation. The following are a few examples:

- taking into consideration the uneven age distribution in the model community
- bio-geographic planting (Kingsbury, 2004), i.e. aiming at as complete a representation of a natural plant community as possible
- achieving organic distribution for plant palettes of commercial (non-native) cultivars

Method	Number of samples	Number of variables	Automatic data processing
Regression	low	any	required
Least Difference	high	any	required
Variable Classification	high	low	not required

Table 5. Recommended methods for different number of sample plots and topo-edaphic variables and modes of data processing.

For the reader who would like to try the methods, a few recommendations are presented. For the Least Difference Method, adjust the elevations from the project site so that the minimum elevation is the same as the model community. For both the Regression Method and the Least Difference Method, convert aspects into Annual Direct Incident Radiation and Heat Load, using the procedure proposed by McCune (2002). Apply Equation 2 to all the three methods, except for the cases where TS for any of the sample plots in the model community is zero. In that case the equation should be modified as follows:

$$N = \frac{EP \times GA \times DF}{100 \times CA \times \cos a} \tag{3}$$

In which, DF is the density factor calculated as the average of TA divided by the average of TS (last two columns of Table 3). It is used to account for the fact that not all the species in the original sample plots were included in the design.

2.3 The role of scientific research

The present trend in design emphasizes issues such as social responsibility, sustainability, environmental responsiveness and human health (Milburn et. al., 2003). In the systematic approach to design, the quality of the outcome is determined by the reliability of the parameters and variables chosen. Such issues require incorporation of research into the landscape design process. The research, basically a literature review, taps into the available body of knowledge acquired through scientific research. Some advances in the field of Landscape Architecture in areas such as irrigation, soil amendments, new cultivars and computer aided design have been due to the scientific and technical achievements of other disciplines. However some information, especially for the implementation of the concept of design with nature, might require studies which are not of common interest to the other disciplines. These have to be dealt with by scientific research within the profession. For instance, more work could be done to increase the accuracy of the planting design methods mentioned above. In the Regression Method, interaction (statistical) between topo-edaphic variables could be considered, or transformations (e.g. logarithmic) on the variables could be tried. With respect to the Least Difference Method, other analysis for determining the relative importance of the different topo-edaphic variables in species distribution should be attempted. More work on combining some of the topo-edaphic variables into ecologically relevant predictors, such as soil fertility, would help to reduce the number of variables.

Some of the problems might be unique to specific projects. In a green belt project around the 3000 ha Mobarekeh Steel Mill Complex in Iran, the author was faced with a number of questions regarding the choice of the species suitable for the calcareous soil of the site, tolerant to the mill's pollutants and their irrigation requirements under the site's arid climate. Also of interest was the relative efficiency of the species in removing the pollutants from the environment. Thus the author in cooperation with two colleagues, an irrigation and a soil scientist, started a long term experiment involving 648 treatments (factorial combination of 72 species, 3 levels and 3 intervals of irrigation). The actual mass planting at the mill was to be carried out over a number of years. As new results were obtained from the experiment, modifications were made in the upcoming plantings.

The above passage pivots on the phrase 'scientific research'. One can get into a long discussion regarding the definition of scientific research, including basic and applied research. For our purpose, let us adopt a practical definition and consider scientific research as one which is accepted by scientific societies for publication. Such societies require that the results of experiments be accompanied by a statement of the probability of the validity of the results (denoted by * or **, indicating 95 or 99% levels of confidence). Such a statement can only be made if there is an estimate of the experimental error, obtained through the application of statistical analysis to the data. The analysis in turn, can only be made on data obtained based on a statistical design. Such an approach to research is imperative when a number of variables (controllable, uncontrollable or unknown) besides the variables in question, affect the results of the study. This necessity was first felt at the Rothamsted Agriculture Research Station in the 1920s which led to the invention of the science of Statistics by Sir R.A. Fisher. Although initially developed for agricultural research, it became widely used in almost all other fields of research.

To bestow research in the field of Landscape Architecture with such quality, it is necessary to incorporate the required skills in the curriculum. To this end the Masters Program should be divided into two sections:

1. Urban Design – the extension of the undergraduate program
2. Rural Design – with focus on projects involving nature at a large scale

The distinction between the two sections will be based on the relative emphasis placed on *innovative creativity* versus the *artistic creativity*. In the Rural Design, emphasis should shift more towards the innovative creativity which enables the future designers to deal with the various problems that nature with all its complexities presents, to devise new techniques and carry out research projects that add to the body of knowledge in the profession.

The modification of the curriculum would involve incorporation of two courses in statistics. In the last term of the undergraduate program, the students should take an introductory course in statistics offered at the Departments of Statistics. This course can be elective, for those who expect to continue their post graduate studies in Rural Design. In the first term of the graduate program, the students of Rural Design should take a course in design and analysis of the experiments, known as Statistical Design. The statistical design course covers the application of statistics to the different types of experiments or surveys pertinent to the specific field of study and is usually offered within the respective department. Of course, the students would also take the necessary courses pertinent to nature, depending on their theses projects.

In the undergraduate program, being bold and daring (appreciated for urban designs) is encouraged. Exposure to scientific research methods leads to a more cautious, sensitive, respectful and humble attitude towards nature and controls the overuse of the so called artistic license. It would be helpful, if a reference manual on the application of statistical designs to Landscape Architecture research is put together.

The application of computers to scientific research is well known. In fact thanks to the computer, research workers (including the author) have been applying the statistical analyses to their data without remembering the details of the calculations involved. With respect to landscape design, once a process is based on laws, the logical next step would be to computerize it. However, the computer is not just to speed up the process; rather it is an integral part of the decision support systems. This is due to the fact that in the case of planting design, for instance, a number of variables (more than the three conceivable Cartesian coordinates) are involved, which mathematically can be considered. Moreover, challenging as it is to envisage the combined effects of a number of variables for placing individual plants from different species, applying it manually to thousands of individual plants would be a monumental task

3. Conclusion

Nature is already under heavy pressure from over population and cannot endure more hammering by designs incompatible with nature. Availability of a systematic approach reduces the occurrence of such designs. There are also a number of advantages in the systematic approach for the designers themselves:

- Confidence in the outcome - Since the best possible locations are determined for the design elements, there is no uncertainty, doubt and indecisiveness during the design process.
- Automation – The approach can be computerized, resulting in accuracy and speed
- Defendability - The questions raised by juries, regarding large scale projects, are mainly with respect to issues such as placement of the design elements, impact on nature, ease of implementation and cost, rather than the aesthetic aspects. These are all the issues

based on which the parameters and variables were initially determined. Placement based on topo-edaphic variables, by minimizing land leveling would address such issues.

4. Acknowledgement

The author wishes to acknowledge the Resource Inventory Branch of the Ministry of Forestry and Range, Victoria, BC, Canada for providing the data.

5. References

Eagle Point (2005a). *Surface Modeling Manual*, Eagle Point, Retrieved from <http://www.eaglepoint.com>

Eagle Point (2005b). *Site Analysis Manual*, Eagle Point, Retrieved from <www.eaglepoint.com>

Harding Hooper, V.; Endter-Wada, J. & Johnson, C. W. (2008). Theory and Practice Related to Native Plants, a Case Study of Urban Landscape Professionals. Landscape Journal, Vol. 27, (January 2008), No. 1, pp. 127-141, ISSN 0277-2426

Kingsbury, N. (2004). Contemporary Overview of Naturalistic Planting Design. In: *The dynamic landscape; design, ecology and management of naturalistic urban planting,* N. Dunnett, & J. Hitchmough, (Eds), 58-96 , Spon Press, Taylor & Francis Group. N.Y., ISBN 97804 15256209

Kirnbauer, M. C.; Kenney, W. A., Churchill, C. J. & Baetz, B. W. (2009). A Prototype Decision Support System for Sustainable Urban Tree Planting Programs. *Urban Forestry & Urban Planning,* Vol. 8, (January 2009), 3-11, ISSN 1618-8667

McCune, B. & Keon, D. (2002). Equations for Potential Annual Direct Incident Radiation and Heat Load. *Journal of Vegetation Science,* Vol. 13, (August 2002), pp. 603-606, ISSN 1100-9233

Milburn, L. S. & Brown, R. D. (2003). The Relationship between Research and Design In Landscape Architecture. *Landscape and Urban Planning,* Vol. 64 (June 2003), pp. 47-66, ISSN 0169-2046

Sepahi, A. (2000) Nature as a Model for Large-scale Planting Design. *Landscape Research,* Vol. 25, No. 1, (January 2000), pp. 63-77, ISSN 0142-6397

Sepahi, A. (2005). Nature as a Model for Large-scale Planting Design II, *Proceedings of the CSLA (Canada) Congress "Landscape Architecture - EXPOSED!"* , Winnipeg, 17-20 August 2005 Winnipeg, Manitoba

Sepahi, A. (2009). Nature as a Model for Large Scale Planting Design, Least Difference Method. *Urban Forestry and Urban Greening,* Vol. 8, No. 3, (June 2009), pp. 97-205, ISSN 1618-8667

Sepahi, A. (2012). Nature as a Model for Large Scale Planting Design, Variable Classification Method. *Landscape Research,* DOI 10.1080/01426397.2011.638740

Thompson, J.H. (1998). Environmental Ethics and the Development of Landscape Architectural Theory. *Landscape Research,* Vol. 23, (July 1998), pp. 175-194, ISSN 0142-6397

Urban, D. L.; Miller, C., Halpin, P. N., and Stephenson, N. L. (2000), Forest Gradient Response in Sierran Landscapes: the Physical Template, *Landscape Ecology*, Vol. 15, (October 2000), pp. 603-620, ISSN 0921-2973

3

Seasonal Reflectance Courses of Forests

Tiit Nilson[1], Miina Rautiainen[2], Jan Pisek[1] and Urmas Peterson[1]
[1]*Tartu Observatory*
[2]*University of Helsinki, Department of Forest Sciences*
[1]*Estonia*
[2]*Finland*

1. Introduction

Studies on the seasonal changes in plant morphology and productivity, also known as phenology, have applied a multitude of methods for detecting changes in the start of spring or length of growing period. Global-scale plant phenology observation is needed, for example, to understand the seasonality of biosphere-atmosphere interactions (e.g. Moulin et al., 1997). Traditional species-specific *in situ* records of leafing and fruiting (e.g. Menzel et al., 2006), monitoring of atmospheric carbon dioxide as an indicator of photosynthesis (e.g. Keeling et al., 1996) and remote sensing of vegetation reflectance (e.g. Zhou et al., 2001) have all been used in attempts to understand the consequences of the changing climate on primary production and global vegetation cover.

Satellite remote sensing enables continuous monitoring of vegetation status (and potentially also physiological processes of plants), and thus, it is not limited to conventional, single-date phenological metrics such as budburst or flowering. Using remote sensing also enables a wider perspective to vegetation dynamics: at its best it can reveal large-scale phenological trends that would not be possible to detect from the ground. Monitoring phenological phases of vegetation using remote sensing has made significant progress during the past decade due to an increased global coverage of satellite images, development of reflectance metrics describing seasonality, and evolution of methods for calibrating time series of remotely sensed data sets. In addition, operational satellite products for monitoring phenology from space have been designed (e.g. Ganguly et al., 2010; Zhang et al., 2003).

Despite rapid progress, remote sensing also faces a multitude of both scientific and technical challenges before reliable phenological statistics with global coverage can be obtained. Firstly, compared to traditional phenological observations, a satellite pixel typically covers a larger area. Therefore, phenological information obtained from remote sensing (especially from satellite images) is aggregated data covering a heterogeneous landscape and requires new interpretation of phenological metrics. Hence, the concept of "land surface phenology" (Friedl et al., 2006) is often used to describe the seasonal pattern of variation in vegetated land surfaces observed from remote sensing. Secondly, coarse scale remotely sensed data sets cannot necessarily differentiate all phenological processes occurring in a fragmented landscape. Thus, the validation of phenological metrics has taken a new meaning: conventional ground measurements (typically made from a few individual plants) need to be upscaled to the level of satellite pixels. Finally, the physical processes driving the seasonal reflectance dynamics observed in satellite images are not yet well known; until now only a few pioneering studies linking forest reflectance seasonality to seasonal changes in forest structure and biochemistry have been carried out.

In this chapter, we review remote sensing of the seasonal dynamics of forest reflectance as well as highlight future challenges in the development of methodologies needed for interpreting seasonal reflectance courses at forest stand level (Section 2: Literature review). In addition, using empirical data from European hemiboreal and boreal forests, we demonstrate how seasonal reflectance courses of forests can be obtained from satellite remote sensing and how they can be used to examine the cyclic changes taking place in the vegetation (Section 3: Case study).

2. Literature review

In the following section, we will briefly outline the most common seasonal reflectance changes observed in remotely sensed datasets and discuss their driving factors in forests. Next, we will describe forest reflectance models as a tool for linking reflectance data to stand structure. Finally, we will explain how calibrated time series of satellite data can be created from multi-year image archives. The literature review focuses on forest stand-level phenomena i.e. global or large-scale monitoring of vegetation phenology is not covered.

2.1 Seasonal reflectance changes in satellite images

It is commonly understood that the seasonal reflectance course of deciduous forests is more pronounced than that of coniferous forests (e.g. Peterson, 1992). The phenology of evergreen coniferous species is more difficult to monitor from satellite images than that of deciduous species due to the smaller changes in foliage i.e. because new shoots account for only a small part of the green biomass. Jönsson et al. (2010) have estimated that seasonal changes in typically used spectral vegetation indices are actually more related to snow dynamics than to changes in needle biomass in coniferous forests.

The differences in seasonal reflectance dynamics between forests ultimately arise from the differences in their canopy structure and understory vegetation. In the same tree species, reflectance differences may be due to, for example, site fertility (e.g. Nilson et al., 2008) or topographical elevation (e.g. Guyon et al., 2011). Currently, there is only a limited understanding of reflectance seasonality and its driving factors at forest stand level: seasonal reflectance variation has preliminarily been linked to changes in stand structure (e.g. Kobayashi et al., 2007; Kodani et al., 2002; Nilson et al., 2008; Rautiainen et al., 2009; Suviste et al., 2007) or variation in leaf spectra resulting from, for example, climatic-elevational factors and site quality (e.g. Richardson & Berlyn, 2002; Richardson et al., 2003). The seasonal reflectance course of a forest results from the temporal reflectance cycles of the tree canopy and understory (also occasionally called forest floor or background) layers and the presence of snow. Seasonal reflectance changes of the two layers, in turn, are explained by changes in biochemical properties of plant elements and geometrical structure (e.g. crown shape, leaf area index, canopy cover) of the different plant species. In addition, the seasonal and diurnal variation in solar illumination influences forest reflectance (e.g. Sims et al., 2011): often a forest may exhibit a clear seasonal reflectance course even if there were no phenological changes occurring in the vegetation cover (Nilson et al., 2007).

The main drivers of the seasonal course of forest reflectance are found from the tree canopy layer. The amount of foliage, also characterized by leaf area index (LAI) (Chen & Black, 1992; Watson, 1947), undergoes significant seasonal changes in deciduous species and much smaller changes in coniferous species throughout the growing period (Rautiainen et al., 2011a). Phenological greening (an increase in LAI) or senescence (a decrease in LAI) influence both the scattering process in the tree canopy volume as well as have an indirect impact on canopy cover (and hence, the visibility of the understory layer or forest floor to the satellite sensor). The relationship of LAI and forest reflectance is complicated and somewhat species-specific: an increase in LAI may result both in an increase or decrease in forest reflectance, depending on the geometric arrangement of the canopy, optical properties of single leaves and optical properties of the understory layer. Theoretical studies have shown that the relationship of LAI and canopy reflectance depends on viewing and illumination angles as well as the absolute value of LAI and canopy cover (e.g. Gastellu-Etchegorry et al., 1999; Knyazikhin et al., 1998; Myneni et al., 1992; Rautiainen et al., 2004). In empirical studies, shortwave spectral bands typically have a decreasing trend when canopy LAI increases (e.g. Brown et al., 2000; Nilson et al., 1999; Stenberg et al., 2004; Stenberg et al., 2008a).

The contribution of leaf optical properties to stand reflectance is the greatest in dense stands with closed canopies (e.g. Eklundh et al., 2001). The optical properties of single leaves (or coniferous needles) change through the growing season due to physiological processes (e.g. Middleton et al., 1997). For example, Demarez et al. (1999) reported for temperate tree species that the seasonal course of leaf albedos (especially in the green and SWIR bands) is characterized by a decrease at the beginning of the growing season, a relatively constant value during summer and finally an increase towards the end of the growing season. Currently, there is no public database covering the seasonality of leaf spectra. Leaf optical properties models, such as the widely used PROSPECT model (e.g. Jacquemoud & Baret, 1990), enable us to predict quantitatively the seasonal changes in leaf reflectance and transmittance spectra, if the changes in leaf biochemical parameters (contents of various leaf pigments, water, organic compounds, etc.) are known. However, there are too few systematic investigations to prove the validity of leaf reflectance models in the seasonal context and to describe quantitatively changes in leaf biochemical parameters throughout the growing season. Further, the leaf optical properties are also affected by less predictable attacks of insects and diseases, typically in the second half of the vegetation.

Understory vegetation contributes significantly to forest reflectance (e.g. Eriksson et al., 2006; Spanner et al., 1990); the contribution of understory in sparse canopies can overrule the contribution from the tree canopy layer. In addition, if the temporal cycles of the tree canopy and understory layers differ considerably (Richardson et al., 2009), the greening of the understory prior to the tree canopy may complicate interpreting the budburst date for the region from satellite images (Ahl et al., 2006). The seasonal variability of understory reflectance is biome- and site type-specific (e.g. Miller et al., 1997; Shibayama et al., 1999). For example, most boreal forest understory layers have strong seasonal dynamics during their short snow-free growing period. Their seasonal course of understory reflectance is very characteristic of stand type, and the reflectance differences between stands tend to become more pronounced as the growing period progresses (Miller et al., 1997). Furthermore, Rautiainen et al. (2011b) reported that the spectral differences between and within boreal understory types are the largest at the peak of the growing season whereas in the beginning and end of the growing season the differences between the understory types are marginal. They also noted that understory vegetation growing on fertile sites had the brightest near infrared (NIR) spectra throughout the growing season whereas infertile sites appeared darker in NIR. It is very likely that the relative contributions of the understory and tree canopy layers to forest reflectance change as the growing season proceeds. Thus, separating the biophysical properties of the tree canopy layer from remote sensing data is further complicated.

Finally, the presence of snow significantly influences the seasonal course of forest reflectance, especially in the visible part of the spectrum. According to Manninen & Stenberg (2009), the presence of snow on the forest floor significantly increases forest albedo in sparse canopies (i.e. tree canopy LAI < 3). Furthermore, they reported that snow or hoar frost on trees alters the red band forest albedo completely. In addition, forest reflectance may vary when snow melts even though there are no changes in canopy leaf area index (e.g. Dye & Tucker, 2003).

The basic idea how to link a seasonal series of satellite images to phenology is to determine key moments of the season from the image series. Different physical quantities derived from the satellite images could be used to plot the time series; however, the most feasible is to use top-of-canopy (TOC) reflectance factors or multispectral indices derived from the TOC reflectance factors. Thus, pre-processing of the images that includes the absolute calibration and atmospheric correction of the images is needed. In addition, if the images in the series are acquired at variable view angles, the so-called BRDF (bidirectional reflectance distribution function) correction should be applied to reduce the reflectance factors to a fixed (e.g. nadir) direction. For instance, such image processing is included in the MODIS nadir reflectance product chain. The most popular multispectral index used to link to the phenology is NDVI, which is defined as

$$NDVI = (R_{NIR}-R_{RED})/(R_{NIR}+R_{RED}), \quad (1)$$

where R_{NIR} and R_{RED} are TOC reflectance factors in the NIR and red spectral region.

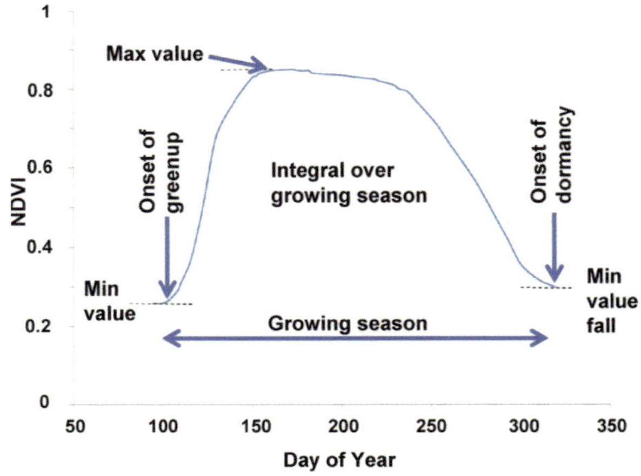

Fig. 1. A typical seasonal course of NDVI during the snow-free period with key phenology moments

Several authors have defined many parameters defining the shape of the seasonal curve (Crist, 1984; Jönsson & Eklundh, 2002, among others). A typical NDVI seasonal profile of green vegetation with indicated key moments of the season is shown in Fig. 1. A very important quantity is the integral of NDVI over the growing season. This integral is assumed to be proportional to the fraction of absorbed PAR over the season and, according to Kumar & Monteith (1981), linearly related to annual net primary production (NPP). The beginning of the active season could be defined as the moment when the first derivative of the NDVI differs reliably from zero. The moment when the maximum NDVI value is reached could be interpreted as the moment when the maximum leaf area is attained. The end of the growing season, also known as the beginning of dormancy, is assumed to start when a stable minimum NDVI value is measured, e.g. when the NDVI derivative has reached the stable value of zero. It is important to note that the spring minimum NDVI value and the minimum reached in fall after the active vegetation period is over does not necessarily have to be the same. According to our own experience from the hemiboreal region, the fall minimum is typically higher than in the spring for deciduous dominated stands. Several reasons could be responsible for this: there are still some leaves left on trees in the fall; on average the ground is wetter in spring compared to conditions in fall; and the layer of fresh leaf litter in the fall is brighter than the old decomposed litter after snowmelt in spring.

In practice, instead of a smooth NDVI curve we have a scattered series of points, because the image calibration and atmospheric procedures inevitably cause errors in the TOC reflectance values and vegetation indices. Thus the first task is to smooth the series. Several methods have been proposed for smoothing (e.g. TIMESAT, Jönsson & Eklundh, 2002). The problem is further complicated by the fact that most of the key parameters should be determined from an analysis of the first (sometimes second) derivative of the time course. The problem of estimating the derivatives from an empirical curve has been long known as a difficult task.

Another question is how are the important moments of phenological development of plants related to the seasonal curves of reflectance or vegetation indices? Determining the beginning or end of the green development is more or less clear. However, it is not as evident for another important phenological phase, the flowering.

The detection of flowering is an important phenology-related application of remote sensing. For instance, estimates of phenological phases derived from satellite images can be used in honeybee management to detect blooming of major honey plants (e.g. Nightingale et al., 2008), to estimate crop pollination by honeybees, to estimate honeybee nectar flow, etc. An option is to use an indirect way to detect the beginning of the vegetation period and then predict the flowering dates from the known meteorological data.

The possibility to detect flowering in visible spectral bands depends on the colour of flowers and on the cover fraction of flowers exposed in the view (nadir) direction. To detect flowering, a rather frequent temporal reflectance sampling is needed. In the NIR region, the pigments causing the coloration of flowers do not have a primary effect on reflectance. However, the scattering properties of leaves and flowers could be different in the NIR band as well, mostly because of their different structure. Floral reflectance databases are available on the Web, such as the FReD http://www.reflectance.co.uk//. Typically, these databases do not contain data in the NIR region or for the main tree species of the boreal and temperate region.

Detection of flowering by coarse resolution satellite scanners can be successful, if the cover fraction of flowers in the pixel is sufficiently large, which seems to be a rather hopeless task. On a subpixel scale, it is also possible to detect flowering if a linear unmixing method (Singer and McCord, 1979) is used and the spectral signatures of the flowers are known to define the respective endmember.

2.2 Seasonal forest reflectance modeling

Analyzing the role of each of the contributing factors (see Section 2.1) can only be achieved by linking stand-level forest reflectance modelling to empirical reflectance and forest inventory data sets. Forest reflectance models are parameterized using mathematical descriptions of tree canopy and understory structure and optical properties of plant elements to produce spectral signatures of the canopy leaving radiation field (see Stenberg et al., 2008b or Widlowski et al., 2007 for an overview of canopy reflectance models). Forest reflectance simulations can be applied, for example, (1) to understand how the spectral signature of a stand is formed, (2) to investigate quantitative relationships between remotely sensed reflectance data and forest attributes, and (3) to simulate seasonal and age courses in forest reflectances (Nilson et al., 2003). Even though forest reflectance models have existed for three decades, there is a lack of detailed *in situ* ground reference data sets for the validation of the seasonal simulation results. Thus, many quantitative links between seasonal changes in canopy structural and biochemical properties and forest reflectance are yet to be established.

Currently, there are only a handful of published studies, which have applied forest reflectance models to analyze the driving factors of reflectance seasonality. Based on empirical reflectance data and radiative transfer modelling, Kobayashi et al. (2007) reported that seasonal changes in a larch forest understory (or forest floor) influence both the absolute values of reflectance and seasonal reflectance trajectories even if the LAI of the tree layer remains constant. Next, Nilson et al. (2008) and Suviste et al. (2007) reported that in sparse stands (e.g. peatlands), the seasonal course of Sun elevation can be a significant contributing factor to reflectance seasonality i.e. the continuous change in Sun angle typically results in a bell-shaped seasonal reflectance trajectory for a forest. In addition, they reported that the typical seasonal course of canopy LAI results in a bowl-shaped seasonal reflectance trajectory in the visible bands. Most recently, Rautiainen et al. (2009) studied hemiboreal birch forests and concluded that the main driving factors in forest reflectance seasonality in the red and green bands were stand LAI and leaf chlorophyll content, in the NIR band stand LAI, and in the shortwave infrared (SWIR) band LAI and general water content. Furthermore, they suggested that future work on detecting boreal forest phenological phases from satellite images should be focused on quantifying the role of debris and forest floor water on stand reflectance during the early phases of leaf development in the spring, and better characterizing the highly variable surface roughness of the understory layer.

A general challenge in studies which link forest reflectance modeling to seasonal reflectance changes is that expert guesses and interpolation often need to be used to determine many of the input parameters (e.g. biochemical changes in leaves during the growing season, structural changes in forest understory plants), and thus only preliminary conclusions can be drawn from the simulations.

2.2.1 Boreal and temperate forests as multi-component plant communities

Boreal and temperate forests should be treated as multi-component plant communities in forest reflectance modelling. To understand how the seasonal changes in reflectance of forest communities are formed one should be able to describe the seasonal changes occurring in all of the community components. A complete structural model of a typical forest community as an input for a forest reflectance model should include a quantitative description of the following components:

1. The first (dominating) tree layer, separately for all (main) tree species;
2. The second (and possibly third) tree layer, separately for all species;
3. Understory vegetation, including the tree regeneration, shrub and ground layer vegetation (different herbs, dwarf shrubs, lichens, mosses);
4. Soil surface including the layer of litter on the surface;

Each of the vegetative components changes its appearance through the growing season. From the optical point-of-view, the changes in phenology induce the following structural and optical changes:

1. Growth of new leaf area, stems and branches, emergence of flowers and fruits, as well as senescence and dying of leaves and other components or whole plants in overstory and understory. It results mainly in changes of cover fraction of different components (through the effects of mutual shadowing) as seen from the view direction of the satellite sensor;
2. Changes in leaf inclination and orientation along with the growth of new leaves and branches. This has an effect on redistribution of radiation within the plant community and on scattering properties of canopy elements;
3. Changes in colour and surface structure of leaves and other organs mainly due to changes in pigmentation or in other optically active chemical compounds (water, organic compounds);
4. Changes in soil optical properties mainly due to changes in ground surface wetness, appearance of new litter on soil surface and decay of old litter;
5. Changes in spatial distribution of plants and their elements, clumping or surface roughness in overstory, understory and soil.

This is a complex set of changes. Although all these changes are known in principle, reliable data for their quantitative description are typically lacking. In addition, successional transitions and disturbance-caused changes contribute to the seasonal course as well.

2.3 Creating seasonal time series of higher resolution satellite images

Interpreting seasonal reflectance dynamics of forests from satellite images requires processing multitemporal satellite images to create a smoothed time series of reflectance factors. Several coarse resolution satellite scanners, such as MODIS, MERIS, etc. provide multispectral reflectance data that can be arranged into seasonal series and analysed in terms of changes in phenology and view geometry. Because of coarse spatial and temporal resolution, these data are well fitted to analyse seasonal changes at global and regional scales. However, we also need to study seasonal changes at much finer scales, e.g. at a forest stand level. Due to relatively low revisit capability of most popular scanners (Landsat TM and ETM+, SPOT) and frequent occurrence of clouds, the reconstruction of reliable seasonal courses of reflectance in different spectral bands in boreal and temperate regions is practically impossible during a particular growing season. New emerging higher-resolution satellite-borne scanner systems have higher revisit frequency, and thus, could provide high spatial resolution seasonal reflectance time series in the near future.

Currently, the only feasible alternative is to derive the seasonal reflectance series averaged over a certain period of years, and to produce a seasonal series called the "climatological" mean seasonal reflectance course. It is possible to accumulate cloudless images over several years covering the whole vegetation period. Using this method, it is possible to create a time series for, for example, middle-aged and mature forests, since there is a long period of stand age when the reflectances are stable (Nilson & Peterson, 1994). A natural way to produce the time series is to calibrate all the images into the units of TOC reflectance factors by making use of the absolute calibration and atmospheric correction of images. In spite of apparent success in the atmospheric correction procedures, the resulting time series is typically scattered and may need some additional smoothing.

Nilson et al. (2007) proposed a method for smoothing such multi-year time series using the assumption that some forest types have smooth average seasonal reflectance curves during the snowless period of the year. A smoothing procedure was used to correct the two calibration coefficients defining the linear relation between TOC reflectance and the digital number (DN) representation for each image in each spectral band. The corrected calibration coefficients were selected so that the squared sum of residuals between the smoothed seasonal curve and initial TOC reflectance values for a selected sample set of stands was minimized. The selected stands in Nilson et al. (2007) included middle-aged and mature forests growing on fertile sites as dark targets and a selection of pine bogs as a bright target. Next, the corrected calibration constants were applied to the time series of images for all the study stands.

When images from multiple years are used, a problem of phenology differences between the years arises. To account for the interannual differences in meteorological conditions, a concept of accumulated or growing degree-days (GDD) or temperature (also called as physiological or phenological) time has often been used. There are several methods to calculate GDD. Here we use a simple method where daily mean air temperatures from the nearest meteorological station exceeding a predetermined level (5°C in our case) are cumulated starting from the day when the daily mean air temperature permanently exceeds the 5°C level. Using GDD as the time axis reduces the air temperature differences between years; however, the differences in precipitation and air humidity remain.

To test which time axis should be preferred in analyzing the resulting time series, Nilson et al. (2008) applied the smoothing procedure in two different ways: one version when the GDD time axis was used, and the second version with the day of year (DOY) used as the seasonal time axis. When describing the quality of smoothing by the residual standard deviation of measured reflectance factors with respect to the smoothed curve, it appeared that both methods gave nearly the same value of standard deviation. However, the GDD time axis seemed to perform somewhat better in the first half of the vegetation period. According to phenological observations of the Estonian Plant Breeding Institute (L. Keppart, pers. comm.) in Jõgeva (60 km away from the main study site in Järvselja, Estonia), a considerable variation in arrival dates of main phenophases of tree species have been observed during 1986-2003. For instance, on average DOY=129 was the beginning of flowering in Silver birch and the standard deviation of that date was 7.4 days. When recalculated into GDD, the average flowering date was at GDD=87 degree days, the standard deviation being 27. Since the mean daily contribution into GDD for that period of time is 5°C, the standard deviation of the flowering date is about 5.4 days. This confirms that the use of GDD axis to plot the seasonal reflectance series is preferable at least in the first half of the vegetation period.

After the smoothed calibration coefficients have been determined for all images in the seasonal series, it is possible to analyse the "average" seasonal reflectance courses of different forest types. Here we can make use of potential forestry databases over the study area. By queries from these databases it is possible to extract forests with similar site type and site index, age class, species composition, etc. Using a digital map of the forest area, the pixels corresponding to the query results can be extracted from the images. Next, the pixel values can be converted from the original DN format into physical radiance units, and finally, via atmospheric correction into the TOC reflectance factors. Aggregation of pixels leads to smoother seasonal curves. If a sufficiently large total area of forests is queried from the database, one can expect smooth seasonal curves (Nilson et al., 2008). However, as can be seen in our case study (Section 3), more or less smooth seasonal curves can be obtained even to describe the average ("climatological") seasonal course of an individual forest stand.

3. Case study

The aim of the case study is to demonstrate how seasonal reflectance courses of forests can be obtained from high and coarse resolution satellite images and how they can be used to examine the phenological changes taking place in hemiboreal and boreal forests. First, we examine stand-level seasonal courses of reflectance and spectral vegetation indices for several forest types in Estonia during the snow-free period (typically from April/May until October). Next, we look at seasonal reflectance changes of typical hemiboreal understory communities, interpret the changes in the community by changes of aspecting species and discuss the problems of flowering detection. Finally, we examine landscape-level dynamics of forest reflectance seasonality using MODIS data from Estonia and Finland.

3.1 Study areas

In this case study, we have two study areas. The first study area is located in Järvselja Training and Experimental Forestry District (Estonia, 58.15°N, 27.28°E), a representative of the lush hemiboreal zone forest. The dominating tree species are Silver birch, Scots pine, and Norway spruce. The site types vary over a large range from fertile types to infertile transitional and raised bogs and lowland mires (Nilson et al., 2007). The Järvselja area is used in all parts of the case study. In addition, ground-based radiometric measurement data from Peterson (1989-1990) is on grass communities of clearcut forests in the Konguta Forest District, Estonia (relatively close to Järvselja) are included.

The second study area is located in Hyytiälä (Finland, 61.83° N, 24.28° E) and represents typical southern boreal vegetation. The stands range from herb-rich Silver birch forests to xeric Scots pine forests with understory dominated by lichens and heather. A more detailed description of the Hyytiälä forest is available in Rautiainen et al. (2011b). The Hyytiälä area is used only in the landscape-level analysis of MODIS data (Section 3.5).

3.2 Stand-level seasonal reflectance courses of forests

3.2.1 Creating a time series of Landsat and SPOT images

From the time period extending from 1986 to 2007, a total of 34 Landsat and SPOT images over the Järvselja area covering practically the whole growing season were collected. The aim was to produce a smooth time series of reflectances, to apply the time series in change detection and to study the primary productivity of different forest types by applying the Monteith hypothesis (Kumar & Monteith, 1981). The data set is described in detail by Nilson et al. (2007).

Scanner and channel	Wavelength region	Wavelength range, nm
TM1	blue	TM: 450-520
TM2 and XS1	green	TM: 520-600, XS 500-590
TM3 and XS2	red	TM: 630-690, XS 610-680
TM4 and XS3	near infrared (NIR)	TM: 780-900, XS 780-890
TM5 and XS4	shortwave infrared (SWIR1)	TM: 1550-1750, XS 1580-1750
TM7	shortwave infrared (SWIR2)	TM: 2080-2350

Table 1. Satellite sensors and spectral bands used to form a time series of forest reflectance for the Järvselja site. [TM = Landsat channels, XS = SPOT channels].

Using the previously described (Section 2.3.1) method of smoothing the image calibration coefficients, seasonal courses of reflectance in six spectral bands corresponding to Landsat TM and ETM+ scanner were produced (Table 1).

After the smoothed calibration coefficients for the whole set of 34 images had been identified, it was possible to study the seasonal course of reflectance and various multispectral indices at different levels of pixel aggregation starting from sample plots covering 1 ha up to areas corresponding to MODIS pixels. Alternatively, queries from the forestry database were used to select forests of the same dominating species, age, site quality, etc.

We demonstrate the seasonal changes in two commonly used multispectral indices (normalized difference vegetation index NDVI (Eq.1) and normalized difference water index NDWI), and two linear indices (Greenness, Wetness). NDWI was calculated as

$$NDWI = (R_{NIR} - R_{SWIR1})/(R_{NIR}+R_{SWIR1}). \qquad (2)$$

The linear indices were defined similar to the tasselled cap transformation by Kauth & Thomas (1976); however, the coefficients of the transformation were calculated for the TOC reflectance factor in all six reflective bands of Landsat TM or four bands of SPOT. The formulas to calculate the indices were for Landsat TM and ETM+ as follows:

$$Greenness = -0.0912\, R_1 - 0.1614\, R_2 - 0.4214\, R_3 + 0.8045\, R_4 - 0.2242\, R_5 - 0.3009\, R_7 \qquad (3)$$

$$Wetness = 0.5078\ R_1 + 0.421\ R_2 + 0.4407\ R_3 + 0.1019\ R_4 - 0.4682\ R_5 - 0.3755\ R_7 + 0.1, \quad (4)$$

where R_i is the TOC reflectance factor in Landsat band *i*. The formulas to calculate the indices are scanner-specific. For the SPOT4 images they were defined as

$$Greenness = -0.2100\ R_1 - 0.5031\ R_2 + 0.7745\ R_3 - 0.3209\ R_4 \quad (5)$$

$$Wetness = 0.4984\ R_1 + 0.4807\ R_2 + 0.1555\ R_3 - 0.7045\ R_4 + 0.1, \quad (6)$$

where Ri is the TOC reflectance factor in SPOT4 band *i*. For SPOT3 images, these indices were not calculated. The value 0.1 was added to the Wetness index to obtain typically positive values of the index.

3.2.2 A comparison of reflectance seasonality in coniferous and deciduous stands

We show the climatological average seasonal course of reflectance in six Landsat bands as functions of the GGD (temperature time) for three different forest types: a Scots pine, Norway spruce and Silver birch forest representing typical hemiboreal vegetation in Järvselja (Fig. 2). A description of the stands is provided in Table 2. The pine and birch stands also serve as test stands within the RAMI-IV (RAdiation transfer Model Intercomparison) effort (Widlowski et al., 2007) (http://rami-benchmark.jrc.ec.europa.eu/ HTML/RAMI-IV/RAMI-IV.php).

According to previous studies (see Section 2 for review), the seasonal course of forest reflectance in the visible bands is primarily controlled by changes in the amount of pigments (mainly chlorophyll) per ground area in the tree layer and understory. In other words, the main driving factor of reflectance seasonality in the visible bands is the green LAI. NIR region is considered to be sensitive to the total amount of leaf tissue or LAI, and SWIR bands to the total amount of water in leaves per ground area.

In general the seasonal curves in Fig. 2 are in agreement with the theory. The seasonal reflectance changes in the two evergreen forest plots, spruce and pine, are less pronounced than in the birch plot. However, the reflectance factors of the two coniferous stands are not constant during the season. There are deciduous species present in the spruce stand (approximately 25 % of trees), so some features of their seasonal course are evident. In general, the seasonal changes of reflectance in pure spruce forests should be smaller than what was observed for the spruce-dominated stand in this case study (Fig. 2). This highlights how the species composition, especially the presence of deciduous species in a coniferous forest, influences stand reflectance and its seasonal trends.

In the deciduous birch stand, the most notable seasonal courses of reflectance are related to emergence of new leaves in spring. In the visible bands a decreasing trend of reflectance is seen mainly caused by growing of green LAI. The seasonal course in the NIR region follows the development of total LAI (canopy + understory). The expected loss of chlorophyll during the second half of the vegetation period is hardly notable. The scatter of points in Fig. 2 is larger during late summer and fall if compared with midsummer. This possibly refers to problems with the atmospheric correction of the images during the period or to the fact that the GDD time axis fails to represent adequately the phenological time. The seasonal trends in the SWIR1 and SWIR2 regions should be explained by changes in the total amount of water in leaves of canopy and understory, and to some extent by moisture conditions of soil surface and litter. In addition, these spectral regions are somewhat sensitive to organic compounds (cellulose, proteins, lignin) in leaves and litter (Asner, 1998; Jacquemoud & Baret, 1990). As the effect of variable atmospheric conditions on the blue band is probably not too reliably eliminated by the atmospheric correction of the images, we have not further analyzed the seasonal course in the blue band.

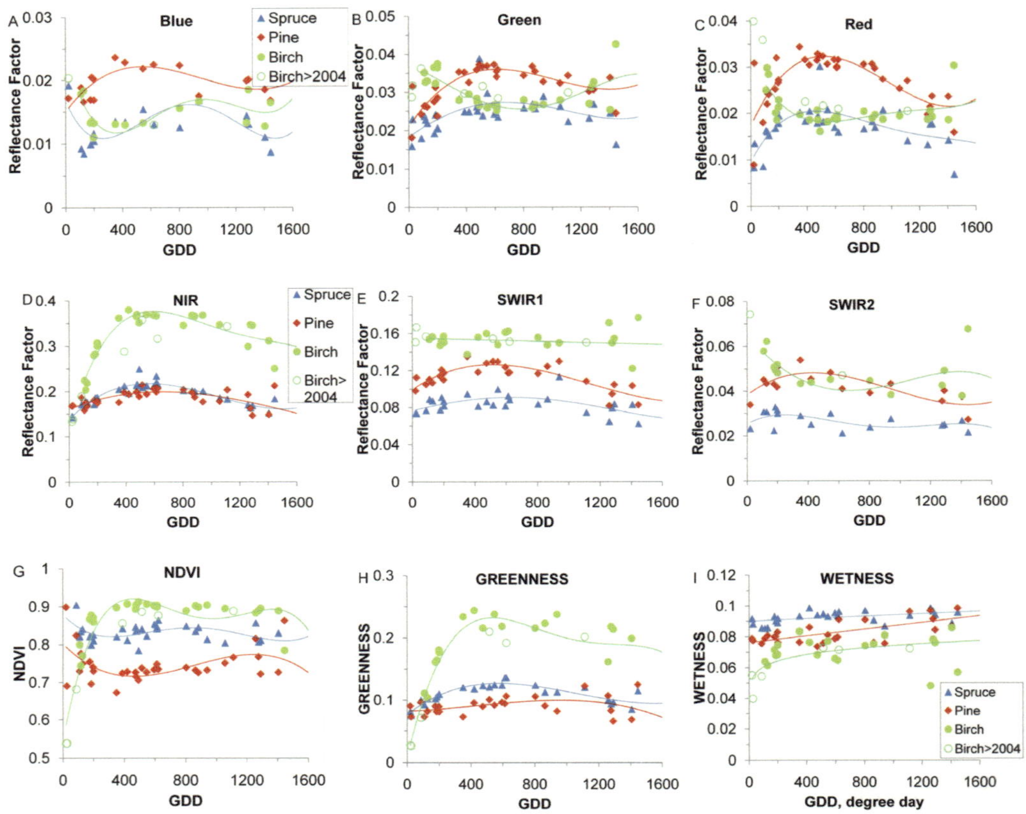

Fig. 2. Seasonal course of reflectance in six spectral bands and three multispectral vegetation indices for three 100x100m sample plots of different dominating tree species (Norway spruce, Scots pine, Silver birch, Table 2) in Järvselja, Estonia. The time axis is in growing degree days (GDD): A - blue band, B - green, C - red, D - NIR, E - SWIR1, F - SWIR2, G - NDVI, H - Greenness, I - Wetness index. Reflectance data for the birch stand after the thinning in 2004 are shown by empty symbols

In the birch-dominated stand, we can see some changes in reflectance caused by thinning during the following years and a stabilization of reflectance after 2-3 years. After the thinning reflectance increased in the visible bands (especially the red band) and decreased in the NIR band. The thinning has changed the absolute values of reflectance in a few spectral bands, but the overall seasonal shape of the curve remained the same.

Some of the seasonal reflectance changes are not easy to explain. What is the reason for the rather pronounced bell-shaped seasonal changes of reflectance in the visible and SWIR bands of the two coniferous stands? This could be partly explained by the dependence of forest reflectance on solar elevation and by seasonal change of the Sun angle during the image acquisition. Why is there a reflectance decrease in the SWIR2 band and almost no change in the SWIR1 band in spring? SWIR bands, especially SWIR1 should be less sensitive to the forest understory than the visible and NIR band (Brown et al., 2000). Leaf water is added along with the development of new leaves, so the reflectance decrease in the springtime is expected. Ground surface moisture is expected to decrease after snowmelt in spring and this effect should be better seen in leafless deciduous forests. Soil surface is drying while new leaves containing water are added, and to some extent these factors cancel out. It is also somewhat problematic to explain why the changes in SWIR2 are more pronounced than in SWIR1. The linear wetness index has a slight increasing trend, and the reason for such a trend is difficult to explain. The autumn coloration of birch leaves is not clearly observed in this dataset, yet in the coarser MODIS data (see below Section 3.5) and in the selection of birch forests (Fig. 3 below) it is clearly seen. Presently, it is difficult to conclude if these seasonal changes are real or if they are partly results of the applied calibration and smoothing procedure.

Stand name	**Dominating tree species in upper layer**	**Tree species in lower layer**	**Stand age (years) in 2007**	**Site type (Paal, 1997) and fertility**	**Understory and soil layer**	**Other remarks**
Pine site	*Pinus sylvestris*	none	124	transitional bog infertile (H100 = 10.8m)	*Ledum palustre, Eriophorum vaginatum, Sphagnum* ssp.	
Spruce site	*Picea abies*	*Betula pendula*	59	*Oxalis-Vaccinium myrtillus,* fertile (H100 = 28.1m)	mosses (*Hylocomium splendens, Pleurozium schreberi*)	High canopy cover (0.89), thus sparse understory
Birch site	*Betula pendula*	*Tilia Cordata, Picea abies*	49	*Aegopodium* fertile (H100 = 28.7m)	several grass species	Thinned in 2004 fall

Table 2. Characteristics of three study sites (100-m x 100-m) in the Järvselja area used to examine climatological average seasonal courses of reflectance (Fig. 2). For more details on these stands see Kuusk et al. (2008, 2009).

The seasonal courses of reflectance and multispectral indices for these forests are similar to those previously averaged over a large number of stands from the Järvselja area (Nilson et al., 2008; Rautiainen et al., 2009). In other words the selected three stands represent well the forest types over the whole study area. The scatter of points in the seasonal curves for these sample plots is somewhat higher than for the respective selection of forest of the same type from the database due to considerable differences in the number of pixels used in averaging. There are minor differences, but these are rather systematic differences in the absolute reflectance values than in the general shape of the seasonal reflectance curves. No time lags in the major developmental phases between the different stands of the same species and site fertility class are notable. For instance, the spruce stand seems to have more chlorophyll and water in tree leaves compared to the selection of spruce forests of the same site index from a local forestry database, but the shape of the seasonal reflectance curve is similar.

3.2.3 Seasonal reflectance courses of birch forests

Next, we used the same time series of Landsat and SPOT images (see Section 3.2.1) to study the pronounced seasonal reflectance courses of deciduous forests in more detail. Our aim was to evaluate the influence of site fertility on reflectance seasonality of birch forests. We selected 145 birch (*Betula* sp.) dominated, approximately 50 year-old stands from the Järvselja area from a database collected by a local forest inventory company (Metsaekspert OÜ in 2001). We grouped the stands into seven classes based on the fertility (so-called Orlov's index) of the site type. More details about the stands can be obtained from Rautiainen et al. (2009).

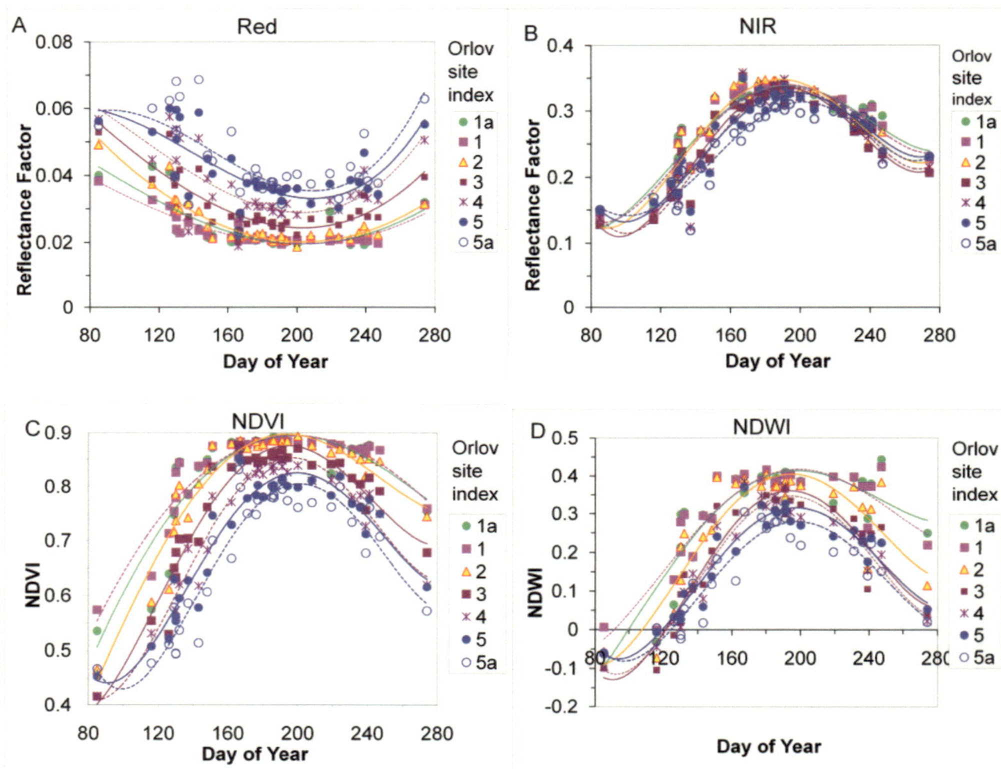

Fig. 3. Seasonal course of reflectance in the red (A) and NIR (B) bands and of the indices NDVI (C) and NDWI (D) for the selection of birch-dominated forests growing on sites of different fertility from Järvselja, Estonia. Site fertility is shown by the Orlov site index: 1a - the most fertile, 5a - most infertile

Our results showed that the more fertile the site type, the lower the reflectance of the stand in the red band and the higher in the NIR band, with the relative differences being larger in the red band (Fig. 3). Among the birch stands of different fertility, the shapes of the NDVI curves are similar, yet some time lag in the green development in the poorest sites can be observed in the spring. Also, the maximum NDVI value is reached later among the forests growing on poor sites. The seasonal integrals of NDVI (see Fig. 1) over the vegetation period can be interpreted as being proportional to the yearly net primary production (NPP), thus confirming the hypothesis by Kumar & Monteith (1981).

3.3 Seasonal reflectance changes in understory vegetation

In a community consisting of several species (e.g. the understory layer of a forest), the seasonal course of a reflectance factor or a multispectral index is formed by the reflectance spectra of all components. Due to extremely high variability of irradiance at the ground level it is difficult to obtain reliable measured reflectance spectra and their seasonal trends for understory vegetation. Typically, pure patches of understory vegetation have been measured (Miller et al., 1997; Peltoniemi et al., 2005; Rautiainen et al., 2011b). As a first approximation, the linear mixture approach can be used with cover fractions of community components serving as weights of the components. According to theory, linear approximation should work better in the visible and SWIR bands, where the first-order scattering with its linear nature is dominating. In the NIR region, because of non-linear effects in reflectance, mostly due to multiple scattering of radiation within the plant community, the reflectance spectrum of a community is not exactly a linear mixture of the spectra of its components.

If the aim of the analysis is to estimate the roles of different components in the community in forming the spectrum of reflected radiation, it is logical to use different linear multispectral indices, such as the Brightness, Greenness, Wetness etc. indices suggested by Kauth & Thomas (1976). Linear multispectral indices preserve all linear features if they exist at a single band level.

To demonstrate this we present results of ground-based spectral reflectance measurements conducted by Peterson (1989) with a four-channel field radiometer in a species-rich grass ground-elder (*Aegopodium podagraria*) dominated community growing on a fresh boreo-nemoral (Paal, 1997) site in a four-year-old clearcut in Konguta, south-central Estonia.

The same sample plots (ca $2m^2$) were measured throughout the whole growing season. The cover fractions of individual species were visually estimated three times during the season. Following the ideas by Kauth & Thomas (1976) the index Greenness was defined for the four-channel field radiometer used in these measurements as follows:

$$\text{Greenness} = -0.2923\,R_1 - 0.2773\,R_2 - 0.5775\,R_3 + 0.7101\,R_4, \quad (7)$$

where R_1, R_2, R_3 and R_4 are the reflectance factors in bands 1 - blue (480 nm), 2 - green (553 nm), 3 - red (674 nm) and 4 - near infrared (780 nm). The cover fractions were used as weights to linearly mix the Greenness values of different aspecting species as measured from the pure plots.

The contribution of different aspecting species to the seasonal profile of the Greenness index is shown in Fig. 4 (reproduced from Peterson, 1989). This example shows us that the whole-community Greenness course may be more or less a smooth curve, in spite of the gradual transition from one aspecting species to another. So, for this particular community, it is very difficult to identify the domination of one or another aspecting species by measuring the seasonal profiles of reflectance of the whole community. The approximate linear mixing of Greenness values of different community species yields an acceptable agreement with the Greenness values of the whole community. This has little practical importance, since it is extremely laborious to measure cover values of different species in a community. However, it shows that linear mixture method can also be applied at the tree canopy level and with spatially aggregated pixels. For example, we can create the spectral signatures and seasonal reflectance courses of mixed forests using the spectral data of monospecific forests and inventory data on species composition. However, there are likely to exist specific community-level structural effects, caused by the fact that a mixed stand is not just a mechanical mixture of different species. This probably results in deviations from the linear mixture. Another question is how well the species composition given in a forestry database reflects the crown coverage (which is important in defining the proper weights of the components) of different species seen by remote sensing instrument.

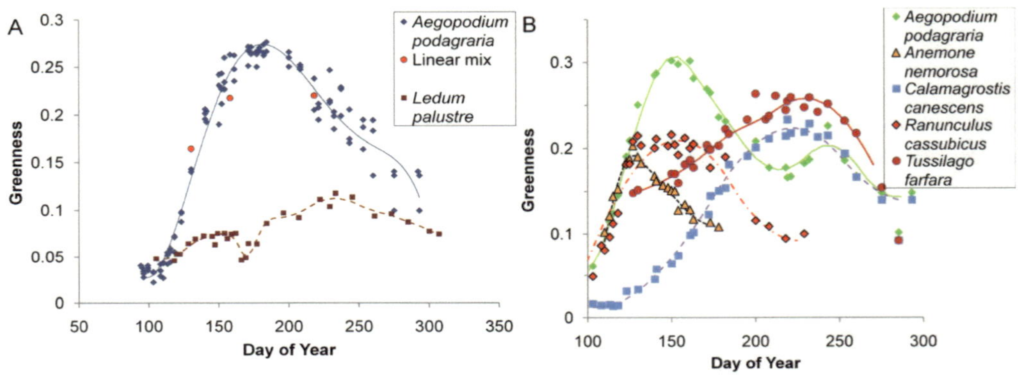

Fig. 4. A: Seasonal course of Greenness of a ground-elder (*Aegopodium podagraria*) dominated clearcut community growing on a fresh boreo-nemoral site compared to a Labrador tea (*Ledum palustre*) dominated community growing on a infertile transitional bog site. Red dots have been calculated by linearly mixing Greenness of different aspecting species in the ground-elder community. B: Greenness course of different aspecting species in the *Aegopodium* community. The pure patches of the respective species were maintained by periodic clipping of other species. Results of ground-based measurements by Peterson (1989)

3.4 Detection of phenophases from reflectance data

So far, the majority of studies related to phenology deal with the onset of greening. Can other key phenophases, such as flowering, emergence of new shoots in conifers and beginning of dormancy be detected? For instance, new shoots in spruce have a lower chlorophyll content and are well exposed when viewed from above, so the moment when they have fully emerged should be, in principle, detectable. However, from the multiyear Landsat and SPOT data set, this moment is very difficult to identify (Fig. 2). Even if the series contained images exactly corresponding to the period where new shoots with less chlorophyll emerged (or with tree flowers exposed), these effects may be partly lost as a result of the present smoothing procedure.

As detecting flowering of tree species from satellite images has been shown to be difficult, we will examine next if it would be possible to identify the moment of flowering of smaller plants from high resolution data collected at the ground level. A radiometer was used to measure the reflectance of a Labrador tea (*Ledum palustre*) dominated grass community in four spectral bands (Peterson, 1989). The community was located in a forest clear cut in a wet oligotrophic site in Konguta, Estonia. The size of the measured plot was ca 2 m^2, and thus, the detection of flowering was conducted at an extremely high spatial resolution.

At DOY≈170 a significant peak is clearly seen in all the three visible bands confirming the visual impression of mass flowering (Fig. 5). The flowers of Labrador tea are white, so the peak in all visible bands was expected. In the NIR band a simultaneous slight reflectance decrease during intensive flowering can be noticed; it is likely that Labrador tea flowers scatter less radiation in the NIR band than the leaves. In this particular case, the estimated fraction of flowers at the peak time of flowering was 21% while the total coverage by Labrador tea was 66%. Taking into account that reflectance of white flowers can exceed that of green leaves up to ten times in the visible bands, a detectable signal from flowers can happen at rather low cover values of flowers. However, the temporal frequency of acquiring images must be high enough to catch the peak of flowering.

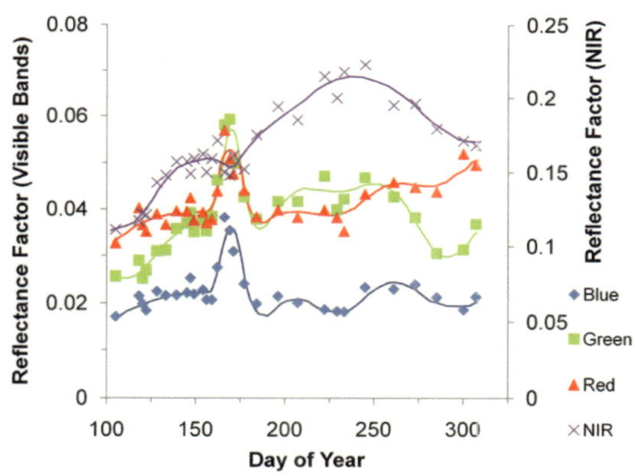

Fig. 5. The seasonal reflectance course of a *Ledum palustre* dominated clearcut community. The effect of flowering on is clearly seen at DOY≈170 as a significant peak in the visible bands (Peterson, 1989)

Is it possible to detect the mass flowering of the same Labrador tea community through the canopy of pine trees (as it could be for our pine stand discussed previously in Fig. 2, Table 2)? This is extremely difficult to achieve when the "climatological" seasonal series are used, as it is the case in this study. From model simulations by the forest reflectance model FRT (Kuusk & Nilson, 2000), we can conclude that even if the understory reflectance in the red band increased by 50% (like in Fig. 5) because of flowering of Labrador tea, the nadir reflectance of the stand would increase only by 22%. Although the measured canopy cover of the stand was rather high - (0.74), the pine crowns were relatively transparent, and the total gap fraction in the vertical direction in the canopy was 0.46. So, the understory contributes considerably to the reflected signal from the nadir in the stand. If the cover of Labrador tea were 100%, its flowering could be detectable through the pine canopy. However, in reality, although Labrador tea is the dominating species in the understory of the pine stand, its cover fraction is less than 0.20. So, even its mass flowering is extremely difficult to detect from satellite data, since the expected change is of the same order of magnitude with the reflectance measurement error.

3.5 Landscape-level seasonal reflectance courses of forests

A landscape-level analysis of reflectance seasonality was carried out with MODIS data. The MODIS BRDF/Albedo Product (MCD43A1, version 5) (Lucht et al., 2000; Schaaf et al., 2002) was used to reconstruct the nadir reflectances every 8 days at 500 m resolution over Järvselja and Hyytiälä areas in 2006 (dry summer) and 2010 (normal summer). The sun zenith angle was determined for each day for 6:15 GMT i.e. roughly corresponding to the Landsat overpass time. Only the best quality retrievals (QA=0 in the associated MCD43A2 product) were used. Additionally, phenology dates (greenness increase, maximum, decrease, and minimum) were obtained from MODIS MCD12Q2 product for year 2006. Our aim was to use these data to compare the seasonal reflectance curves in two different locations with supposed time lags in phenology and to compare seasonal curves in two years with different precipitation. The idea of comparing the seasonal reflectance courses from MODIS data and from Landsat/SPOT series was to show the reliability of image calibration and to examine the effects of spatial averaging in a fragmented landscape on the seasonal reflectance profiles.

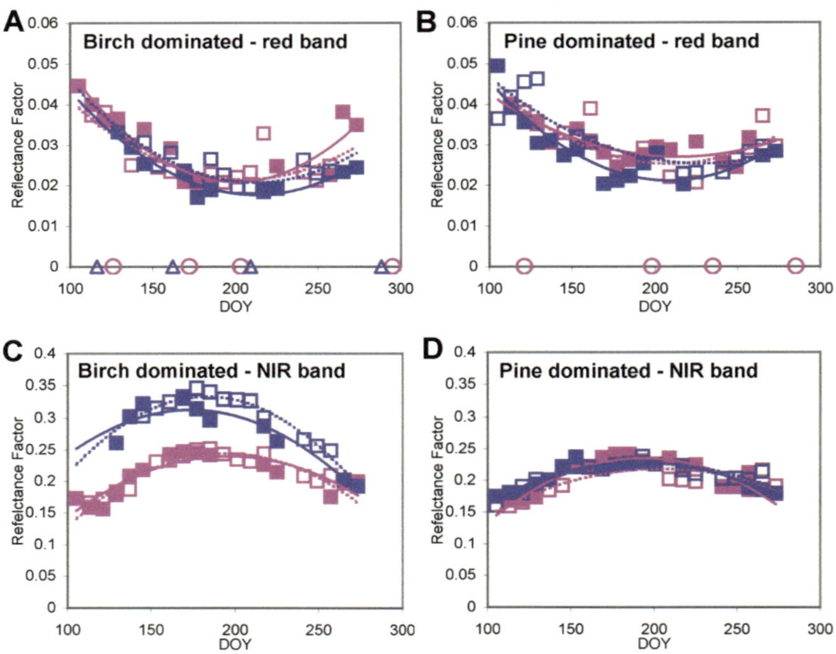

Fig. 6. Comparison of Järvselja (blue squares) and Hyytiälä (purple squares) seasonal reflectance curves for MODIS pixels with birch (A, C) and pine dominated stands (B,D) for years 2006 (dry summer-empty symbols) and 2010 (typical summer-filled symbols). If available, the phenology dates (greenness increase, maximum, decrease, and minimum) from MODIS MCD12Q2 product for year 2006 are also marked (Järvselja - empty triangles; Hyytiälä - empty circles)

First, no clear time lag in the green development can be observed from MODIS reconstructed retrievals between Järvselja and Hyytiälä (Fig. 6). This may be partly due to rather long compositing period of the MODIS BRDF product (16 days). The green development at Järvselja preceded Hyytiälä by around 10 days in 2006 according to the MODIS MCD12Q2 phenology product. Järvselja stands also show more rapid reflectance change in early spring, and greater differences in the absolute values of reflectance between the year with dry (2006) and typical (2010) summer (Fig. 6). The more southerly Järvselja stands thus appear to be more sensitive to the variation in available moisture than Hyytiälä. The pronounced difference in amplitude between the birch dominated stands in Järvselja and Hyytiälä in the NIR band (Fig. 6C) is due to the higher share of coniferous stands in the Hyytiälä MODIS pixel. Finally, the differences in the red band (Fig. 6A,B) indicate the general higher vegetation abundance in the Järvselja stands.

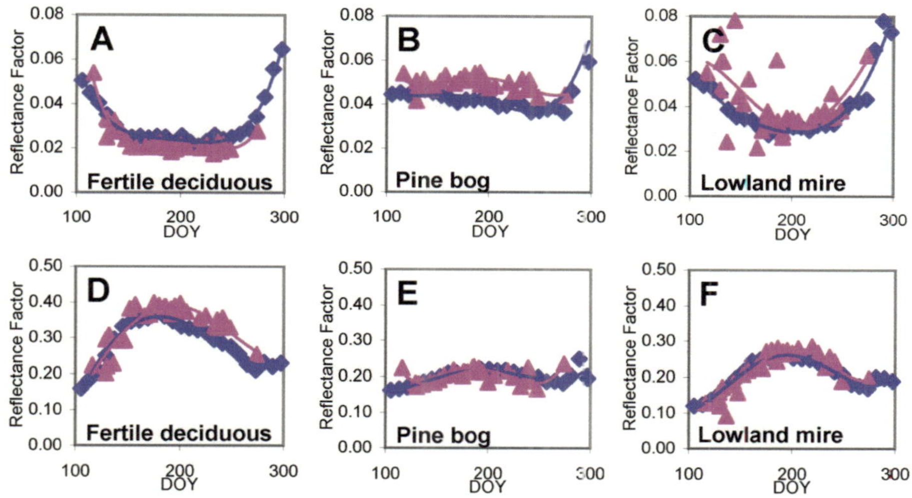

Fig. 7. Comparison of Landsat-SPOT red and NIR band time series (purple triangles) and average MODIS seasonal curves for period 2000-2010 (blue diamonds) for a few typical MODIS pixels from Järvselja. A, B, C - red band; D, E, F - NIR band

A comparison of MODIS seasonal curves and Landsat-SPOT time series for a few typical MODIS pixels from Järvselja is shown in Fig. 7. The course of seasonal curves may agree quite well, as in case of the fertile deciduous black alder and birch dominated stand (Fig. 7A, D). The rapid reflectance increase in red band (Fig. 7A) during the fall (DOY 280-300) is due to the loss of chlorophyll in broadleaf trees and understory. A similar increase can also be observed for averaged MODIS reflectances over a pine bog (Fig. 7B). Most of the MODIS retrievals over the pine bog during the fall period have been flagged to be of low quality and were subsequently left out of the averaging procedure, while the remaining retrievals sometimes indicated even early presence of snow. The effect of spatial averaging can be illustrated over the pine bog stand as well. MODIS does not sample from the same footprint every time it passes over, which increases the uncertainty quite considerably (Tan et al., 2006). In fact, a single pixel value may have been obtained from up to 4 different pixels. This complicates the situation particularly in heterogeneous areas in visible bands (Fig. 7B). In the NIR band, the differences are usually smaller due to the weaker contrast and stronger multiple scattering effect

(Fig. 7E,F) (Pinty et al., 2002). As for the lowland mires, the large scattering in the red band of Landsat-SPOT time series (Fig. 7C) is caused by water level, which varies a lot in Järvselja from year to year, especially in the first half of the vegetation period.

The heterogeneity of the landscape seems to be the main reason why, for some other MODIS pixels in Järvselja such as a treeless bog (results not shown here), the comparison between the MODIS-retrieved seasonal reflectance courses and that derived from the Landsat/SPOT series showed systematic differences. Pixels with typically high reflectance (especially in the visible bands, such as treeless bogs) are extremely sensitive with respect to possible contribution by neighbouring dark forest areas. Such contaminating effects can influence not only the absolute reflectance values at different moments of the season, but can modify the seasonal curves as well and even cause shifts in the key phenology dates determined from the seasonal reflectance curves.

Thus, at the landscape level, the seasonal course of a forest pixel depends much on the share of coniferous and deciduous forests and of treeless (e.g. clear-cut areas, meadows, agricultural land) areas within the pixel. Once again, to explain the seasonal course of reflectance in mixed pixels a linear mixture approach is possible.

4. Future perspectives and conclusions

Until now, most remote sensing studies have concentrated on detecting the onset of greening from satellite data. Now is the time to move beyond the spring-summer time period i.e. to include the full growing period of forests in the analyses, as shown in our case study. In this way, we can develop satellite-based monitoring methods for forest phenology to link the observed changes in reflectance and phenological development of different species to, for example, carbon stocking and photosynthesis.

The main future challenges in monitoring seasonal changes of forests using reflectance data are related to the methods used for interpreting satellite data. As we have shown in this chapter, the theoretical understanding of the physical phenomena is not yet mature enough even though the key driving factors of forest reflectance seasonality have been qualitatively identified. Detecting short phenological phases (e.g. flowering) or phenological phases that are fragmented in the landscape (e.g. yellowing of leaves) is not trivial. Currently, the onset of greening seems to be the easiest, yet still not routinely successful, to detect.

Another challenge is related to data sets. We need to generate spatially and temporally continuous and consistent time series of satellite data for monitoring the seasonal changes. However, the temporal resolution of satellite images is often poor due to the presence of clouds. This has consequences in understanding the seasonality of primary production in regions with a persistent cloud cover, such as boreal and tropical forests. In addition, we need to develop a method for validating the seasonal changes observed in satellite images. Currently, there are no (global) landscape-level ground reference data for validating remotely sensed seasonal courses - maybe the RGB-analysis of landscape-level photos (e.g. automatically taken by webcams in flux towers) will be a step forward?

5. Acknowledgment

The support by Estonian Science Foundation grants no ETF8290 and ETF7725, ESA/ESTEC contract number 4000103213/11/NL/KML, Academy of Finland, University of Helsinki

Research Funds and Emil Aaltonen Foundation is acknowledged. MODIS MCD43 and MCD12Q2 product tiles were acquired from Land Processes Distributed Active Archive Center (LP DAAC).

6. References

Ahl, D.; Gower, S.; Burrows, S.; Shabanov, N.; Myneni, R. & Knyazikhin, Y. (2006). Monitoring spring canopy phenology of a deciduous broadleaf forest using MODIS. *Remote Sensing of Environment*, Vol. 104, pp. 88–95, ISSN 0034-4257

Asner, G.P. (1998) Biophysical and biochemical sources of variability in canopy reflectance. *Remote Sensing of Environment*, Vol. 64, pp. 234-253.

Brown, L.; Chen, J.M.; Leblanc, S.G. & Cihlar, J. (2000). A Shortwave Infrared Modification to the Simple Ratio for LAI Retrieval in Boreal Forests: An Image and Model Analysis. *Remote Sensing of Environment*, Vol. 71, pp. 16–25, ISSN 0034-4257

Chen, J.M. & Black, T. (1992). Defining leaf area index for non-flat leaves. *Plant, Cell and Environment*, Vol. 15, pp. 421-429, ISSN 0140-7791

Crist, E.P. (1984). Effects of cultural and environmental factors on corn and soybean spectral development patterns. *Remote Sensing of Environment*, Vol. 14, pp. 3-13, ISSN 0034-4257

Demarez, V.; Gastellu-Etchegorry, J.P.; Mougin, E.; Marty, G.; Proisy, C.; Dufrene, E. et al. (1999). Seasonal variation of leaf chlorophyll content of a temperate forest. Inversion of the PROSPECT model. *International Journal of Remote Sensing*, Vol. 20, pp. 879–894, ISSN 0143-1161

Dye, D. & Tucker, C. (2003). Seasonality and trends of snow-cover, vegetation index, and temperature in northern Eurasia. *Geophysical Research Letters*, Vol. 30, No. 7, pp. doi:10.1029/2002GL016384, ISSN 0094-8276

Eklundh, L.; Harrie, L. & Kuusk, A. (2001). Investigating relationships between Landsat ETM+ sensor data and leaf area index in a boreal conifer forest. *Remote Sensing of Environment*, Vol. 78, pp. 239-251, ISSN 0034-4257

Eriksson, H.; Eklundh, L.; Kuusk, A. & Nilson, T. (2006). Impact of understory vegetation on forest canopy reflectance and remotely sensed LAI estimates. *Remote Sensing of Environment*, Vol. 103, pp. 408–418, ISSN 0034-4257

Friedl, M. et al. (2006). Land surface phenology NASA white paper. In: *NASA documents*, 15.8.2011, Available from : http://landportal.gsfc.nasa.gov/Documents/ESDR/Phenology_Friedl_whitepaper.pdf.

Ganguly, S. et al. (2010). Land surface phenology from MODIS : Characterization of the Collection 5 global land cover dynamics product. *Remote Sensing of Environment*, Vol. 114, pp. 1805-1816, ISSN 0034-4257

Gastellu-Etchegorry, P.; Guillevic, P.; Zagolski, F.; Demarez, V.; Trichon, V.; Deering, D. & Leroy, M. (1999). Modeling BRF and radiation regime of boreal and tropical forests : I. BRF. *Remote Sensing of Environment*, Vol. 68, 281-316, ISSN 0034-4257

Guyon, D.; Guillot, M.; Vitasse, Y.; Cardot, H.; Hagolle, O.; Delzon, S. & Wigneron, J.-P. (2011) Monitoring elevation variations in leaf phenology of deciduous broadleaf forests from SPOT/VEGETATION time-series. *Remote Sensing of Environment*, Vol. 115, pp. 615-627, ISSN 0034-4257

Jacquemoud, S. & Baret, F. (1990). PROSPECT: a model of leaf optical properties spectra. *Remote Sensing of Environment*, Vol. 34, pp. 75-91, ISSN 0034-4257

Jönsson, P. & Eklundh, L. (2002). Seasonality extraction by function fitting to time series of satellite sensor data. *IEEE Transactions on Geoscience and Remote Sensing*, Vol. 40, pp. 1824–1832, ISSN 0196–2892

Jönsson, A.; Eklundh, L.; Hellström, M.; Bärring, L. & Jönsson, P. (2010) Annual changes in MODIS vegetation indices of Swedish coniferous forests in relation to snow dynamics and tree phenology. *Remote Sensing of Environment*, Vol. 114, pp. 2719-2730, ISSN 0034-4257

Kauth, R.J. & Thomas, G.S. (1976). The tasseled cap --a graphic description of the spectral-temporal development of agricultural crops as seen in Landsat. *Proceedings on the Symposium on Machine Processing of Remotely Sensed Data*, West Lafayette IN, June 29 -- July 1, 1976, pp. 41-51.

Keeling, C. et al. (1996). Increased activity of northern vegetation inferred from atmospheric CO2 measurements. *Nature*, Vol. 382, pp. 146-149, ISSN 0028-0836

Knyazikhin, Y.; Martonchik, J.; Myneni, R.; Diner, D. & Running, S. (1998). Synergistic alogorithm for estimating vegetation canopy leaf area index and fraction of absorbed photosynthetically active radiation from MODIS and MISR data. *Journal of Geophysical Research*, Vol. D103, pp. 32 257-32 275, ISSN 0148-0227

Kobayashi, H.; Suzuki, R. & Kobayashi, S. (2007). Reflectance seasonality and its relation to the canopy leaf area index in an eastern Siberian larch forest: Multi-satellite data and radiative transfer analysis. *Remote Sensing of Environment*, Vol. 106, pp. 238-252, ISSN 0034-4257

Kodani, E.; Awaya, Y.; Tanaka, K. & Matsumura, N. (2002). Seasonal patterns of canopy structure, biochemistry and spectral reflectance in a broad-leaved deciduous Fagus crenata canopy. *Forest Ecology and Management*, Vol. 167, pp. 233-249, ISSN 0378-1127

Kumar, M. & Monteith, J.L. (1981). Remote sensing of crop growth. In: *Plants and the daylight spectrum*, Smith (ed), pp. 133–144, Academic Press, London

Kuusk, A.; Lang, M.; Kuusk, J.; Lükk, T.; Nilson, T.; Mõttus, M. et al. (2008). Database of optical and structural data for the validation of radiative transfer models. Technical Report. Tartu Observatory, 52 pp, Retrieved from http://www.aai.ee/bgf/jarvselja db/jarvselja db.pdf.

Kuusk, A.; Kuusk, J. & Lang, M. (2009). A dataset for the validation of reflectance models. *Remote Sensing of Environment*, Vol. 113, pp. 889–892, ISSN 0034-4257

Kuusk, A. & Nilson, T. (2000). A directional multispectral forest reflectance model. *Remote Sensing of Environment*, Vol. 72, pp. 244–252, ISSN 0034-4257

Lucht, W.; Schaaf, C.B. & Strahler, A. H. (2000). An algorithm for the retrieval of albedo from space using semiempirical BRDF models. *IEEE Transactions on Geoscience and Remote Sensing*, Vol. 38, pp. 977–998, ISSN 0196–2892

Manninen, T. & Stenberg, P. (2009). Simulation of the effect of snow covered forest floor on the total forest albedo. *Agricultural and Forest Meteorology*, Vol. 149, pp. 303-319, ISSN 0168-1923

Menzel et al. (2006). European phenological response to climate change matches the warming pattern. *Global Change Biology*, Vol. 12, pp. 1969-1976, ISSN 1354-1013

Middleton, E.; Sullivan, J.; Bovard, B.; Deluca, A.; Chan, S. & Cannon, T. (1997). Seasonal variability in foliar characteristics and physiology for boreal forest species at the five Saskatchewan tower sites during the 1994 Boreal Ecosystem-Atmosphere Study. *Journal of Geophysical Research*, Vol. 102, No. D24, (December 1997), pp. 28831-28844, ISSN 0148-0227

Miller, J.; White, P.; Chen, J.M.; Peddle, D.; McDemid, G.; Fournier, R.; Shepherd, P.; Rubinstein, I.; Freemantle, J.; Soffer, R. & LeDrew, E. (1997). Seasonal change in the understory reflectance of boreal forests and influence on canopy vegetation indices. *Journal of Geophysical Research*, Vol. 102, No. D24, (December 1997), pp. 29475 – 29482, ISSN 0148-0227

Moulin, S.; Kergoat, L.; Viovy, N. & Dedieu, G. (1997). Global-Scale Assessment of Vegetation Phenology Using NOAA/AVHRR Satellite Measurements. *Journal of Climate*, Vol. 10, pp. 1154-1170, ISSN 0894-8755

Myneni, R.; Asrar, G.; Tanré, D. & Choudhury, B. (1992). Remote sensing of solar radiation absorbed and reflected by vegetated land surfaces. *IEEE Transactions on Geoscience and Remote Sensing*, Vol. 30, No. 2, pp. 302-314, ISSN 0196-2892

Nightingale, J.; Esaias, W.; Wolfe, R.; Nickeson, J. & Ma, P. (2008). Assessing honey bee equilibrium range and forage supply using satellite-derived phenology. *Proceedings of IEEE International Geoscience and Remote Sensing Symposium*, Boston, MA, July 7-11, 2008, pp. III-763-766, doi:10.1109/IGARSS.2008.4779460.

Nilson, T.; Anniste, J.; Lang, M. & Praks, J. (1999). Determination of needle area indices of coniferous forest canopies in the NOPEX region by ground-based optical measurements and satellite images. *Agricultural and Forest Meteorology*, Vol. 98-99, pp. 449-462, ISSN 0168-1923

Nilson, T.; Kuusk, A.; Lang, M. & Lükk, T. (2003). Forest reflectance modeling : theoretical aspects and applications. *Ambio*, Vol. 33, No. 8, (December 2003), pp. 534-540, ISSN 0044-7447

Nilson, T.; Lükk, T.; Suviste, S.; Kadarik, H. & Eenmäe, A. (2007). Calibration of time series of satellite images to study the seasonal course of forest reflectance. *Proceedings of the Estonian Academy of Sciences, Biology / Ecology*. Vol. 56, pp. 5-18, ISSN 1406-0914

Nilson, T. & Peterson, U. (1994). Age dependence of forest reflectance: Analysis of main driving factors. *Remote Sensing Environment*, Vol. 48, pp. 319-331, ISSN 0034-4257

Nilson, T.; Suviste, S.; Lükk, T. & Eenmäe, A. (2008). Seasonal reflectance course of some forest types in Estonia from a series of Landsat TM and SPOT images and via simulation. *International Journal of Remote Sensing*, Vol. 29, No. 17-18, pp. 5073-5091, ISSN 0143-1161

Paal, J. (1997). *Classification of the Estonian vegetation site types*, Estonian Environment Information Centre, ISBN 9985-9072-8-0, Tallinn (in Estonian)

Peltoniemi, J.I.; Kaasalainen, S.; Näränen, J.; Rautiainen, M.; Stenberg, P.; Smolander, H.; et al. (2005). BRDF measurement of understory vegetation in pine forests: Dwarf shrubs, lichen, and moss. *Remote Sensing of Environment*, Vol. 94, pp. 343–354, ISSN 0034-4257

Peterson, U. (1989). *Seasonal reflectance profiles for forest clearcut communities at early stages of secondary succession.* Academy of Sciences of the Estonian SSR, Section of Physics and Astronomy, Preprint A-5 (1989), Tartu

Peterson, U. (1992). Seasonal reflectance factor dynamics in boreal forest clear-cut communities. *International Journal of Remote Sensing*, Vol. 13, pp. 753-772, ISSN 0143-1161

Pinty, B.; Widlowski, J.L.; Gobron, N.; Verstraete, M.M. & Diner, D.J. (2002). Uniqueness of multiangular measurements - Part 1: An indicator of subpixel surface heterogeneity from MISR. *IEEE Transactions on Geoscience and Remote Sensing*, Vol. 40, pp. 1560-1573, ISSN 0196-2892

Rautiainen, M.; Heiskanen, J. & Korhonen. L. (2011a). Seasonal changes in canopy leaf area index and MODIS vegetation products for a boreal forest site in central Finland. *Boreal Environment Research*, in press, ISSN 1239-6095

Rautiainen, M.; Mõttus, M.; Heiskanen, J.; Akujärvi, A.; Majasalmi, T. & Stenberg, P. (2011b). Seasonal reflectance dynamics of common understory types in a northern European boreal forest. *Remote Sensing of Environment*, Vol. 115, pp. 3020-3028, ISSN 0034-4257

Rautiainen, M.; Stenberg, P.; Nilson, T. & Kuusk, A. (2004). The effect of crown shape on the reflectance of coniferous stands. *Remote Sensing of Environment*, Vol. 89, pp. 41-52, ISSN 0034-4257

Rautiainen, M.; Nilson, T. & Lükk, T. (2009). Seasonal reflectance trends of hemiboreal birch forests. *Remote Sensing of Environment*, Vol. 113, pp. 805–815, ISSN 0034-4257

Richardson, A. & Berlyn, G. (2002). Spectral reflectance and photosynthetic properties of Betula papyrifera (Betulaceae) leaves along an elevational gradient on Mt. Mansfield, Vermont, USA. *American Journal of Botany*, Vol. 89, pp. 88–94, ISSN 0002-9122

Richardson, A.; Berlyn, G. & Duigan, S. (2003). Reflectance of Alaskan black spruce and white spruce foliage in relation to elevation and latitude. *Tree Physiology*, Vol. 23, pp. 537-544, ISSN 0829-318X

Richardson A. & O'Keefe J. (2009). Phenological differences between understory and overstory: a case study using the long-term Harvard Forest records, In: *Phenology of Ecosystem Processes,* A. Noormets (Ed.), 87–117, Springer Science + Business Media, ISBN 978-0-4419-0025-8, New York, USA.

Schaaf, C.B.; Gao, F.; Strahler, A.H.; Lucht, W.; Li, X.; Tsang, T. et al. (2002). First operational BRDF albedo, nadir reflectance products from MODIS. *Remote Sensing of Environment*, Vol. 83, pp. 135–148, ISSN 0034-4257.

Shibayama, M.; Salli, A.; Häme, T.; Iso-Iivari, L.; Heino, S.; Alanen, M.; Morinaga, S.; Inoue, Y. & Akiyama, T. (1999). Detecting phenophases of subarctic shrub canopies by using automated reflectance measurements. *Remote Sensing of Environment*, Vol. 67, pp. 160–180, ISSN 0034-4257

Singer, R.B. & McCord, T.B. (1979). Mars: Large scale mixing of bright and dark surface materials and implications for analysis of spectral reflectance. *Proceedings of the*

10th Lunar Planetary Science Conference, Houston TX, March 19-23, 1979, pp. 1835–1848.

Sims, D.; Rahman, A.; Vermote, E. & Jiang, Z. (2011). Seasonal and inter-annual variation in view angle effects on MODIS vegetation indices at three forest sites. *Remote Sensing of Environment*, in press, ISSN 0034-4257

Spanner, M.; Pierce, L.; Peterson, D. & Running, S. (1990). Remote sensing of temperate coniferous forest leaf area index: the influence of canopy closure, understory vegetation and background reflectance. *International Journal of Remote Sensing*, Vol. 11, No. 1, pp. 95-111, ISSN 0143-1161

Stenberg, P.; Mõttus, M. & Rautiainen, M. (2008b). Modeling the spectral signature of forests: application of remote sensing models to coniferous canopies, In: *Advances in Land remote Sensing: System, Modeling, Inversion and Application*, S. Liang, (Ed.), 147-171, Springer-Verlag, ISBN 978-1-4020-6449-4, New York, USA.

Stenberg, P.; Rautiainen, M.; Manninen, T.; Voipio, P. & Mõttus, M. (2008a). Boreal forest leaf area index from optical satellite images: model simulations and empirical analyses using data from central Finland. *Boreal Environment Research*, Vol. 13, pp. 433-443, ISSN 1239-6095

Stenberg, P.; Rautiainen, M.; Manninen, T.; Voipio, P. & Smolander, H. (2004). Reduced simple ratio better than NDVI for estimating LAI in Finnish pine and spruce stands. *Silva Fennica*, Vol. 38, pp. 3-14, ISSN 0037-5330

Suviste, S.; Nilson, T.; Lükk, T. & Eenmäe, A. (2007). Seasonal reflectance course of some forest types in Estonia from multi-year Landsat TM and SPOT images and via simulation. *Proceedings of the International Workshop on the Analysis of Multi-temporal Remote Sensing Images* (MultiTemp 2007), 5 p. doi:10.1109 /MULTITEMP.2007.4293068, ISSN 1-4244-0846-6, Leuven, Belgium, July 18-20, 2007.

Tan, B.; Woodcock, C.E.; Hu, J.; Zhang, P.; Ozdogan, M.; Huang, D.; Yang, W.; Knyazikhin, Y. & Myneni, R.B. (2006). The impact of gridding artifacts on the local spatial properties of MODIS data: Implications for validation, compositing, and band-to-band registration across resolutions. *Remote Sensing of Environment*, vol.105, pp. 98–114, ISSN 0034-4257

Watson, D. (1947). Comparative physiological studies in the growth of field crops. I. Variation in net assimilation rate and leaf area between species and varieties, and within and between years. *Annales Botanicales*, Vol. 11, pp. 41-76.

Widlowski, J-L.; Taberner, M.; Pinty, B.; Bruniquel-Pinel, V.; Disney, M.; Fernandes, R.; Gastellu-Etchegorry, J-P.; Gobron, N.; Kuusk, A.; Lavergne, T.; Leblanc, S.; Lewis, P.; Martin, E.; Mõttus, M.; North, P.R.J.; Qin, W.; Robustelli, M.; Rochdi, N.; Ruiloba, R.; Soler, C.; Thompson, R.; Verhoef, W.; Verstraete, M. & Xie, D. (2007). The third RAdiation transfer Model Intercomparison (RAMI) exercise: Documenting progress in canopy reflectance models. *Journal of Geophysical Research - Atmospheres*, Vol. 112, No, D09111, pp : doi:10.1029/2006JD007821, ISSN 0148-0227

Zhang, X. et al. (2003). Monitoring vegetation phenology using MODIS. *Remote Sensing of Environment*, Vol. 84, pp. 471-475, ISSN 0034-4257

Zhou, L. et al. (2001). Variations in northern vegetation activity inferred from satellite data of vegetation index during 1981 to 1999. *Journal of Geophysical Research - Atmospheres*, Vol. 106, pp. 20069-20083, ISSN 0148-0227

4

Assessment and Mitigation of Nutrients Losses from Forest Harvesting on Upland Blanket Peat – A Case Study in the Burrishoole Catchment

Liwen Xiao*, Michael Rodgers, Mark O'Connor
Connie O'Driscoll and Zaki-ul-Zaman Asam
Civil Engineering, National University of Ireland, Galway
Republic of Ireland

1. Introductions

Since the 1950s, large areas of upland peat were afforested in northern European countries. In Ireland, it was estimated that in 1990 about 200,000 ha of forest were on peatland (Farrell, 1990) and between 1990 and 2000, about 98,000 ha of peat soils were afforested (EEA, 2004). Before the 1980s, most of the Irish peatland forests were planted without riparian buffer strips in upland areas that contain the headwaters of rivers, many of them salmonid. These forests are now reaching harvestable age. Due to the sensitive of the upland water and blanket peat to the disturbance, concerns have been raised about the possible impacts of harvesting these forests and associated activities on the receiving aquatic systems (Coillte Teo, 2007). In order to minimize the possible negative impact of forest harvesting on water quality, good management practices were introduced in the UK (Forestry Commission, 1988) and in Ireland (Forest Service, 2000b, 2000c and 2000d). These practices targeted the process of soil erosion, and included proper harvesting methods and the use of thick brash mats to limit surface disturbance. The findings of earlier harvesting studies in the UK and Ireland were not relevant for the impact assessment of forestry operations carried out under the new forest and water guidelines (Stott et al., 2001). To date, few studies have focused on the impact of post-guideline harvesting on water quality (Nisbet, 2001; Stott et al., 2001). In this study, an assessment of the impact of post-guideline harvesting on the suspended solid and phosphorus release was carried out in an upland blanket peat catchment that had been afforested in the 1970s without buffer strips - typical of most Irish forests now approaching harvestable age. It comprised a control area upstream of an experimental area. We hypothesize that if the best management practice are strictly followed (1) suspended solids release will be low but (2) P release will increase significantly due to a combination of poor P adsorption capacity in blanket peat soil, high rainfall (>2000 mm) and runoff in the study area, and labile P sources being available after harvesting.

Nutrients release to the water body can be minimized by (1) preventing the nutrients transportation from sources to water and (2) reducing nutrient sources. In Ireland and the

* Corresponding Author

UK, many of the earlier afforested upland blanket peat catchments were established without any riparian buffer areas, with trees planted to the stream edge (Ryder et al., 2010). To reduce P release to recipient water courses, buffer strips with a width of 15-20 m are recommended. However, their effect may be limited if (1) most of the P release occurs in storm events, when there would be low residence times for the vegetative uptake of soluble P and (2) most of the P release are dissolved reactive phosphorus. Thus, a specific aim of the study was to investigate the P release pattern in storm events, and to quantify the P release occurring during storm events and base flow conditions. In order to reduce nutrient sources, whole-tree harvesting (WTH) is recommended (Nisbet et al. 1997). In the UK, WTH is usually achieved by removing the whole tree (i.e. all parts of the tree above the ground) from the site in a single operation (Nisbet et al. 1997). In Ireland, in experimental trails conducted by Coillte, an adapted WTH procedure was adopted where the forest harvest residues are bundled and removed from the selected sites after the conventional harvesting of stem wood (personal communication, Dr. Philip O'Dea, Coillte Teoranta, 2010). To increase the understanding of the effect of whole-tree harvesting on P release, a small-scale pilot survey was also performed to investigate if the water extractable P (WEP) contents in soil below windrow/brash material are significantly higher than for areas without windrow/brash material.

Previous studies have indicated that vegetation can retain the available P in situ and reduce P release from forest activities. In Finland, Silvan et al. (2004) demonstrated that plants are effective in retaining P in peatlands. In China and Australia vetiver grass in buffer zones and wetlands has shown a huge potential for removing P from wastewater and polluted water (Wagner et al. 2003). Loach (1968) found that Molinia caerulea could uptake 3.4 kg TP/ha in the wet-heath soils. Sheaffer et al. (2008) reported a P uptake of 30 kg/ha by Phalaris arundinacea in their wastewater treatment sites. In this study we hypothesized that by stimulating the growth of the native grass species in the blanket peat forest area immediately after harvesting, significant amounts of P will be quickly taken up and conserved in situ, which will result in reduced P release sources. To test this hypothesis, a trial experiment was first carried out to identify the successful germination grass species in the blanket peatland. The grass species were then sown in three harvested blanket peat forest plots. The area without grass seeding worked as control. The biomass and P content of the above ground vegetation and the water extractable phosphorus in the study and control plots were tested one year after grass seeding.

2. Study catchment description

The Burrishoole catchment, located in County Mayo, Ireland, in the west of Ireland, consists of important salmonid productive rivers and lakes (Figure 1). About 18% of the catchment is covered by forests that were planted in the 1970s and which are now being, or are about to be, harvested. The study site (9°55'W 35°55'N), which is a sub-catchment of the Burrishoole catchment drained by a small first order stream, was planted with Lodgepole Pine (Pinus contorta) between January and April 1971. The stream is equipped with two flow monitoring stations at stable channel sections, one upstream (US) and the other downstream (DS) of the experimental area. The US measures flows from the control area (area A in Figure 1) of 7.2 ha and the DS covers the control coupe and the experimental coupe (coupes B in Figure 1) with a total combined area of 17.7 ha. Before the start of this study, road drainage into the channel near the US gauge was diverted into an adjoining sub-catchment. In August 2005, a wind-blown tree blocked one of the collector drains, resulting in an increase of the upstream forest control area (coupe D), to about 10.8 ha (coupes A plus D in Figure 1). Meanwhile the downstream harvested area increased to about 14.5 ha due to the blockage of a drain by brash mat during the harvesting, incorporating another part of the total harvested area (coupe C). Fortunately, in both cases the additional area had the same characteristics of vegetation and soils, and the relative sizes of US and DS remained unchanged - US increasing only marginally from 41% of the total area to DS before harvesting and 43% afterwards. All unit area depths in this paper have been calculated using these values. The blanket upland peat soil in all four areas A - D had been double mouldboard ploughed by a Fiat tractor on tracks creating furrows and ribbons (overturned turf ridges) with a 2 m spacing, aligned down the main slope, together with several collector drains aligned close to the contour. The trees were planted on the ribbons at 1.5 m intervals, giving an approximate soil area of 3 m^2 per tree. The initial stand density was about 2800 trees per ha but was reduced to about half by thinning and natural die-off before harvesting. The area was fertilized manually immediately after planting at a rate of 80 kg ground mineral phosphate (GMP) per ha - equivalent to 12 kg P per ha. This rate is low comparing with the normal rate of 250 kg GMP per ha. The catchment had an average peat depth of more than 2 m above the bedrock of quartzite, schist and volcanic rock, and the peat typically had a gravimetric water content of more than 80%. In the catchments, the mean annual rainfall is more than 2000 mm and the mean air temperature is about 11 ^{o}C. Hill slope gradients in areas B and C average 8^0 and range between 0° – 16°. Bole-only harvesting was conducted in area B and C from July 25th to September 22nd 2005. The volume of lodgepole pine upon harvesting in area B (Figure 1) was about 400 m^3 ha^{-1}. The timber was harvested using a Valmet 941 harvester, and the residues (i.e. needles, twigs and branches) were left on the soil surface and collected together to form windrows. During harvesting, the boles were stacked beside the windrow for collection. A Valmet 840 forwarder delivered the boles to truck collection points beside the forest service road. To minimise soil damage, the clearfelling and harvesting were conducted only in dry weather conditions during the period from July to September 2005. That time period is recommended for harvesting in the Irish Forest Harvesting and the Environment Guidelines since ground conditions tend to be drier (Forest Service, 2000b). Mechanised operations were suspended during and immediately after periods of particularly heavy rainfall. Another important good management practice used during the harvesting operation was the proper use of brash mats for machine travelling. Tree residues (i.e. needles, twigs and branches) were collected together to form brash mats on which the harvesting machines travelled, thus protecting the soil surface, and reducing erosion. The width of the windrows/brash mats is about 4 m. The distance between two adjacent brash mats/windrows mats is about 12 m. In the lowest part of the site where the stream is deeply incised, the trees were cut with a chain saw and left behind. The non-harvested upstream area of A and D, was used as a control area in this study as it had the same type and age of trees, similar soil, hydrologic characteristics and size, as the harvested experimental area of B and C. In the experimental area, the furrows and windrows/brash mats - formed from the harvest residues - are, in general, parallel with the study stream, which is at right angles to the contours. The surface water flows along the furrows, is collected by collector drains (arrows in Figure 1c) and joins the study stream.

The second rotation of lodgepole pine was planted in December, 2005 at a density of 2,800 per ha with no cultivation and no new drainage. No fertilizer was applied in the replanting operation. A buffer zone was established by replanting birch, rowan, alder and willow (instead of pine), in a 15-20 m-wide strip on each side of the stream. Furrows, ribbons, drains and brash/windrows were left in situ. Very little re-vegetation was observed in the harvested area until late Summer, 2008.

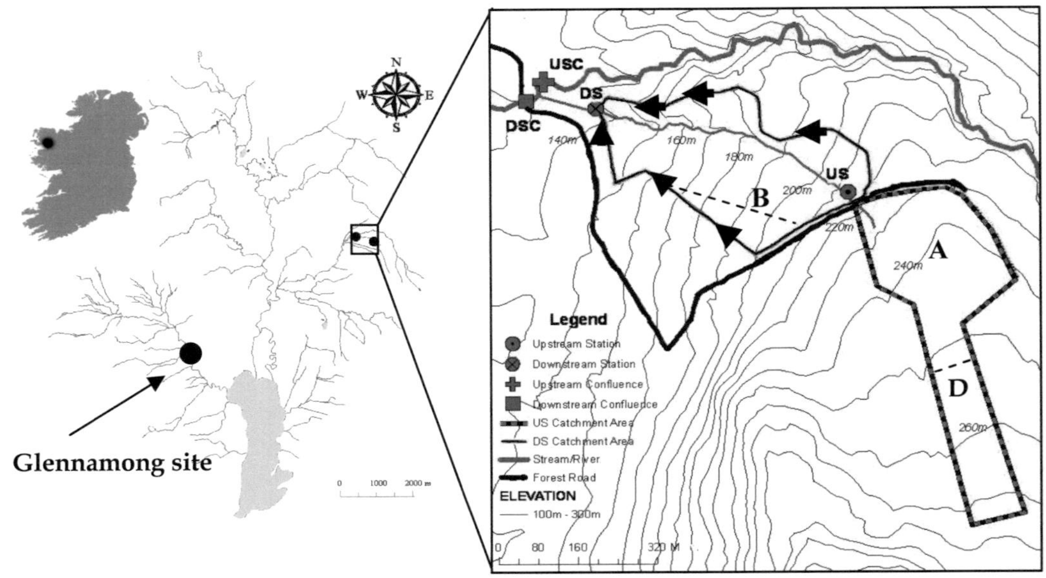

Fig. 1. Study sites

3. Measurement and methods

3.1 sampling and measurement

From April 2004 - March 2005, continuous water levels in the study stream were recorded at both the upstream station (US) and downstream station (DS), and converted to flows by a rating equation based on dilution gauging and current meter measurements. In April 2005, H-flume flow gauges were installed at the sites for flow measurement. At US and DS, water samples were taken: (i) manually every 20 minutes from April 2004 to March 2005 during flood events; (ii) hourly from April 2005 to September 2009 using ISCO automatic water samplers and (iii) manually in base flow conditions through the study period. Rainfall water samples were also collected by placing an open and clean plastic container near the DS station during storm events for P analysis. Suspended solid concentrations of the water samples were measured at the Marine Institute in Newport, Co. Mayo in accordance with the Standard Methods (APHA, 1995) using Whatman GF/C (pore size 1.2 μm) filter papers. Water samples collected May 2005 to September 2009 were analysed for phosphorus content. Water samples were frozen at -20 °C in accordance with standard methods (APHA,

1995) until water quality analyses were conducted. The following analyses were carried out on the water samples: total reactive phosphorus (TRP), dissolved reactive phosphorus (DRP) – filtered using Whatman Cellulose Nitrate Membrane Filters (pore size 0.45 μm) - and total phosphorus (TP) - after digestion with acid persulfate – using a Konelab 20 Analyser (Konelab Ltd., Finland). Sites of about 1 ha in areas A and B were chosen for soil sampling. Forty and thirty-eight 100-mm-deep soil cores consisting of the humic and upper peat layers were collected using a 30-mm-diameter gouge auger from the ribbons in A and B in May 2005, April 2006, March 2007, April 2008 and March 2009. 15, 26, 25 and 28 more soil cores were taken under the windrow/brash in the DS harvested area in April 2006, March 2007, April 2008 and March 2009, respectively. Since the brash mats/windrows - formed from the harvest residues – are parallel to the study stream and furrows, and along the slope, P from the brash mats/windrows didn't enrich the brash-free soil. Soil samples were analyzed for gravimetric water content and water extractable P (WEP). The core samples were placed in bags, hand mixed until visually homogenized, and subsamples of approximately 0.5 g (dry weight) were removed and extracted in 30 ml of distilled deionized water, and measured for P using a Konelab 20 Analyser. The remaining core samples were dried to determine their gravimetric moisture contents (Macrae et al., 2005).

In the Glennamong site, an area of about 1 ha was clearfelled in August 2009 and three plots of 100 m^2 (plot 1), 360 m^2 (plot 2) and 660 m^2 (plot 3) were identified for the grass seeding plot-scale study. Each plot received the same sowing treatment, which comprised of a 50:50 ratio of *Holcus lanatus* and *Agrostis capillaris*. The ground was undisturbed and the seed was distributed evenly by hand at an initial rate of 36 kg ha^{-1} on top of the old forest residue layer in October 2009. December 2009 and January 2010 were exceptionally cold months and a layer of snow measuring 30 cm in depth was recorded above the seeded area. To eliminate the risk of seed establishment failure the plots were seeded again in February 2010 at the same rate of 36 kg ha-1. The area which was not seeded was used as control. 100-mm-deep soil cores consisting of the humic and upper peat layers were collected using a 30-mm-diameter gouge auger in the Glennamong site one year after grass seeding. 4, 8 and 14 soil samples were taken from plot 1, 2 and 3, respectively. Soil samples were analyzed for gravimetric water content and water extractable P (WEP). To estimate the aboveground vegetation biomass in the study and control plots, thirty two 0.25 m x 0.25 m quadrats were randomly sampled (Moore and Chapman 1986) in each site in August 2010. All vegetation lying within the quadrat was harvested to within 1 cm and dried at 80 °C in the laboratory on the day of collection for 48 hours. Samples were then weighed and the biomass was calculated by using Equation 1. Total phosphorus (TP) content of the vegetation was measured in accordance with Ryan et al. (2001). About 1 g of dry matter from each sample was weighed, ground and put into a furnace at a temperature of 550oC overnight, then 5 ml of 2 N HCl was added to extract the P and subsequently diluted to 50 ml with deionised water. P in the solution was analyzed using a Konelab 20 Analyser (Konelab Ltd.).

$$B_p = \frac{Wt}{St} \times 10000 \quad (1)$$

Where Bp is the biomass production (kg/ha); Wt is the total dry weight of the samples (kg) and St is the total area (m2).

3.2 Analysis methods

Storm flow was defined as the total flow (including the base flow) from the time where stream flow begins to increase on the rising limb to the time when the flow on the falling limb intercepts the separation line with a constant slope of 0.0055 L s^{-1} ha^{-1} $hour^{-1}$ (Yusop et al., 2006). Monthly TRP loading was calculated in base flow and storm flow periods as follows (Yusop et al., 2006):

$$Q_{mTRP} = CQ_m \tag{2}$$

where Q_{mTRP} is monthly TRP load (μg $month^{-1}$); C is the discharge-weighted mean concentration (μg L^{-1}) and Q_m is the total flow (L $month^{-1}$). For each month, C (μg L^{-1}) values at base flows and storm flows were calculated separately, using the following equation (Fergusson, 1987):

$$C = \sum_{n=1}^{n} c_i q_i \, / \sum_{i=1}^{n} q_i \tag{3}$$

where c is the instantaneous concentration (μg L^{-1}), q the corresponding discharge during sampling (L s^{-1}) and n is the number of low flow or storm flow samples in the respective month. Finally, the annual loading is calculated as the summation of monthly loadings during both low and storm flow periods.

The TRP loads were calculated using the following linear equation:

$$Q_{TRP} = \alpha Q + \beta \tag{4}$$

where QTRP represents the TRP yield (μg), Q is the water discharge (L), and α (μg L^{-1}) and β (μg) are obtained by the least squares method using observed TRP yield and water discharge data.

At the DS station, the values of α and β in the base flow and storm flow were calculated for the following periods: August 2005 - July 2006, August 2006 - July 2007, August 2007 - July 2008, and August 2008 - July 2009. At the US station, because there was no significant change during the study period, the values of α and β in the base flow and storm flow were calculated from August, 2005 to July, 2009.

The differences in WEP in soil in kg ha^{-1} between the areas without windrow and with windrow were calculated by assuming that windrow comprises 25 % of the harvested area and that soil density is similar in areas below windrow and without windrow.

The difference between the daily total reactive phosphorus (TRP) concentrations at the US and DS stations in the first four years after harvesting was analysed using a paired samples t-test at the 95% significance level (P=0.05). The difference between the soil WEP in A and B before harvesting was analysed using an independent samples t-test at the 95% significance level (P=0.05). After harvesting, the differences between the soil WEP in (i) area A and the brash/windrow-free area in B and (ii) under the brash/windrow and in the brash/windrow-free area in B were also analysed using an independent samples t-test at the 95% significance level (P=0.05). All the t-test was done with the SPSS statistical tool (http://www.spss.com).

4. Results and discussions

4.1 Suspended solids release

During base flow conditions, suspended solid concentrations at the US and DS stations were generally low before and after harvesting and ranged from 0.1 to 5 mg/l. Stream suspended solid are usually episodic - most solid is carried in high flows - so this study focused on the storm events. A rainfall event was defined as a block of rainfall that was preceded and followed by at least 12-hours of no rainfall (Hotta et al., 2007). A total of 23 events were studied in this paper: 8 before and 15 after harvesting. 114 and 394 water samples were collected at both stations before and after harvesting, respectively. Figure 2 shows the suspended solid concentrations and flows in some storm events before and after the harvesting period. As expected, variations in suspended solid concentration roughly correlate to the temporal profile of water discharge, and bigger storm events generally result in higher suspended solid concentrations. In most of the studied storms, suspended solid increased quickly at the beginning of the water discharge and reached the maximum prior to the water discharge peak, which could be due to the build-up of the soil fraction available for release and erosion prior to rainfall. Similar phenomena were also observed by Drewry et al. (2008) and Baca (2002).

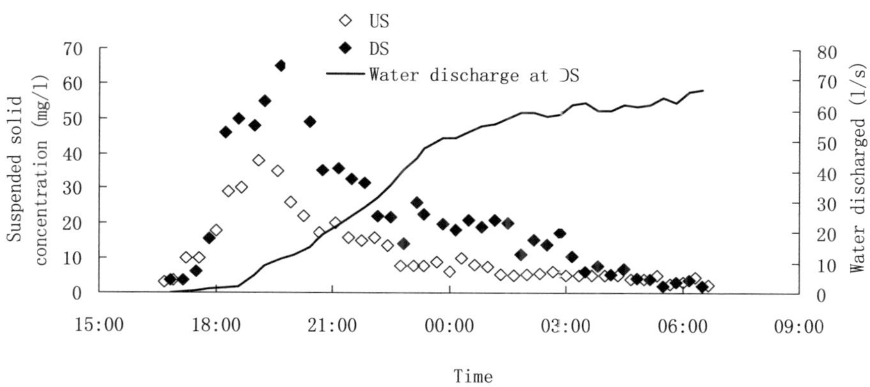

Fig. 2. a. Pre-harvesting (22/06/2004)

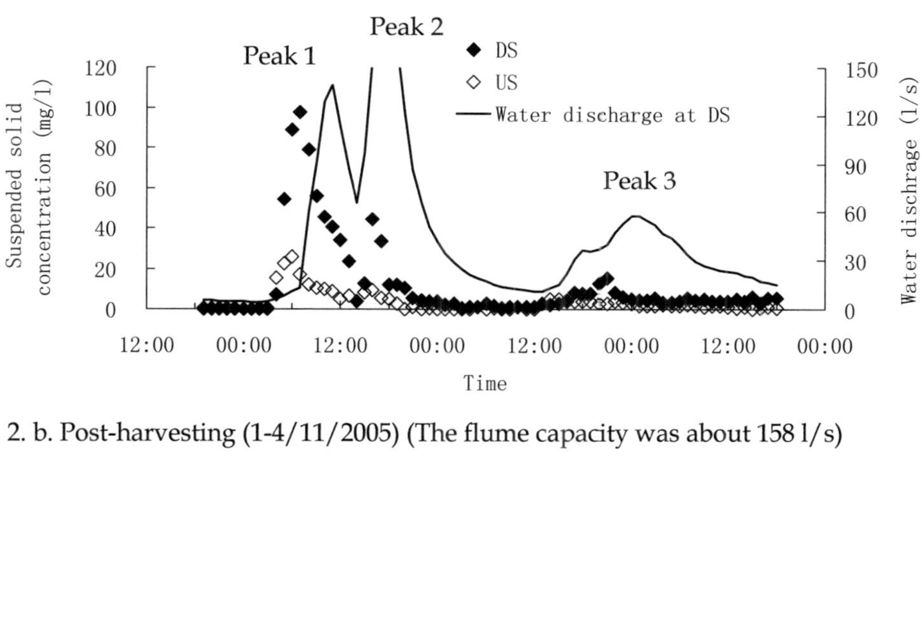

Fig. 2. b. Post-harvesting (1-4/11/2005) (The flume capacity was about 158 l/s)

Figures 3a and 3b show the relationships between suspended solid concentrations of the US and DS before and after harvesting, respectively. Larger scatter was found in the correlation of US and DS suspended solid concentrations after harvesting. In most of the storm events the peak flows passed US earlier than DS with the time difference of less than 30 minutes. Simple power equations were used to describe the solid relationships between the two stations:

$$C_{DS} = a.C_{US}^{\ b} \tag{5}$$

Where C_{DS} and C_{US} are the suspended solid concentrations at DS and US stations, and a and b were obtained by the least squares method.

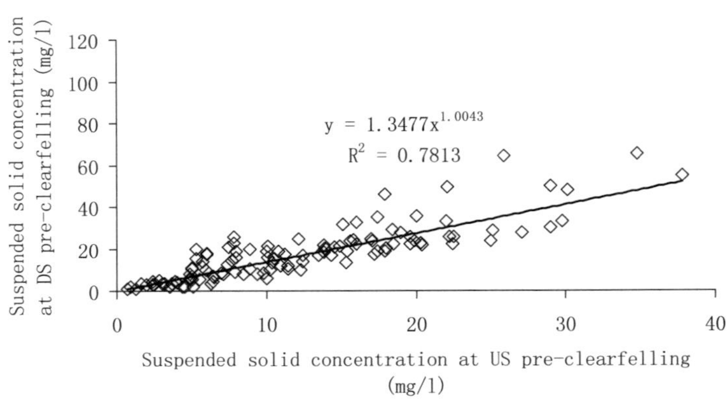

Fig. 3. a.

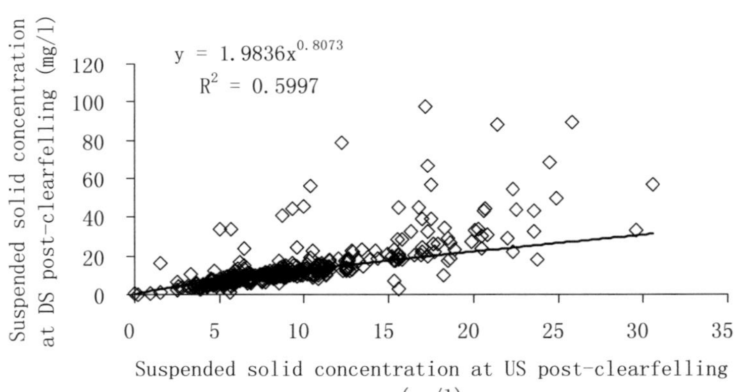

Fig. 3. b.

Fig. 3. The relationship between the suspended solid concentrations at US and DS stations (Figure 3a before harvesting; 3b after harvesting)

Parameter a increased from about 1.35 before harvesting to about 1.98 after harvesting and b decreased from 1.01 to 0.81. In order to examine the impact of the harvesting activities on the sediment release, the solid at DS was estimated as the dependent variable by using the pre-harvesting power function equation (a = 1.35 and b = 1.0) and the observed post-harvesting solid at US as the independent variable. The estimated and measured solid concentrations at DS were compared using a paired samples t-test at the 95% significance level (P=0.05) (http://www.spss.com), which indicated that there was no statistically significant difference between the estimated and measured concentrations.

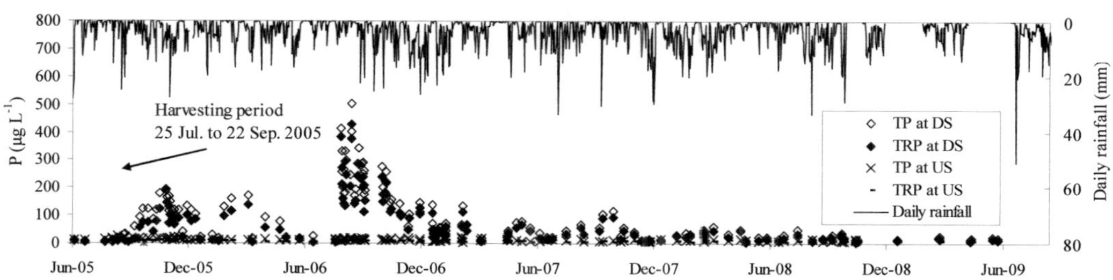

Fig. 4. The daily rainfall and daily discharge-weighted mean total phosphorus (TP) and total reactive phosphorus (TRP) concentrations at downstream station (DS) and upstream station (US) during the study period

4.2 Phosphorus release

4.2.1 General trends

The average P concentrations in the rainfall were 13 ± 6 µg L^{-1} of TP and 4 ± 3 µg L^{-1} of TRP. Figure 4 shows the daily discharge-weighted mean TP and TRP concentrations at US and DS stations during the study period. The release pattern of P concentrations - increasing to a clear peak after harvesting, experiencing a distinct declining tail, and then increasing to the maximum peak in the next summer - was also observed by Cummins and Farrell (2003) in a study carried out in a blanket peatland forest in the west of Ireland. The maximum peak in the next summer after harvesting was also observed by Nieminen (2003) in a Scots pine-dominated peatland in southern Finland. The daily discharge-weighted mean P concentrations at the DS station reduced to less than 15 µg L-1 of TRP and 20 µg L-1 of TP in July 2009, four years after harvesting. Statistical analysis indicated that P concentrations at the DS station were significantly higher than that at the US station (P=0.05) in the 4-year period following harvesting. Figure 5 shows the relationship between the DRP, TRP and TP at the DS station during the study period. Linear regressions were established for DRP and TRP versus TP. TRP and DRP were about 87 % and 77 % of TP, respectively, which indicated that: (1) the majority of TP was reactive and (2) particulate P concentrations were low. Renou-Wilson and Farrell (2007) found that in water samples with high organic matter content, TRP may be equal to TP.

4.2.2 Effect of storm flow events

Over 120 storm events were analysed in this study. Along with being influenced by the elapsed time after harvesting, P concentrations were also affected by the flow rates (Figure 6).

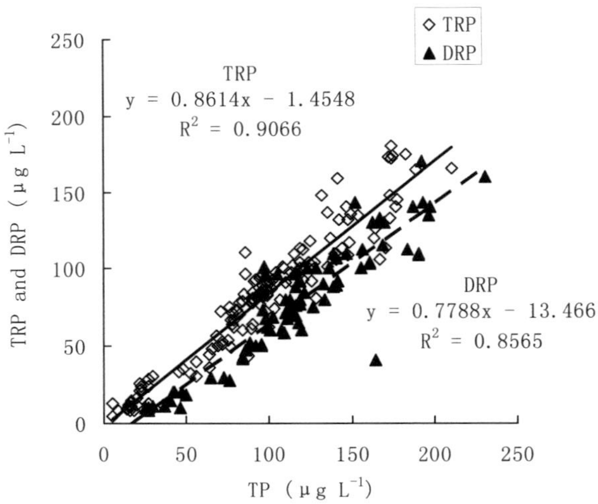

Fig. 5. The relationship between the instantaneous concentrations of dissolved reactive phosphorus (DRP), total reactive phosphorus (TRP) and their linked total phosphorus (TP) concentration at downstream station (DS) during the study period

In over 80 % of the monitored storm events, P concentrations increased at the discharge rising stage, reached the maximum prior to the peak flow rate, and then reduced to a relatively stable value. The major part of the P loading in receiving waters after harvesting activities was derived from the P movement from the topsoil to the stream during overland flow events (McDowell and Wilcock, 2004; Monaghan et al., 2007). Shigaki et al. (2006) and Quinton et al. (2001) found that high rainfall intensity resulted in a greater degree and depth of interaction between runoff and surface soil, including high runoff DRP concentrations, compared to what occurs during low rainfall intensities. The P concentrations were also affected by antecedent weather conditions. In the storm event of November 2nd 2005, peak TRP concentrations were 197 μg L^{-1}, 106 μg L^{-1} and 113 μg L^{-1} in Events 1, 2 and 3 (Figure 6). The peak TRP concentration in Event 2 was lower than in Event 1, although the flow rate was higher, which could be due to less labile P sources being available for release in Event 2. When a storm event follows immediately after a previous storm event, much of the labile P has already been removed by the previous flood (Bowes et al., 2005). Similar phenomena were also observed in other storm events.

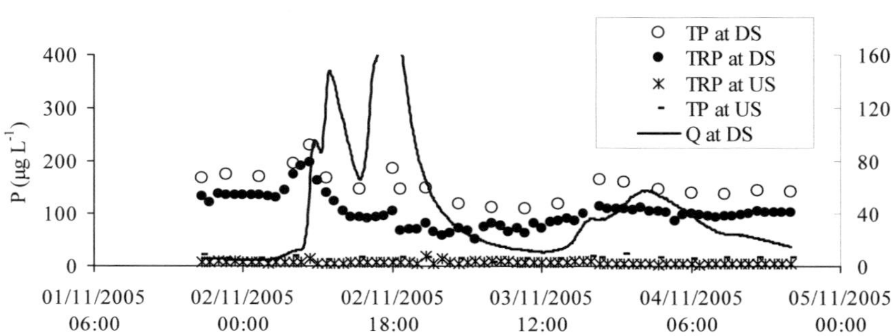

Fig. 6. The instantaneous P concentrations at upstream station (US) and downstream station (DS) with the instantaneous DS flow rate (Q) in a storm event

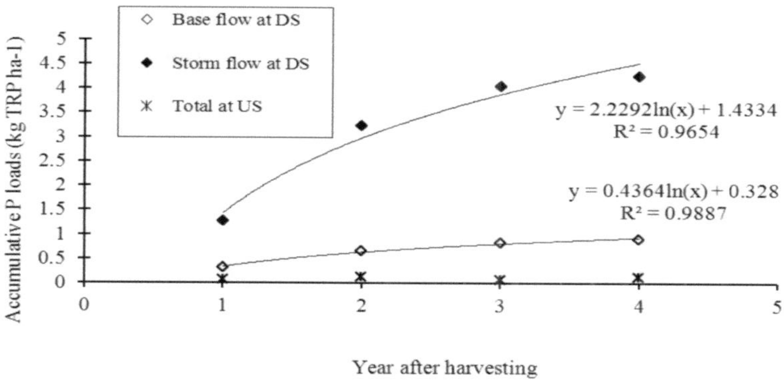

Fig. 7. Accumulative total reactive phosphorus (TRP) loads from the control site (US) and from the harvested area (DS) in base flow and storm flow after harvesting

4.2.3 Phosphorus loads

Annual TRP loads from the control area were steady and low during the study period, with values of less than 60 g ha^{-1}. Figures 7 shows the TRP loads from the harvested area in base flow and storm flow in the first 4 years after harvesting. A total of about 5.15 kg ha-1 of TRP was released from the harvested area in the four years after harvesting, and mainly occurred in the first three years. The highest TRP load of 2303 g ha^{-1} was recorded in the second year after harvesting. Most of the TRP was released in storm events.

4.2.4 P concentrations in downstream river

Phased felling is recommended in the UK (Forest Commission 1988) and Ireland (Forest Service 2000a) to diminish the negative impact of harvesting on water quality. Harvesting appropriately sized coupes in a catchment at any one time can minimise the nutrient concentrations in the main rivers (Rodgers et al. 2010). In their study, Cummins and Farrell (2003) found that the study streams had P concentrations well above critical levels for eutrophication, but they didn't know what implications these pollutions had for downstream river-water quality in larger channels. This study found that the P concentration at the DS station in the study stream did not have a large impact on the P concentration in the main river, which covers an area of 200 ha above its confluence with the study stream and should have a dilution factor of about 8 for the study stream. These preliminary results indicated that catchment-based selection of the harvesting coupe size could limit the P concentrations in the receiving waters after harvesting.

4.2.5 Water extractable P concentrations of the soil after harvesting

Figure 8 shows the WEP of the soil between the windrows and areas under the windrows in the DS harvested area and the US control area in May 2005, April 2006, March 2007, April 2008 and March 2009. The independent samples t-test indicated that (i) before harvesting (in May, 2005), the difference between the WEP concentrations in area A and area B was not significant; (ii) after harvesting (in April, 2006 and March, 2007), WEP concentrations were significantly higher in the brash/windrow-free soils in area B than in area A (P=0.05); (iii) in the harvested area B, the WEP concentrations under the windrows/brash were significantly higher than those in the windrow/brash-free area in April 2006, March 2007, April 2008 and March 2009 (P=0.05). The high WEP value under the windrows/brash material lasted longer than for the windrow-free areas, which could be due to the relatively low decomposition rates of bark and branches (Ganjegunte et al., 2004).

4.3 Possible mitigation methods

4.3.1 Whole tree harvesting, buffer zone and phased felling

In order to reduce nutrient sources after harvesting, whole-tree harvesting is recommended (Nisbet et al., 1997). Needles and branches have much higher nutrient concentrations than stem wood. Whole-tree harvesting may reduce nutrient sources by 2 to 3 times more than bole-only harvesting (Nisbet et al., 1997). This study found higher WEP contents in harvested areas below windrow/brash material than for the brash-free sites, indicating that whole-tree harvesting could be used as a means to decrease P release.

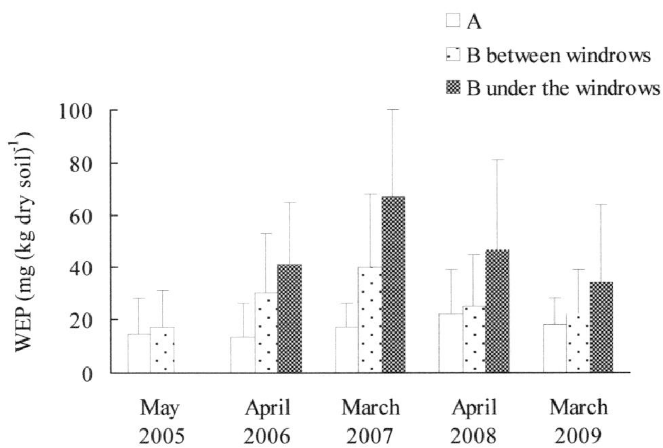

Fig. 8. Soil water extractable phosphorus (WEP) in non-harvested (A) and harvested areas (B) between the windrow and under the windrow in May 2005, April 2006, March 2007, April 2008 and March 2009 (The bars indicate the standard deviation)

A buffer zone is an area adjacent to an aquatic zone and managed for the protection of water quality (Forest Service, 2000a). Within buffer zones, natural vegetation and/or planted suitable tree species are allowed to develop. Buffer zone has been widely used by forestry practitioners in the protection of freshwater aquatic systems (Newbold et al., 2010). However, this study shows that traditional buffer zones may not be an efficient method to mitigate the P release from all harvested areas, since, in this study, about 80 % of TP in the study stream was soluble and more than 70 % of the P release occurred in storm events when there would have been low residence times for the uptake of soluble P in the buffer zones. In fact, using buffer zones to reduce P release could do the opposite, as the buffer zones which usually adjoin to the water body could become P release sources (Uusi-Kämppä, 2005).

Phased felling and limiting size to minimise negative effects have been recommended in the UK (Forestry Commission, 1988) and Ireland (Forest Service, 2000a). Harvesting proper proportion of a catchment at any one time can reduce the nutrient concentrations on aquatic systems. This study found that due to the dilution capacity of the main river, the P concentrations in the river were low after harvesting, indicating that catchment-based selection of the harvesting coupe size could limit the P concentrations in the receiving waters after harvesting. However, the management strategy does not reduce the total P load leaving the harvested catchment.

4.3.2 A possible novel practice – Grass seeding

Figures 9 shows the P content of the above ground biomass in the sown and control plots. Seeding of *Holcus lanatus* and *Agrostis capillaris* increased the above ground vegetation P content one year after grass seeding. While there was very little vegetation growth in the control plots (22 kg biomass ha-1 with P content of 0.02kg TP ha-1) , vegetation biomass of 2753 kg ha-1, 723 kg ha-1 and 2050 kg ha-1 were observed in the three study plots, giving the TP content of 2.83 kg ha-1, 0.65 kg ha-1 and 3.07 kg ha-1, respectively (Figure 9). The P content of above ground biomass in the sown plots was significantly higher than in the control plots (t test, $p < 0.01$). The vegetation collected for testing was cut to 1cm aboveground level so these estimates could in fact be higher when taken below ground biomass production into account which has been estimated at 30% of the total plant biomass (Scholes and Hall 1996). In the UK, Goodwin et al. (1998) found that *Holcus lanatus* produced biomass of 3405 kg ha-1 with P concentrations of 1.64 mg TP (g biomass)-1, giving the total P content of 5.58 kg P ha-1. Figure 10 shows the water extractable phosphorus (WEP) concentrations in the sown plots and the control plots. The WEP in the three study plots were 9 mg P (kg dry soil)$^{-1}$, 12 mg P (kg dry soil)$^{-1}$ and 6 mg P (kg dry soil)-1, respectively, which was significantly lower than the value of 27 mg P (kg dry soil)-1 in the control areas (Figure 9) (t-test, $p < 0.01$).

Future research on the potential of grass seeding as forestry BMP should measure stream chemistry to assess the success of the practice at protection water quality. It is expected that the P measured in the grass would render a corresponding reduction in the P exported by the stream after harvesting. However this has not been addressed by this study. Sowing grass immediately after harvesting may affect forest regeneration.

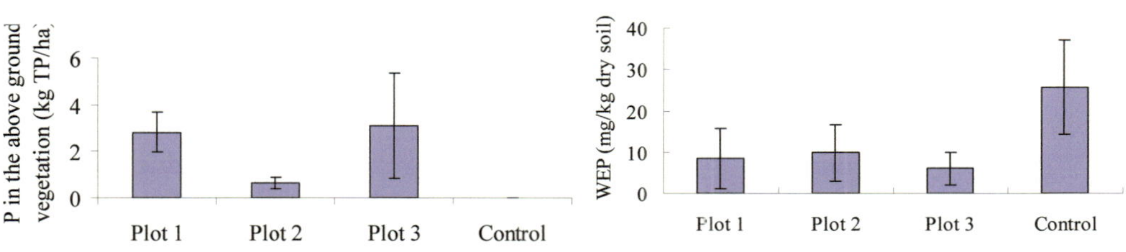

Fig. 9. P content of above ground vegetation and soil WEP in the study plots and control in the Glennamong

5. Conclusions

The results of this study indicated that post-guideline harvesting (1) did not have long-term impact on the suspended solid concentrations and did not change the erosion characteristics of the catchment, but (2) increased the TRP export in the study stream, and this impact could last for more than four years. In the first three years following harvesting, up to 5.15 kg ha-1 of TRP were released from the catchment to the receiving water; in the second year alone after harvesting, 2.3 kg ha-1 were released.

About 80 % of TP in the study stream was soluble and more than 70 % of the P release occurred in storm events, indicating that traditional buffer zones may not be an efficient method to mitigate the P release from all harvested areas. Due to the dilution capacity of the main river, the P concentrations in the river were low during the study period, indicating that rational sizing of the harvesting coupe could be an efficient practice to limit the P concentration in the receiving waters following harvesting. However, the study comprised only one experimental catchment. In future research, more paired sites should be investigated. Higher WEP concentrations found under the windrows/brash material, were due to P release from decomposing logging residues. This indicates that whole-tree harvesting could, at least to some extent, be used as a means to decrease P release from blanket peats. The results of this study also indicate that *Holcus lanatus* and *Agrostis capillaris* can be quickly established in blanket peat forest areas after harvesting and has the potential to immobilize the P that would otherwise be available for leaching. Further research into the feasibility of grass seeding as a potential new best management practice is clearly warranted.

6. References

APHA. (1995). Standard Methods for Examination of Water and Wastewater, 19th edition. American Public Health Association, Washington.

Baca, P. (2002). Temporal variability of suspended sediment availability during rainfall-runoff events in a small agricultural basin. ERB and Northern European FRIEND Project 5 Conference, Demänovská dolina, Slovakia.

Bowes M.J., House A.W., Hodgkinson A.R and Leach V.D. (2005). Phosphorus - discharge hysteresis during storm events along a river catchment: the River Swale, UK. Water Res. 39, 751-762.

Coillte Teo. (2007). Code of Best practice for the management of water run off during forest operations. Coillte, Newtownmountkennedy, Co. Wicklow, Ireland

Cummins, T., and E. P. Farrell. (2003). Biogeochemical impacts of clearfelling and reforestation on blanket peatland streams I. phosphorus. Forest Ecology and Management, 180(1-3), 545-555.

Drewry, J. J., Cameron, K. C. and Buchan, G. D. (2008). Pasture yields and soil physical property responses to soil compaction from treading and grazing: a review. Australian Journal of Soil Research 46: 237-256

EEA. (2004). Revision of the assessment of forest creation and afforestation in Ireland. Forest Network Newsletter Issue 150, European Environmental Agency's Spatial Analysis Group

Farrell, E.P. (1990). Peatland forestry in the Republic of Ireland. In: Biomass Production and Element Fluxes in Forested Peatland Ecosystems, Hånell, B., Ed., Umea, Sweden, 13, 1990.

Fergusson, R.I. (1987). Accuracy and precision of methods for estimating river loads. Earth Surf. Proc. Land 12, 95–104.

Forestry Commission. (1988). Forests and water guidelines. 1st ed. London: HMSO. Revised (2nd ed. 1991, 3rd ed. 1993, 4th ed. 2003)

Forest Service. (2000a). Forest Harvesting and the Environment Guidelines. Irish National Forest Standard. Forest Service, Department of the Marine and Natural Resources, Dublin.

Forest Service. (2000b). Forestry and Water Quality Guidelines. Forest Service Guideline. Forest Service, Department of the Marine and Natural Resources, Dublin.

Forest Service. (2000c). Code of Best Forest Practice - Ireland. Irish National Forest Standard. Forest Service, Department of the Marine and Natural Resources, Dublin.

Forest Service. (2000d). Forest Harvesting and the Environment Guidelines. Irish National Forest Standard. Forest Service, Department of the Marine and Natural Resources, Dublin.

Ganjegunte K.G., Condron M.L, Clinton W.P., Davis R.M. and Mahieu N. (2004). Decomposition and nutrient release from radiate (Pinus radiata) coarse woody debris. Forest Ecol. Manag. 187: 197–211.

Goodwin, M. J., R. J. Parkinson, E. N. D. Williams, and J. R. B. Tallowin. (1998). Soil phosphorus extractability and uptake in a Cirsio-Molinietum fen-meadow and an adjacent Holcus lanatus pasture on the culm measures, north Devon, UK. Agriculture, Ecosystems & Environment, 70(2-3), 169-179.

Hotta N., Kayama T. and Suzuki M (2007). Analysis of suspended sediment yields after low impact forest harvesting. Hydrological processes 21: 3565-3575

Loach, K. (1968). Seasonal Growth and Nutrient Uptake in a Molinietum. Journal of Ecology, 56(2), 433-444.

Macrae M.L., Redding T.E., Creed I.F., Bell W.R. and Devito K. J. (2005). Soil, surface water and ground water phosphorus relationships in a partially harvested Boreal Plain aspen catchment. Forest Ecol. Manag. 206, 315-329.

McDowell, R.W., Wilcock, R.J. (2004). Particulate phosphorus transport within stream flow of an agricultural catchment. J. Environ. Qual. 33, 2111–2121.

Monaghan R.M., Wilcock R.J., Smith L.C., Tikkisetty B., Thorrold B. S., and Costall D. (2007). Linkages between land management activities and water quality in an intensively farmed catchment in southern New Zealand. Agr Ecosyst Environ 118, 211–222.

Moore, P. D. and S.B. Chapman. (1986). Methods in Plant Ecology, Alden Press, Oxford.

Newbold D. J., Herbert S., Sweeney W. B., Kiry P. and Alberts J.S. (2010). Water quality functions of a 15-year-old riparian forest buffer system. J Am Water Resour As. 46(2):299-310

Nieminen, M. (2003). Effects of clear-cutting and site preparation on water quality from a drained Scots pine mire in southern Finland. Boreal Env. Res. 8: 53-59.

Nisbet T, J. Dutch and A.J. Moffat (1997). Whole-tree harvesting: a guide to good practice. Forestry practice guide, London, HMSO.

Nisbet T.R. (2001). The role of forest management in controlling diffuse pollution in UK forestry. Forest Ecol. Manag. 143, 215-226.

Quinton J.N., Catt J.A., Hess T.M. (2001). The selective removal of phosphorus from soil: is event size important? J Environ Qual 30, 538-545

Renou-Wilson. F. and Farrell P. E., 2007. Phosphorus in surface runoff and soil water following fertilization of afforested cutaway peatlands. Boreal Env. Res 12, 693-709.

Rodgers, M., M. O'Connor, M. G. Healy, C. O'Driscoll, Z. Asam, M. Nieminen, R. Poole, M. Muller, and L. Xiao. (2010). Phosphorus release from forest harvesting on an upland blanket peat catchment. Forest Ecology and management , doi:10.1016/j.foreco.2010.09.037

Ryan J., E. George, and R. Abdul. (2001). Soil and plant analysis laboratory manual. Second edition. Jointly published by the International Center for Agricultural Research in the Dry Areas and the National Agricultural Research Center. Available from ICARDA, Aleppo, Syria.

Ryder, L., E. de Eyto, M. Gormally, M. Sheehy-Skeffington, M. Dillane and R. Poole. (2010). Riparian zone creation in established coniferous forests in Irish upland peat catchments: Physical, Chemical and Biological Implications. Biology and Environment: Royal Irish Academy. (In press).

Scholes, R.J. and D. O. Hall. (1996). The carbon budget of tropical savannas, woodlands and grasslands. In: Breymeyer A.I., D. O. Hall, J. M. Melillo, and G. I. Agren. (Eds.) Global Change: Effects on Coniferous Forests and Grasslands, SCOPE Volume 56. Wiley, Chichester.

Sheaffer C.C., Rosen C.J. and Gupta S.C. (2008). Reed Canarygrass Forage Yield and Nutrient Uptake on a Year-round Wastewater Application Site. Journal of Agronomy and Crop Science, 194 (6), 465-469

Shigaki F., Sharpley A.N. and Prochnow I. L., 2006. Rainfall intensity and phosphorus source effects on phosphorus transport in surface runoff from soil trays. Sci Total Environ, doi:10.1016/j.scitotenv.2006.10.048

Silvan N., H. Vasander, and J Laine. (2004). Vegetation is the main factor in nutrient retention in a constructed wetland buffer. Plant soil, 25, 179-187

Stott T., Leeks G., Marks S. and Sawyer A. 2001. Environmentally sensitive plot-scale timber harvesting: impacts on suspended sediment, bedload and bank erosion dynamics. Journal of Environmental Management 63: 3-25.

Uusi-Kämppä, J. (2005). Phosphorus purification in buffer zones in cold climates. Ecological Engineering, 24(5), 491-502.

Wagner, S., P. Truong, A. Vieritz, and C. Smeal. (2003). Response of Vetiver Grass to Extreme Nitrogen and Phosphorus Supply. In: Proceedings of Third International Vetiver Conference, Guangzhou, China. October 2003.

Yusop Z., Douglas I., Nik R.A. (2006). Export of dissolved and undissolved nutrients from forested catchments in Peninsular Malaysia. Forest Ecol. Manag. 224, 26-44.

5

Evaluation for the UMA's of Diversified Breeding in the Mixteca Poblana, México

Oscar Agustín Villarreal Espino Barros[1], José Alfredo Galicia Domínguez[1], Francisco Javier Franco Guerra[1], Julio Cesar Camacho Ronquillo[1] and Raúl Guevara Viera[2]
[1]Benemérita Universidad Autónoma de Puebla
[2]Universidad de Camagüey
[1]México
[2]Cuba

1. Introduction

The ethnic zone named Mixteca in the state of Puebla, Mexico; is a region with rough topography, arid and semiarid climate, critical poverty, and isolated of the development. In that area, the white tail deer (*Odocoileus virginianus*) from the "*mexicanus*" subspecies is used, in Units for the Management and Wildlife Conservation or UMA's, by means of the model called Diversified Breeding (livestock) (Villarreal, 2006). This technology is based on a productive model of Diversified Integrated and Self-sufficient type (Sustainable Farming Systems), where the exploitation of bovine of meat is diversified, by means of rational and sustained utilization of the white tailed deer, other species of the wild fauna and their habitat, in the hunting game and the tourism of nature. These sustainable models are an alternative for the conservation of the natural resources, since they´ll favor the recycling of nutrients, production of biomass and their movement across the ecosystem, achieving to establish schemes that integrate the productive managing, with the exchange of energy and nutrients, with a natural base of coherent performance (Pimentel, 2001). The objective of this work was to realize an ecological and socioeconomic evaluation of the application of the model of Diversified Breeding in the UMA's of white-tailed deer, in the Mixteca region in the south of the Mexican state of Puebla. (Villarreal et al. 2008).

1.1 Study´s setting

The Mixteca poblana belongs to the dry tropic of the depression of the Balsas River (Fig. 1), with habitat whit tropical deciduous forest (Fig. 2), arid brushwood and oaks forest, among other vegetative types. It covers 47 Municipalities, with a principally mountainous surface of 10.565 km^2. Due to its geographical conditions the region shows under agricultural potential, the activities of the primary sector of the economy are the agriculture of temporarily and the extensive ranching of bovine's and goats livestock. The secondary and tertiary activities concentrated in two growth points, the cities of Tehuacán and Izúcar of Matamoros (Villarreal, 2006). Due to the lack of opportunities of development the rural population migrates principally to the New York, California and Texas states in the American Union. In the region the white tailed deer of the *"mexicanus"* subspecies, is distributed in 37 Municipalities by a surface of 547.550 ha. (Villarreal & Guevara, 2002).

Fig. 1. Map from location of Mixteca region, Puebla State, Mexico.

Fig. 2. Tropical deciduous forest in the Mixteca Poblana, México (Picture, Oscar Villarreal).

2. Materials and methods

The application of the Pressure-State-Response framework or PSR was used, which allows to analyze and to quantify the socioeconomic and environmental sustainability of agricultural systems to regional or local levels (De Camino & Muller, 1993; OCDE, 1993 Winograd, 1995). The work was realized by means of technologies of group by the farmers from seven UMA's, as well as the collection of information of all the UMA's in governmental dependences, into federal level as the Environment and Natural Resources Secretary (SEMARNAT), and Puebla State Government as, the Sustainable Environmental and Territorial Organization Secretary (SSAOT), and the Rural Development Secretary (SDR). The information was analyzed attending to a group of variables of the technological model such as: the population density, evaluation of the habitat, carrying capacity and diversity of the diet, the rate of crop and regional development between others. The determined results for the different variables used in the matrix appeared for unit of measure, or according to the type of action quantifiable.

3. Results and discussions

The socioeconomic and environmental evaluation by means of the PSR matrix to the UMA's of Diversified Breeding, threw the following results (Table 1): the increase of the population density of deer like response to the application of the technological model; though they are not spectacular for treating itself about animals in free life, represent a positive progression already brought in hunting ranches in the southwest of the United States, and UMA's the North-East of Mexico (Brown, 2004). On the other hand, the calculations "*in situ*" of the primary productivity of phytomass usable in the order of 0,79 and 0,88 ton,/dray matter per year, has contributed in the decisions of managing of the population density and its relation with the capacity of load, which is between 7,28 and 9,41 has./UA (United Animal), besides 139 vegetable species which have been identified like consumed by the deer. As a consequence, it has produced an optimization itself in the employment of food supplements and waters in critical epochs (Villarreal, 2006). The consumption of herbaceous plants and tree and shrub in the diet constitutes an advantage for the system of corporal reservations in drought (Savory, 2005; Villarreal & Marín, 2005); that produced increases in the rates of crop of the deer. Another undeniable aspect though not quantified, it is the relative to environmental services for the capture of carbon of the biomass, and the recycling nitrogen in the soil (Savory, 2005). In relation to the use of land, has been observed that for the past period from March, 2001 to December, 2009, occurred unexpectedly an increase in the number of properties incorporated as UMA's from 13 to 72, which means the pass from 14.423,92 ha. to 82.522,02 ha., incorporated into this model which respects the biodiversity and takes advantage of the animal resource in rational form; in addition, of six initial Municipalities it has increased to 35. This information relates to the degree of adoption in the time of the technology in the region, where the works of conservation, managing and crop, developed by institutions of top education (Benemérita Universidad Autónoma de Puebla) and lenders of technical services for the UMA's, have had good results recognized and supported by the Federal institutions Environment and Natural Resources Secretary (SEMARNAT) and Government of Puebla State, that are the organisms tracers of the public policies in the rural way.

Variables	Element	Indicator	Period or years	Effects
Deer`s population	Growth	Population density	Before Later	-- + +
Habitat	Components	Evaluation	Before Later	-- +
Deer`s supply	Phytomass Consumption of forage	Capacity of load Variety	Before Later Before Later	-- Evaluated *in situ* Unknown 139 *Spp.*
Big game trophies	Utilization	Rate of crop	Before Later	-- +
Management of land	Extension	Surfaces (has.)	2001 2009	14.423.92 82.522,02
Biodiversity	UMA's	Number of lands	2001 2009	13 Predios 72 Predios
Regional development	Municipalities	Numbers	2001 2009	13 35
Socio-economic Development	Generation of permanents employments	Increase of each 1000 ha., of operation	Before Later	00 2-3
Information and participation	Activities of training and capture of decisions	UMA's Management plan.	Before Later	-- ++
Agreements and events	Agreements big game trophies	Regional Tournament and Mexican "Slam"	Before After	Non include Include
Investigations and consultations	Universities and ONG´s	Researching groups Working groups	2000 2011 2000 2011	1 3 Anyone 5
Conservation	Protected natural areas	Number of Municipalities N° hectares	1998 2011 1998 2011	9 11 145.715,8 161.662,8

Table 1. Analysis of PSR Matrix (framework) to model of Diversified Breeding from Mixteca Poblana region, Mexico

Other measured indicators in the PSR matrix are: the participation of the producers in the accomplishment of socio cultural events such as: the "Thummler Award" (Mexican Deer Super Slam) and the "Regional Tournament of Hunting Game, slams directed in order to obtain big game trophies, which confirms the validity of the technological adoption, in the search of the sustainability of the use of the natural resources, in harmony with the agricultural activities to attack the poverty and social inequality (Fig 3); which they are a reason of emigration of the population in productive age to the United States, and that affect the sustainable regional development (Villarreal, 2002, 2006).

Fig. 3. Arturo Villarreal, young hunter and his trophy of Mexican white tailed deer, taken in the Mixteca Poblana region (Picture from Oscar Villarreal).

In addition, it is necessary to indicate, the labor in favor of the wild fauna with game potential, since there are three groups of investigators that realize functions of researching, transferring of technology, extensions, promoting and advising (Villarreal et al. 2011). On the part of Benemérita Universidad Autónoma de Puebla (BUAP), stand out the groups of the Faculty of Veterinary and Animal Science (FMVZ) and the Biology School, besides the group of the Faculty of Veterinary and Animal Science, from Universidad Autónoma Metroplitana (UAM). The Benemérita Universidad Autónoma de Puebla has come organizing the "Symposiums on Game Animals of Mexico", academic and national event that looks for the conservation of the natural resources and generation of socioeconomic benefits, using as tool the hunting game and the ecotourism.

On the other hand, the sector of the Non-Governmental Organizations (ONG's), is represented in a general form by three groups that are employed at subject matters of forestry development join with wild fauna. Finally only there was only a natural protected area. The Biosphere Tehuacán-Cuicatlán Reserve (RBT-C), dependent organism on the National Commission of Natural Protected Areas (CONANP) from Environment and

Natural Resources Secretary, which includes nine Municipalities of the east of the region, with a surface of 145.715,8 ha., (Villarreal, 2006). On May 2, 2011, the Sustainable Environmental and Territorial Organization Secretary, of the Puebla State Government, had decreed the Natural Protected State Area "Tentzo's Sierra", which includes inside the Mixteca Poblana two north Municipalities with 15.947 ha. from the conservation and sustainable managing in to the region(Fig. 4).

Fig. 4. Dr. Rafael Moreno Valle, Governor of Puebla State and Eng. Juan Elvira Quesada, Minister of Environment and Natural Resources Secretary, during inauguration of Natural Protected State Area "Tentzo's. Sierra". At right show up the Mexican royal (golden) eagle *Aquila chrysaetos canadensis* (Picture Oscar Villarreal).

4. Conclusion

We can conclude that the analysis-summary of the PSR matrix for the conservation and managing of the Mexican white-tailed deer in the UMA's from Mixteca, demonstrated the potentials of the rational utilization of the cervid and its habitat, a regional level as resource of wild life inside the model of Diversified Breeding, to reach the sustainability of this technological model in the region, from the auto management, the empowerment and the community participation, respecting its biodiversity. Therefore, there is advisable the application of PSR matrix for similar valuations in other regions of Mexico.

We recommend to have care in the conservation and the appropriate managing of this biodiversity, since some threats besides the deforestation due to the advance of the urban borders, industrial and agricultural, it is the introduction of plants and species and subspecies (geographical races) exotic animals, it means, foreign to the regional ecosystems, such as: the red deer (*Cervus elaphus*), sika deer (*Cervus nippon*), axis deer (*Axis axis*), fallow deer (*Dama dama*), Texan white tailed deer (*Odocoileus virginianus texanus*) and European wild boar (*Sus scropha*), among other exotic species (Álvarez et al. 2008). The society in general and the government in its three levels (Municipal, State and Federal) are correspondents of the conservation, managing and rational utilization and supported of the Mexican white tailed deer and its habitat for the benefit of the Puebla society, Mexico and the world.

5. References

Álvarez-Romero, J. G.; Medellín R. A.; Oliveras de Ita, A.; Gomes de Silva, H. & Sánchez, O. (2008). *Animales exóticos en México; una amenaza para la biodiversidad.* Comisión Nacional para el Conocimiento y Uso de la Biodiversidad, Instituto de Ecología, Universidad Nacional Autónoma de México, Secretaría del Medio Ambiente y Recursos Naturales; ISBN 978-970-9000-46-7. México, D. F. 518 pp.

Brown, K. (2004). Experiences in Texas Private Properties: Lessons in wildlife coordinated management, regulatory decrease and free market economy. *Proceedings of: 16th National Diversified Livestock (Wildlife Breeders) Congress.* Asociación Nacional de Ganaderos Diversificados Criadores de Fauna, Nuevo Laredo, Tams., México, pp 20-21.

De Camino R. y S. Muller. (1993). *Sostenibilidad de la agricultura y los recursos naturales. Bases para establecer indicadores.* Proyecto IICA/GTZ. San José, Costa Rica, 38 pp.

OCDE (Organization for Economic Co-Operation and Development). (1993). *OECD core set of indicators for environmental performance reviews. A synthesis report by the Group on the State of the Environment. Environment monographs.* N° 83. OCDE/GD (93)179. 39 pp.

Pimentel, D. (2001). Limits of biomass utilization. *Encyclopedia of Physical Science and Technology.* 3a Ed. Vol. 2. Academic Press, New York. USA.

Savory, A. (2005). *Manejo Holístico; Un nuevo marco metodológico para la toma de decisiones.* Secretaría del Medio Ambiente y Recursos Naturales, ISBN: 968-817732-6 México, D. F. pp 115-180.

Villarreal, O. A. (2002). El "Grand Slam" del Venado Cola Blanca Mexicano, una Alternativa Sostenible. *Archivos de Zootecnia,* Vol. 51, N° 193-194. Instituto de Zootecnia; Facultad de Veterinaria. Universidad de Córdoba, España; pp 187-193. ISSN 0004-0592.

Villarreal, O. A. & Guevara, R. 2002. "Distribución Regional del Venado Cola Blanca Mexicano (*Odocoileus virginianus mexicanus*); en la Mixteca Poblana, México". *Producción Animal,* Año 2002. Vol. 14, N° 2. pp 35-40. Facultad de Ciencias Agropecuarias. Universidad de Camagüey, Ministerio de Educación Superior, Cuba. ISSN 0258-6010.

Villarreal, O. A. & Marín. M. (2005). Agua de Origen Vegetal para el Venado Cola Blanca Mexicano. *Archivos de Zootecnia.* Vol. 54 Núm. 206-207, pp 191-196. Instituto de Zootecnia; Facultad de Veterinaria. Universidad de Córdova, España. ISSN 0004-0592.

Villarreal O. A. (2006). *El venado cola blanca en la Mixteca poblana. Conceptos y métodos para su conservación y manejo"*. Benemérita Universidad Autónoma de Puebla, ISBN: 968 863 992-3. Puebla, México. 191 pp.

Villarreal. O. A.; Franco, J.; Hernández, J. E.; Romero, S.; T. & Guevara, R. (2008). Evaluación de las UMAS de venado cola blanca en la región Mixteca, México. *Zootecnia Tropical*, Año 2008, Vol. 26, N° 3: pp 395-398. Instituto Nacional de Investigaciones Agrícolas; Ministerio del Poder Popular para la Agricultura y Tierras, Venezuela: ISSN 0798-7269.

Villarreal, O. A.; Plata, F. X.; Franco, F. J.; Hernández, J. E.; Mendoza, G. D.; Aguilar, B.; Camacho, J. C. (2011). Conservación y manejo del venado cola blanca en México: región Mixteca Poblana. *Ciencia Tecnología e Innovación para el Desarrollo de México (PCTI)*: Año 3 N° 70. ISSN: 2007-1310.

Winograd, M. (1995). *Indicadores ambientales para Latinoamérica y el Caribe. Hacia la sustentabilidad en el uso de tierras*. Proyecto IICA/GTZ, OEA. Instituto de Recursos Mundiales. San José, Costa Rica.

Section 2

Forest and Organisms Interactions

6

The Development of a Port Surrounds Trapping System for the Detection of Exotic Forest Insect Pests in Australia

Richard Bashford
Forestry Tasmania
Australia

1. Introduction

This chapter traces the development of trap equipment and techniques over a decade, leading to an effective and efficient unban monitoring system to detect exotic woodborer insects of importance to Australian plantation forestry.

The interception of exotic forestry insect pests entering Australia through airports and seaports depends on a three-tiered system of zoning of inspection. Primary focus is placed at the port of entry with a detailed examination of goods and passengers by AQIS and Customs staff. The second zone is the port surrounds within a 5 kilometre radius of the port. Except for the Asian Gypsy Moth surveys within this zone there was no mechanism for the detection of exotic forest insect pests that have escaped barrier interception and have spread and possibly established in this zone. The third zone is the forest plantation estate existing beyond the port surrounds area. This zone is subject to regular forest health surveillance designed to detect all damaging forest insect pests and diseases.

The implementation of the GIMP, (Generic Incursion Management Plan for forest pests and diseases), provides a process to enable rapid response following detection of potential exotic incursions. Central to the GIMP is an effective surveillance system based on the three zones of interception.

A committee was established within RWG 7 (Forest Health) in 2000 with the aim of investigating ways to implement port environs surveys to detect the presence of exotic forest insects and diseases.

In 2004 additional money for biosecurity was made available through the 'Securing the Future' program. A portion of this new annual funding was to target the detection of exotic pests and diseases in the transition movement phase between ports of international cargo entry and forest plantations and agricultural crop sites. The value of monitoring the port surround area adjacent to sea or air ports is supported by data from New Zealand where, with 44 years of surveillance experience, Carter (1989) determined that 47% of first records of pests and diseases on living trees had been within the port surround zones.

The Forest Health Surveillance Unit of Queensland Forest Research Institute conducted intensive visual surveys (Blitz surveys) of port environs at five seaports and two airports in

Queensland during 1999 and 2000 (Wylie, *et al.*, 2000). Although no exotic forest pests or pathogens were found during these surveys the exercise demonstrated the necessity for rapid assessment surveys at specific times of the year to detect pathogens. The blitz visual assessment system is expensive to conduct, has only moderate detection rates (Bulman *et al* 1999) and samples only a narrow window of time. The Queensland exercise also emphasised the need to have a system in place for ongoing detection of insect pest species with cryptic life histories to augment visual surveys.

Static trapping techniques have been defined by Speight & Wylie, (2001), as having four functions. (a) detect presence of imported noxious insects (b) determine the spread and range of recently introduced pests in a region (c) determine the seasonal appearance and abundance of insects in a locality and (d) determine the need for application of control measures.

2. Assessment of trap designs and lures combinations

A number of static trapping techniques were evaluated including sticky banding of trees, a range of different trap types, and lure combinations.

2.1 Sticky banding

For two summer seasons sticky band traps were placed on a range of native tree species within 1 km of three major shipping ports and Hobart Airport in Tasmania (Fig. 1). Sticky waxed paper bands 400mm wide were stapled around the trunks of several species of *Eucalypus* and *Acacia* in differing stages of health at each site. The bands were coated with 'Tangle-Trap' (The Tanglefoot Company, Grand Rapids, USA). The bands were removed at monthly intervals during the flight period of native wood borer species. Specimens were removed from the sheets using mineral turpentine and individual specimens cleaned in 'De-Solv-It' (RCR International, Sandringham, Australia). In the 2001-2002 summer the port of Bell Bay on the north coast was targeted and that port and the ports of Burnie, Devonport as well as Hobart Airport were targeted for six summer months during 2002-2003. In all 960 specimens comprising 50 species within the coleopteron families Elateridae, Buprestidse, Scolytidae, Anobiidae and Cerambycidae were collected. The funding report (Bashford, 2002) demonstrates the effectiveness of this technique for a wide range of wood inhabiting insect species. The advantage of sticky traps is that continuous monitoring is achieved at very low cost compared to manual collecting. A comparison of sticky trap sampling of bark dwelling carabids with a hand collection technique at the same site was presented by Bashford (2001). It was estimated that 20 man-hours of manual searching of bark resulted in 14 species of carabids (36 specimens) being found. By comparison 5 man-hours of sticky trap servicing on the same number of trees yielded 15 species (247 specimens). However some larger species of beetle families such as cerambycids were only occasionally trapped. A total of 21 native species of cerambycids were collected using this technique, about one third of the known number at these sites. The technique has been used previously for monitoring a small species of cerambycid, the sugi bark borer *Semanotus japonicus,* by Shibata *et al* (1986). This study was able to correlate emergence hole numbers with adult populations attacking Japanese cedar and cypress trees. In more recent times the sticky band technique has been used in host selection monitoring over a four year period, of the exotic buprestid, *Agrilus planipennis,* in Canada (Lyons *et al.*, 2008).

Fig. 1. Examination of sticky band trap.

2.2 Field trial to evaluate lures

A large scale field trial was run in 2003 to test five commercially produced lures attractive to wood borer insects. Lures were obtained from Advanced Pheromone Technologies, Inc., and Phero Tech Inc. The five lures were ethanol, cineole, alpha-pinene, phellandrene, and a multilure (pinene, phellandrene, cineole, terpinene and cymene). The final selection for field testing was based on two other studies that included eucalypt attractants. (Brockerhoff *et al.*, 2006; Barata *et al.*, 2000). The lures were tested with four replicates of each in panel traps placed equally in a same age *Eucalyptus* plantation, half of which had been commercially thinned, following tree mortality caused by drought stress and subsequent wood-borer attack. As a result of the field trial the ethanol lure was selected to be used as a generalist lure for *Eucalyptus* wood-borers in the Bell Bay quarantine monitoring pilot study (Table 1).

	Thinned Plantation		Unthinned Plantation		
Lure	Species	Specimens	Species	Specimens	Total species
Ethanol	12	173	11	129	17
Pinene	11	102	10	78	16
Cineole	10	80	9	59	14
Multilure	10	68	7	23	11
Control	7	34	6	19	8
Phellandrene	5	21	5	35	6

Table 1. Attractiveness of tested lures on *Eucalyptus* wood-borer beetles

2.3 Portable static traps

Lindgren funnel traps, intercept panel traps, delta traps, chimney traps and bucket traps were tested against each other over a two season period using the same lures. The best performing traps in terms of both numbers of wood borers captured and number of species were the Intercept panel traps closely followed by the Lindgren funnel traps (Fig. 2).

Fig. 2. Intercept panel trap (left) and Lindgren funnel trap (right).

Lindgren funnel traps are currently used in warehouses at several western United States seaports to detect exotic insect pest species while imported goods are in storage. A series of lures within the traps are used to target likely incursions, especially those species with a history of incursion. The traps are quickly serviced and tests show high levels of detection of target species. It is important that the precise combination of lures/release rates/concentrations is used to attract target species/groups.

An evaluation of these traps and lure release mechanisms was warranted since many are now in commercial production and experience using these combinations has accumulated over the past two decades. For example several companies specifically researched the use of ethanol slow release lures within a range of different trap designs for timber insects including cerambycids. This is of special importance with the increasing threat of introduction of many exotic species into Australia. For example the Asian longhorn beetle,

Anoplophora glabripennis, has in recent years entered several western countries including United States with devastating economic and environmental consequences.

In order to further investigate the products available and techniques used in operational trapping in forestry I was awarded a Gottstein Fellowship which enabled me to travel to USA, Canada and UK (Bashford, 2003). The information obtained during this Fellowship has been subsequently applied to the development of urban trapping systems around Australia. The study tour provided information on the methodology and static trapping techniques used in the United States, Canada and United Kingdom for the early detection of exotic forest insects, methods of containment and eradication, and community involvement in those programs. The visit also included visiting a number of commercial trap and lure suppliers to see the latest applications and developments.

2.4 Pilot studies

Some species of exotic timber insects would have a devastating economic impact on the forest industry and other land management agencies if they became established in Australia. The cost of eradication or control of any damaging exotic insect would be millions of dollars added to the loss of resource. It is now recognised in counties such as Canada, United States and New Zealand that early detection is the vital key in preventing huge economic and resource losses. Early detection within the 5-km zone around port entry sites enables eradication to be attempted and containment measures to be initiated under the Generic Incursion Management Plan (GIMP). Placement of traps within plantation and nursery areas would provide a further early detection zone specifically targeting exotics of orchard and forestry trees.

Port area. Entry of cargo into any entry port provides a transport mechanism for the arrival of exotic pests and diseases. In recent times extra emphasis has gone into the examination of pallet wood, packing crates and airport warehouses as pathways and centres of biotic invasion. Despite the high levels of inspection the movement of some exotic insects out of entry port areas is inevitable.

Port surrounds area. An area of 5 km radius of the entry port site is used for trap site selection. This area may consist of high intensity buildings in cities, urban parks and reserves, and residential areas. In urban areas large numbers of trees and shrubs comprising a wide range of species, especially exotic species, are present in parks, street trees and gardens. In rural areas blocks of native bush, hedgerows and planted shelterbelts consist mainly of native tree species.

It is the experience in New Zealand, Canada, and United States it is within this 5 km zone around port of entry sites that initial establishment of exotic pest species occurs. It is also the case that if containment within this area is not achieved within two years of establishment then eradication is not possible. In Australia there is no formal monitoring for the detection of exotic forest insects within this entry zone.

Plantations. In many parts of Australia forestry plantations have been established within or close to the 5 km radius zone of entry ports. Often this is a practical decision for movement of commodities to and from ports. Establishment of a pest species within the port surround area provides a pathway for rapid movement into the plantation estate and severe economic consequences for the forest industry.

Prior to the pilot study trapping had been conducted at the port site and surrounding areas for one flight season as part of the trap testing program. Using funnel, panel and pipe traps a total of 21 wood borer species (267 specimens) were collected. This provided, with the blitz survey collections, a baseline collection against which to assess the pilot trapping results.

2.5 Pittwater pilot study

In a small area adjacent to Hobart International Airport both native forest, mainly *Eucalyptus* and *Acacia* trees and a *Pinus radiata* plantation are present. Panel traps, funnel traps and pipe traps were set up and run for two summer seasons to determine the woodborer species present. This data was added to previous collections made at this site in previous years to form a baseline species list and voucher collection. In all 61 species (1957 specimens) of woodborer insects were included in the voucher collection.

2.6 Rapid visual assessment surveys (Blitz surveys)

Rapid visual search surveys of the five major port entry sites for insect presence on leaves and stems of all trees and shrubs within a 2 km radius of the facility was conducted in 2002 -2003. The surveys followed a protocol established in Queensland by Wylie *et al.* (2000). All trees in twenty 1000 metre transects of native forest were examined for insect pests and pathogens (Fig. 3). The surveys were conducted in the north of the State at Bell Bay and in the south at Hobart International Airport. The collection data for woodborer species combined with the data from tree banding and static trapping has provided baseline data on the species of native woodborer species present at each site. Identification of this material had the establishment of voucher series within the TFIC enables us to rapidly screen subsequent monitoring material for exotics. We were able to show that visual searching was both inefficient in terms of man-hours and also ineffective in detecting many cryptic species. In the blitz surveys 3200 specimens were collected including leaf fungi and cankers as well as herbivore insects.

2.7 Testing the use of individual trees for regular inspection

At the Bell Bay site eleven tree species were obtained in advanced growth forms from a commercial nursery. These were planted out approximately 100 metres from the wharf area with a clear view to the unloading areas. The trees were planted in September 2004. Three blocks, each with two pairs of six tree species were set up with a drip irrigation system and surrounded by a one metre high floppy top fence of plastic trellis supported by star pickets. The trees were examined for insect pests and plant diseases every two months during the summer months. The trees were sprayed with the insecticide Confidor (imidacloprid ai: 0.25g/kg) following each inspection. Several new host records for Tasmania resulted from leaf and lesion samples taken from these trees and opportunistic sampling in parks and reserves. Leaf spot fungi and lesions were identified by specialists at the Tasmanian Department of Agriculture.

Cryptosporiopsis sp. on *Platanus acerifolia.*
Idiocercus australis on *Betula* sp.
Cladosporium orchidearum on *Acer rubrum.*
Phaeophleospora eucalypti on *Eucalyptus viminalis.*
Fairmaniella reposa on *Eucalyptus obliqua.*
Spilocaea sp. on *Alnus incana.*

Fig. 3. Blitz surveys. Use of a beating trap to capture herbivorous insects.

An option to Blitz surveys is the establishment of sentinel planting plots. A sentinel tree plot consists of small plantings of a range of tree species pruned to 3 metres (Fig. 4). These plots enable rapid examination for herbivorous insect species such as aphids and psyllids and observation of disease symptoms. Three or four plots at each site containing 2 - 3 trees of each selected species. Tree selection would include commercial timber trees, urban street trees, and fruit trees. This technique is an adaption of blitz sampling. The plots would be maintained as part of entry port landscape program and be sampled four times a year.

Fig. 4. Planting potted advanced growth trees for sentinel plots (upper) and established plots (lower).

2.8 Bell Bay pilot study

A pilot study was established to determine the logistic and cost requirements of running a practical monitoring system at one major port area. This two year study was funded by the Australian Commonwealth government Department of Agriculture, Fisheries and Forestry, at a cost of $A33, 000. The northern port of Bell Bay is the major entry port for container cargo into Tasmania. Shipping from mainland Australian ports and world-wide international shipping use this port. Static traps were set up within a 5km radius of the wharf area to encompass warehouses industrial work sites, cargo container storage facilities and adjacent native forest, parkland and a commercial *Eucalyptus* plantation. 'Intercept 'panel traps and Lindgren funnel traps were set up in pairs throughout the area. Ethanol lures were used at sites near native tree species and pinene lures near softwood host species. A Gypsy Moth lure delta sticky trap was placed with each pair of traps. Within the plantation traps were placed in areas that were (a) fire damaged (b) drought stressed and (c) apparently healthy. An addition panel trap was included with each pair within the plantation areas charged with a generalist longhorn beetle lure. The three traps were set in a triangle some 5 metres apart. The traps were serviced monthly from September until May for two years. A total of 75 wood-borer species were collected.

The number of traps was dependant on the budget provide for the pilot project and this in turn by the projected future budget for operational monitoring at four port and one airport within Tasmania. Included in the financial assessment was cost of quarantine officers to service the traps, travel costs annual supply of lures and trap equipment, sorting of samples and subsequent diagnostics. The advantage of a baseline collection of woodborers for each site from previous sampling greatly reduced the diagnostic effort required.

2.9 Operational urban surveillance for woodborer insects using static traps

Commonwealth funding was obtained to place traps in urban areas around commercial ports and airports in Tasmania and two mainland ports (Brisbane - Queensland and Fremantle – Western Australia) for the detection of exotic woodborer insects.

2.9.1 2005-2006

In December 2005, the Urban Hazard Site Surveillance Project commenced in Tasmania targeting exotic wood-borer insects. Staff of Quarantine Tasmania, at each port site, was trained in trap servicing, specimen collection and transportation of those samples to the diagnostic laboratory in Hobart (Fig. 5). At each of the four port sites and Hobart International Airport static traps were set up within a 5km radius of the transit site. At each site, two types of static traps were used – panel traps (with ethanol lures) and funnel traps (with alpha-pinene/ethanol lures). In late February Elm Bark Beetle (*Scolytus multistriatus*) lures were added to park sites in Hobart where Elm trees were present . The addition of Elm Bark Beetle lures enhanced the detection of Elm Bark Beetle (which is a vector of Dutch elm disease). The beetle has not been recorded in Tasmania but is widespread in mainland Australia. Other target species were *Arhopalus ferus* (Cerambycidae), and *Ips grandicollis* (Scolytidae), both present in mainland Australia. Exotics of high threat indices for Tasmania which were targeted were the cerambycid species *Monochamus* species, (vectors of the Pine Wilt Nematode *Bursaphelenchus* sp.), *Stromatium barbatum*, *Hylotrupes bajulus* and

Anoplophora glabripennis. To select suitable sites City Councils in each region were consulted. The knowledge of parks and gardens managers proved very useful in identifying parks, gardens and land reserves suitable for the placement of traps. The traps were serviced fortnightly from late December to late April. This servicing involved the collection of the sample jar at the base of the trap (a mixture of dilute antifreeze (propylene glycol) and detergent) and change of lure at the appropriate time. The sample bottles were packed in tote boxes and transported to the diagnostic office by public transport bus cargo.

Fig. 5. Training quarantine officers in static trap assembly and set up.

A combined total of 39 species of woodborer insects were collected (4082 specimens). Four species were known established exotics and new records of establishment in the State were recorded for the two Australian native scolytids *Xyleborus perforans* and *Ficicis varians*.

Cost of the first year of monitoring, which included purchase cost of reusable traps and support pickets, was $A85, 230.

2.9.2 2006-2007

In the second year of operational monitoring the same port and airport areas were utilised but traps placed at different sites within those areas. The monitoring was expanded to include the main importing plant nursery in each port area. Five pairs of traps were placed in each nursery. In addition mature growth potted trees three meters in height were purchased to act as sentinel trees within wharf boundaries at all sites. The trees were a combination of street trees, commercial forestry trees and several orchard species. The aim was the early detection close to unloading wharf areas for insect species and leaf fungi. The bagged trees were placed were placed on wheeled frames so they could be moved easily if required. The trees were to be watered regularly and examined every two weeks for insect or fungal presence. Tree were pruned to 3 meters so all foliage could be examined. The tree species were as follows.

Urban street trees

Tilia cordata (Small-leaved Lime)
Betula alba (Silver Birch)
Ulmus carpinifolia (Elm)
Fraxinus americana (Ash)
Quercus robur (English Oak)
Acer pseudoplatanus (Plane)

Orchard trees

Malus spectabilis (Crab Apple)
Prunus armeniaca (Apricot)

Forestry plantation trees

Eucalyptus globulus (Blue Gum)
Acacia melanoxylon (Blackwood)
Pinus radiata (Radiata Pine)

The trees at Bell Bay were planted as part of the landscaping program on a cleared flat area overlooking the length of the wharf area. Problems were experienced with maintaining the health of the potted trees in the arid environment of the wharf areas. The trees suffered from infrequent watering and dessication due mainly to strong wind movements caused by traffic and the salt in offshore breezes. The result was that very few samples were collected and the general health of the plants resulted in the use of the trees ceased halfway through the summer season. The planted trees at Bell Bay thrived in being just outside the wharf area in a more benign site. The trees established well and many samples were collected.

A new initiative was the production of a leaflet outlining the aims of the project. These leaflets were letterboxed to all businesses and households within the survey areas to inform the public of the reason for the traps and to convey a sense of ownership.

A combined total of 33 species of woodborer beetles was collected (832 specimens).

Additional specimens of the two scolytids collected last year were added to the Tasmanian Forest Insect Collection (TFIC). Of interest were specimens of the anthribid *Euciodes suturalis* collected at two northern sites. This species has been established in New Zealand since 1921 being of European origin. (Penman, 1978). The species is a stem borer pest of cereals and the specimens collected are the first confirmed record for Tasmania.

A collection of exotic woodborer insect species has been established within the Tasmanian Forest Insect Collection. A large number of specimens of important species not yet established in Australia were donated by overseas colleagues has greatly enhanced Tasmania's diagnostic capability.

Cost of second year monitoring was $A63, 000.

2.9.3 2007-2008

The trapping program ran from October until the end of May at the same port and airport sites and nurseries as last year. Trap placement within the nurseries was refined to specifically target container unloading areas. A new initiative this year was to conduct exotic ant surveys at selected nurseries, town park areas and within wharf areas at ports. The methodology was established by Bashford and Pompa (2007) using 'BaitPlate' ant traps. Traps were set up in pairs bated with meat and honey. Two trapping periods in December 2007 and March 2008 resulted in a total of 7140 ants of fifteen species were captured. The established exotic Argentine ant, *Iridomyrmex humilis,* was a common capture at many sites.

A total of 3289 woodborer specimens of 37 species were collected in the routine trap monitoring. The bostrichid bark beetle *Xylotillus lindi* (Blackburn) was recorded for the first time in Tasmania with a total of 15 specimens caught in traps at Hobart International Airport. The scolytid *Cryphalus pilosellus* was collected, confirming the single previous record for the species in Tasmania. The grass anthribid *Euciodes suturalis,* first captured last season was again recorded from several northern locations. The Sirex Woodwasp *Sirex noctilio* and its associated egg parasitoid *Ibalia leucospoides* were captured in traps placed in pine windbreaks at two port surround sites.

The use of potted sentinel trees was deleted from the programs with the problems of tree maintenance being too labour intensive in relation to the results obtained. The planted trees at Bell Bay were pruned and continue to be monitored.

Cost of third year monitoring was $A64, 000.

2.9.4 2008-2009

A decision was made not to continue the trapping program as scheduled for the past three years. The rational being that a continues three year intensive monitoring program with a 5 km radius of each port would have detected the presence of newly established exotic woodborers, all of which have a generational period of at least one year in the climatically temperate island of Tasmania. Part of the annual funding was directed to compiling data on the distribution of three established exotic bark beetles within Tasmania. The other component of funding was directed to monitoring several exotic agricultural pests and diseases within Tasmania. The bark beetle information would give indications of rate of spread and potential distribution of a new exotic. The established bark beetles selected for a detailed trapping program were the anobiid *Ernobius mollis*, and two scolytids *Hylastes ater* and *Hylurgus ligniperda*, Both *Hylastes* and *Hylurgus* are potential vectors of pine pitch canker (*Fusarium subglutinous)*, a very damaging *Pinus* fungal disease not yet recorded in Australia. All are of European origin and have been established for several decades in Australian *Pinus* plantations and are attracted to pinene lures. The Tasmanian Forest Insect Collection has numerous records for the three species selected and this enabled the selection of *Pinus* sites for which there were no records. Traps from the port surveillance program were placed in ten sites using funnel traps set with pinene lures. The traps were run for nine weeks from mid-January. Samples were removed from the traps every three weeks and new lures set. The funnel traps were set within recently logged or pruned coupes with four traps, spaced 100 metres apart in a transect within each coupe. During the trapping season 330 target beetles were captured. *Hylurgus* 81 specimens from 8 sites, *Hylastes* 85 specimens from 8 sites and *Ernobius* 33 specimens from 5 sites.

Cost for the bark beetle trapping program was $A24, 000.

2.9.5 2009-2010

Trapping for the three bark beetles was conducted in the remaining ten non sampled and negative record coupes using the same regime as the previous year. A total of 330 bark beetles were collected comprising *Hylurgus* 77 specimens from 7 sites, *Hylastes* 216 specimens from 6 sites, and *Ernobius* 37 specimens from 1 site. Of interest, the two large Bass Strait islands, which have small old plantings of *Pinus radiata*, both had *Hylastes* and *Hylurgus* present while *Ernobius* was found only on Flinders Island. *Sirex noctilio* was captured on both islands. The data has been incorporated into a national survey for these species (Bashford, unpublished).

The capture of three specimens of the rare platypodid, *Carchesiopygus dentipennis*, previously only recorded from Queensland and New South Wales as single specimens, is a surprising new record for Tasmania (Fig. 6).

Cost of the bark beetle distribution project was $A35, 000.

2.9.6 2010-2011

After a break of two years the wood-borer port surrounds trapping program was re-established at the three northern ports of Burnie, Devonport and Bell Bay. In the south the port of Hobart and Hobart International Airport were targeted. It was felt that any new wood-borer incursions which had established might have populations large enough to be detected by trapping. The radius of the trapping area outside the port environs was extended to 7 km in order to utilise a superior range of trapping sites. The lure combinations were changed to include generalist and specific target lures. The panel traps carried ethanol and *Monochamus* specific lures. The funnel traps carried pinene and ipsdienol lures to specifically target bark beetles and *Arhophalus ferus*. A total of 3208 wood borer specimens were collected comprising 34 species.

Two specimens of the cerambycid, *Atesta bifasciata,* were collected at a northern site, a new record for Tasmania. Numerous specimens of *Tropis oculifera*, an uncommon species of cerambycid, was collected in one trap at Burnie.

Cost of the woodborer urban trapping program this year was $A38, 000.

Fig. 6. Specimen of the rare platypodid *Carchesiopygus dentipennis*.

3. Conclusion

This paper documents the first attempt to monitoring outside port areas for exotic woodborer insects on a regular basis. The introduction pathway, through international cargo port into surrounding urban areas, has been well documented (Brockerhoff and Bain, 2000). Expansion of that initial population into forest areas including plantations is a secondary establishment phase where populations are generally deemed to be permanently established. To achieve successful eradication or containment exotic incursions need to be limited in their distribution and restricted to small defined populations. These populations can only be detected by a routine pattern of monitoring at the site of potential establishment.

A monitoring system within forestry plantations, both hardwood and softwood species has been developed that compliments the documented urban monitoring system (Bashford 2008). This system would allow the early detection within plantations of a new exotic wood-borer previously detected in the urban port surrounds monitoring. This system is vital if eradication or containment is to be achieved. However detection within plantations usually means a degree of establishment by an exotic wood-borer. By combining the two systems then early detection in port surrounds enables species specific monitoring to be conducted in nearby plantations.

In the future, depending on Commonwealth funding, trapping will be conducted every third year for wood borer insects of interest to commercial forestry. Since the initial trapping year in 2005 the post barrier program has been expanded to include monitoring for many agricultural plant pests and diseases, tramp ants, Asian gypsy moth trapping, and orchard pests.

There is scope within the program for the addition of surveys specifically to detect leaf spot fungi and cankers of commercial forestry tree species growing within the urban environment.

The program is an evolving process regulated by financial and manpower limitations. An ideal monitoring program would involve many more traps and the costs for servicing and diagnostics would be considerably increased. In the current program up to 120 traps are utilised which is the maximum that can be serviced in a single day at each port entry site. Although not ideal, having a fixed budget and a flexible system enhances the long term commitment of funding.

Another avenue of research that would enhance the system is the use of multiple lures within a single trap. It may be possible to add a number of lures to specifically targeted single species. The generalist kairomone lures currently used could be augmented by specific pheromone lures to attract a suspected new incursion or other woodborer species not attracted to the generalist lures. An example could be the European house borer, *Hylotrupes bajulus*, which infests seasoned softwood timber (Gove *et al.*, 2007). Some preliminary work has been reported by Reddy & Guerrero (2004), looking at the interactions between insect pheromones and plant semiochemicals. Work done on stored-product beetle species (Athanassiou & Buchelos, 2000) showed that a none pheromone multi- attractant (generalist attractant) could be more efficient if formulated for target species.

Although the program has not detected any new exotic (not Australian native species) wood-borer species, the potential to do so is illustrated by the number of new species records for Tasmania of mainland native species. The establishment of a large voucher series of native woodborer species for each international cargo entry site has greatly enhanced our ability to quickly determine the incursion of a new exotic or mainland species.

The establishment of an aggressive or fungal vector species of wood-borer insect could be devastating to Tasmanian forestry plantation given the transition out of native logging and a reliance on plantation timber production. Without a monitoring system, detection would only occur once a species was established and causing some visual damage. The cost of attempting eradication or containment at this stage would be considerably higher than early control and be of longer duration. (McMaugh, 2005).

The program described in this paper has been adopted by several major international shipping ports on the mainland of Australia and integrated into existing systems at ports in the United States, New Zealand and the pacific islands of Fiji and Vanuatu.

4. Acknowledgements

The pilot study at Bell Bay and the operational monitoring was funded by the Commonwealth Department of Agriculture, Fisheries and Forestry. Michael Cole and Paul Pheloung assisted in obtaining ongoing funding and provided access to information from other jurisdictions.

Peter Brown, Rebecca Boon, Megan Szczerbanik and Ben U'ren (Quarantine Tasmania) assisted with project planning and ensured officers were made available on a regular basis for trap servicing. Tim Wardlaw (Forestry Tasmania) allowed time within my work program to conduct the lure trials, attendance at training workshops and overall supervision of the program. The relevant managers all allowed access to restricted port areas and park areas, and encouraged their staff to cooperate in trap establishment activities.

5. References

Athanassiou, C.G., & Buchelos, C. (2000) Comparison of four methods for the detection of Coleoptera adults infesting stored wheat: efficiency and detection sensitivity. *Anzeiger fur Schadlingskunde Pflanzenschutz Umweltshutz* 73: 129-133.

Barata.; E.N.; Pickett, J.A., Wadhams, L.J., Woodcock, C.M., & Mustaparta, H. (2000) Identification of host and nonhost semiochemicals of Eucalyptus woodborer *Phoracantha semipunctata* by gas chromatography-electroantennography. *Journal of Chemical Ecology* 26(8):1877-1882.

Bashford, R. (2001) Some records of arboreal carabid beetles in Tasmania. *Victorian Entomologist* 31(6):97-100.

Bashford, R. (2002) The use of sticky traps as a long term monitoring tool for exotic saproxylic insects of importance to forestry. *Victorian Entomologist* 32:42-44.

Bashford, R. (2003) The use of static traps for the detection and monitoring of exotic forest insects. *J. W. Gottstein Memorial Trust Fellowship Award Report.* 55pp.

Bashford, R. (2004) Port surrounds monitoring for exotic forest insects and tree pathogens. Report to *Department of Agriculture, Fisheries and Forestry,* Canberra. 28pp.

Bashford, R. (2008) The development of static trapping systems to monitor for wood-boring insects in forestry plantations. *Australian Forestry* v71 (3): 236-241.

Bashford, R., & Pompa, Z. (2007) Development of an Urban Surveillance Protocol for Exotic Ants. *Technical Report* 11/2007. Forestry Tasmania, Hobart, Tasmania. Pgs15.

Brockerhoff, E.G., & Bain, J. (2000) Biosecurity Implications of Exotic Beetles Attacking Trees and Shrubs in New Zealand. *New Zealand Plant Protection,* 53:321-327.

Brockerhoff.; E.G., Jones, D.C., Kimberley, M.O., Suckling, D.M., & Donaldson, T. (2006) Nationwide survey for invasive wood-borer and bark beetles (Coleoptera) using traps baited with pheromones and kairomones. *Forest Ecology and Management* 228:234-240.

Bulman; L.S., Kimberley, M.O., & Gadgil, P.D. (1999) Estimation of the efficiency of pest detection surveys. *New Zealand Journal of Forestry Science* 29(1):102-115.

Carter, P.C.S. (1989) Risk assessment and pest detection surveys for exotic pests and diseases, which threaten commercial forestry in New Zealand. *New Zealand Journal of Forest Science* 19(2/3): 353-374.

Gove, A.D.; Bashford, R., & Brumley, C.J. (2007) Pheromone and volatile lures for detecting the European house borer (*Hylotrupes bajulus*) and a manual sampling method. *Australian Forestry* 70(2):134-136.

Lyons, D.B.; de Grout, P., Jones, G.C., & Scharbach, R. (2008) Host selection by *Agrilus planipennis* (Coleoptera: Buprestidae): inferences from sticky-band trapping. *The Canadian Entomologist* 141(1):40-52.

McMaugh, T. (2005) Guidelines for surveillance for plant pests in Asia and the Pacific. *ACIAR Monograph* No. 119, 192p.

Penman, D.R. (1978) Biology of *Euciodes suturalis* (Coleoptera: Anthribidae) infesting Cocksfoot in Canterbury. *The New Zealand Entomologist* 6(4):421-425.

Reddy, G.V.P., & Guerrero, A. (2004) Interactions of insect pheromones and plant semiochemicals. *Trends in Plant Science* 9(5):253-261.

Shibata, E.; Okuda, K., & Ito, T. (1986) Monitoring and Sampling Adult Populations of the Sugi Bark Borer, *Semanotus japonicus* Lacordaire (Coleoptera: Cerambycidae), by the Sticky Trap Banding Method in Japanese Cedar Stands. *Applied Entomology and Zoology*. 21(4):525-530.

Wylie, R.; Ramsden, M., McDonald, J., Kennedy, J., & De Baar, M. (2000) Port and Airport Environs Surveillance. *Queensland Forest Health Surveillance Report* 1999-2000. 40pp.

7

Changes in the Relative Density of Swamp Wallabies (*Wallabia bicolor*) and Eastern Grey Kangaroos (*Macropus giganteus*) in Response to Timber Harvesting and Wildfire

Kelly Williamson, Helen Doherty and Julian Di Stefano
Department of Forest and Ecosystem Science
University of Melbourne
Australia

1. Introduction

Natural disturbances such as wildfire and storms act as major regulating forces in forest ecosystems (Attiwill, 1994; Lugo, 2000; Ryan, 2002), and in more recent times human disturbances such as urbanisation, land clearing and timber harvesting have also had a marked impact on forest extent and structure, and on the distribution and abundance of forest dwelling organisms (Abrams, 2003; Dale et al., 2000; Gaston et al., 2003; Thompson et al., 2003; Wilson and Friend, 1999). For many animals, disturbance events alter predation risk through changes to forest structure, and effect the distribution and abundance of food resources. Because both predation (Ferguson et al., 1988; Hughes et al., 1994) and food (Geffen et al., 1992; Tufto et al., 1996) can have a strong influence on movement patterns and habitat use, disturbance is predicted to alter the habitat choices of many species.

How disturbances such as fire and timber harvesting alter the habitat use of medium to large mammalian herbivores is variable and species dependant (Fisher and Wilkinson, 2005). For example, woodland caribou (*Rangifer tarandus caribou*) tended to avoid early successional stands, whether these resulted from timber harvesting (Mahoney and Virgl, 2003) or fire (Schaefer and Pruitt, 1991). In contrast, other ungulates responded positively to recently burnt (Archibald and Bond, 2004; Gasaway et al., 1989) or harvested (Cederlund and Okarma, 1988; Sullivan et al., 2007) sites, probably due to increased food resources in these areas. The response of herbivores to fire and timber harvesting also changes over time, with effects being detected across a range of temporal scales (Mahoney and Virgl, 2003; Pearson et al., 1995). Catling et al. (2001) showed that in some cases temporal responses of fauna to fire and habitat structure may be non-linear.

In mainland southeastern Australia, macropodid marsupials (kangaroos and wallabies) are the dominant native medium to large herbivorous fauna in woodland and forest ecosystems, with eastern grey kangaroos (*Macropus giganteus*) and swamp wallabies (*Wallabia bicolor*) two of the most abundant species. Eastern grey kangaroos are predominantly grazers (Taylor, 1983), preferring heterogeneous habitats that provide relatively open, high quality foraging sites close to or interspersed with shelter vegetation (Hill, 1981; Moore et al., 2002; Southwell, 1987; Taylor, 1980). In contrast, swamp wallabies are mixed feeders (Davis et al., 2008; Di Stefano and Newell, 2008; Hollis et al., 1986) who prefer densely vegetated habitat, particularly during the day (Di Stefano et al., 2009; Lunney and O'Connell, 1988; Swan et al., 2008; Troy et al., 1992).

The response of these species to timber harvesting and fire has been rarely studied, particularly for eastern grey kangaroos. In native Eucalypt forest, swamp wallabies used recently harvested (1-5 year old) areas more than older regenerating and unharvested sites (Di Stefano et al., 2007; Di Stefano et al., 2009; Lunney and O'Connell, 1988), although factors such as topographical position (Lunney and O'Connell, 1988), sex and diel period (Di Stefano et al., 2009) were important factors influencing habitat use at finer spatial and temporal scales. In a plantation forestry environment, Floyd (1980) found that swamp wallaby density was uniformly high at two year old sites and uniformly low at one year old sites. Density in unharvested forest and ten year old sites was higher at the periphery than in the centre, an effect probably influenced by the close proximity of food resources to the edges of these more sheltered environments. Fewer data are available about the response of eastern grey kangaroos to timber harvesting. Hill (1981) found that kangaroo density was higher in partially harvested forest than in either unharvested forest or open woodland, and suggested this was related to the spatial interspersion of both food and shelter within harvested stands.

There appear to be no published data quantifying the response of swamp wallabies to fire, although Catling et al. (2001) presented an analysis for a pooled sample of swamp and red-necked wallabies (*Macropus rufogriseus*). In this case, time since fire had little effect on wallaby abundance in structurally simple habitats but there was a strong negative relationship in structurally complex habitats. The use of burnt habitat by kangaroos is likely to increase after fire (Catling et al., 2001; Southwell and Jarman, 1987), probably as a result of increased nutrient content of regenerating forage (Murphy and Bowman, 2007) and the removal of dense movement-impeding vegetation (Taylor, 1980). In a dry sclerophyll landscape, Catling et al. (2001) observed a non-linear relationship between kangaroo density and time since fire, with density increasing for about a decade before declining.

Our objective was to quantify changes in the relative density of swamp wallabies and eastern grey kangaroos in response to disturbance. Timber harvesting and wildfire in a commercially managed mixed species eucalypt forest provided four alternative habitat types, and we expected the density of both species to differ between them. On the basis of the work cited above we predicted that swamp wallabies would use densely vegetated sites more than open ones, and that wallaby density would be positively related to the abundance of lateral cover. In contrast, we expected kangaroo density to be positively related to food availability (grass) (Taylor, 1980, 1984), but for this effect to be moderated by vegetation density, as both food and shelter are likely to influence habitat choices by this species (Hill, 1981; Moore et al., 2002; Southwell, 1987).

In addition, the co-occurrence of a wildfire and one of the harvesting events provided a rare opportunity for direct comparison between harvested and burnt sites of the same age, both with respect to the structure of the regenerating forest and the response of the focal species. This comparison is important from a conservation perspective, as congruence between harvesting effects and those of natural disturbances such as fire are likely to result in better conservation outcomes in managed landscapes (Hunter, 1993; Lindenmayer et al., 2006).

2. Methods

2.1 Study site

The Pyrenees State Forest in west central Victoria (37°05′ latitude, 143°28′ longitude) is dominated by a single range running approximately east-west and rising to 750 m above sea level (Figure 1). This dry sclerophyll landscape is dominated by two Ecological Vegetation Classes: Grassy Dry Forest on the northerly aspects and Herb Rich Foothill Forest on the southerly slopes. Messmate/blue gum (*Eucalyptus obliqua*/*E. globulus bicostata*) associations dominate the overstorey and constitute a forest type often referred to as Low Elevation Mixed Species. A sparse shrub layer includes silver wattle (*Acacia dealbata*), common heath (*Epacris impressa*), gorse bitter-pea (*Davisia ulicifolia*) and austral bracken (*Pteridium esculentum*). A more detailed description of vegetation and other site characteristics is given in Di Stefano et al. (2007).

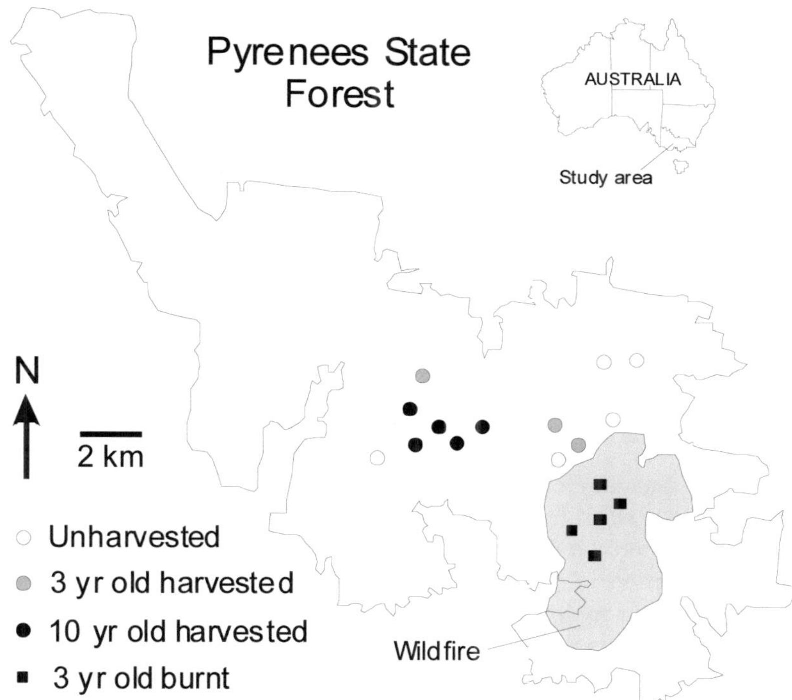

Fig. 1. Map of the Pyrenees State Forest, southeastern Australia, showing the location of the sampling plots.

Since 1990, timber harvesting in the Pyrenees has created 25 regenerating blocks (coupes), 10-30 ha in size and of various ages, surrounded by essentially unharvested forest. The harvesting technique is referred to as the seed tree system, and involves the retention of four to nine mature trees per hectare to provide seed for the next crop and habitat for arboreal animals. Harvesting generally takes place between late spring and autumn (October to April) after which logging debris is burnt to prepare a seedbed and stimulate seed fall (Lutze et al., 1999).

In addition to unharvested areas we used three year old and ten year old regenerating sites to represent three age classes within the harvested landscape. Three year old sites were dominated by relatively homogeneous stands of 1-2 m tall, densely regenerating *Eucalyptus* seedlings, and contained substantial quantities of silver wattle, austral bracken, grass and forbs. Ten year old sites supported patches of dense, closed stands of 3-6 m tall eucalypt regeneration and had variable levels of forb, grass and shrub cover. The cover of bracken, however, remained relatively high.

In March 2001 a wildfire burnt about 2000 ha of forest at the eastern end of the study area (Figure 1). The wildfire and the post-harvest burns at the three year old regenerating sites occurred within a few months of each other, thus presenting the opportunity to compare the impact of the two disturbance types within the same forest ecosystem. In contrast to three year old harvested sites, the burnt area retained many live mature trees regenerating via lignotubers at the base of the trunk and epicormic shoots. It was also relatively heterogeneous, with some sites devoid of regenerating plants, but others densely populated by regenerating eucalypt seedlings, silver wattle and austral bracken. Some patches within the burnt area supported stands of eucalypt seedlings at even higher densities than the harvested areas.

2.2 Experimental design and monitoring

We used all the available three year old sites (n = 3) and selected a random sample of ten year old sites (n = 5) from a pool of eight that were available. In the unharvested and burnt areas we used the road network as a sampling frame and randomly selected sites (n = 5 in each area) from a larger pool of potential areas. On most occasions sites were more than one kilometre apart, although limitations imposed by the road network and the location of harvested areas meant that a few were somewhat closer together. In addition, the location of the burnt and harvested areas prevented the spatial interspersion of sites from each of the four habitats. The locations of the sites within the study area are shown in Figure 1.

At each site, ten 15 m^2 circular plots (radius 2.18 m) were established in a 5 × 2 arrangement on a randomly positioned 40 m grid and within these the accumulation of kangaroo and wallaby faecal pellets were monitored over time as a surrogate for herbivore abundance (Southwell, 1989). In the interpretation of results we make the assumption that relative differences between pellet numbers in each habitat are an accurate reflection of relative differences in animal density. Plots were cleared of faecal pellets in June 2004, and then pellets were counted approximately two and four months later during late winter (July/August) and spring (September/October). Raw data were converted to pellets $ha^{-1}day^{-1}$ prior to analysis. We were able to differentiate between wallaby and kangaroo faecal pellets relatively easily on the basis of size, shape, colour and internal texture (Triggs, 2004).

During the spring monitoring time we used the plots described above to quantify shelter and food resources at each site. For both species shelter (lateral cover) was measured as the vertically projected cover of any live or dead vegetation between 0.5-3 m. Food resources were measured as the cover of grass for kangaroos and the cover of forbs for wallabies, as eastern grey kangaroos eat mainly grass (Taylor, 1983) and swamp wallabies consume substantial quantities of forbs (Hollis et al., 1986), particularly in the Pyrenees (Di Stefano and Newell, 2008). In addition, we quantified the percentage cover of tree canopy, live or dead vegetation <0.5 m, and a variable representing the combined cover of litter and woody debris. Cover values were estimated by a single observer (K.W.) to the nearest 5%. The cover of vegetation <0.5 m was discarded prior to analysis as it was highly correlated with the cover of grass.

2.3 Data analysis

Principle components analysis (PCA) and an associated vector fitting procedure were used to assess how the 18 sites differed with respect to the measured habitat variables. The routine was run in PRIMER 6 and each variable was standardised by its maximum value prior to analysis. Analysis of similarities (ANOSIM), also run in PRIMER 6 on a Euclidian resemblance matrix derived from the standardised data, was used to test for multivariate difference among habitat types.

We used repeated measures ANOVA in GenStat 13 to analyse the effects of habitat type and season (winter, spring) on relative macropod density. Differences between particular habitat types were then assessed by constructing a table of contrasts including mean differences and their associated 95% confidence intervals. Assumptions of normality and homogeneity of variance were assessed using half-normal and fitted value plots respectively. For both data sets the variance was somewhat heterogeneous but as transformations had little effect on the output we ran the analysis on the raw data.

Generalised linear mixed models were used to assess the relationship between relative density and three predictor variables: season, shelter and food. In addition to its role as a predictor, season was used as a random factor in the statistical model to account for the nesting of season within site. Assumptions of normality and homogeneity of variance were assessed for the top ranked models using half-normal and fitted value plots respectively, and no transformations were deemed necessary. Models were run in R v. 2.13 using the lme4 package (Bates and Maechler, 2009).

We developed a candidate set of eight models designed to test the consistency of the data with our a priori expectations about the relationships between relative density and resource abundance. First, we built three models examining the effect of food, shelter, and the interaction between the two. We then build three additional models examining the interaction between season and the two primary resources. Finally, we added a global model which included all three variables and their interactions, and a null model (containing only the intercept and the random factor) to provide a baseline against which potentially more informative models could be compared. Models including interactions always contained the main effects of the interacting variables. In addition, food and shelter values were centred before computing the interaction between them to avoid colinearity between the original variables and their multiplicative effect (Quinn and Keough, 2002).

Models were compared using an information theoretic approach (Burnham and Anderson, 2002). Akaike's Information Criteria (AIC) and Akaike weights were used to rank models and assess their relative fit to the data. The importance of each variable (season, shelter and food) was assessed by calculating predictor weights, the sum of the Akaike weight for each model containing the focal variable (Burnham and Anderson, 2002).

3. Results

3.1 Disturbance and resource availability

The amount of shelter vegetation, forbs and grass in each of the four habitat types is shown in Figure 2. Unharvested sites were characterised by low levels of shelter vegetation and somewhat higher levels of grass and forb cover. All three year old harvested sites contained substantial amounts of shelter while three year old burnt sites had highly variable levels of shelter and grass. Shelter at three year old harvested and burnt sites was composed mostly of regenerating eucalyptus seedlings, austral bracken (*Pteridium esculentum*) and silver wattle (*Acacia dealbata*). Grass cover at ten year old harvested sites was also highly variable, with moderate levels of shelter also appearing in this habitat.

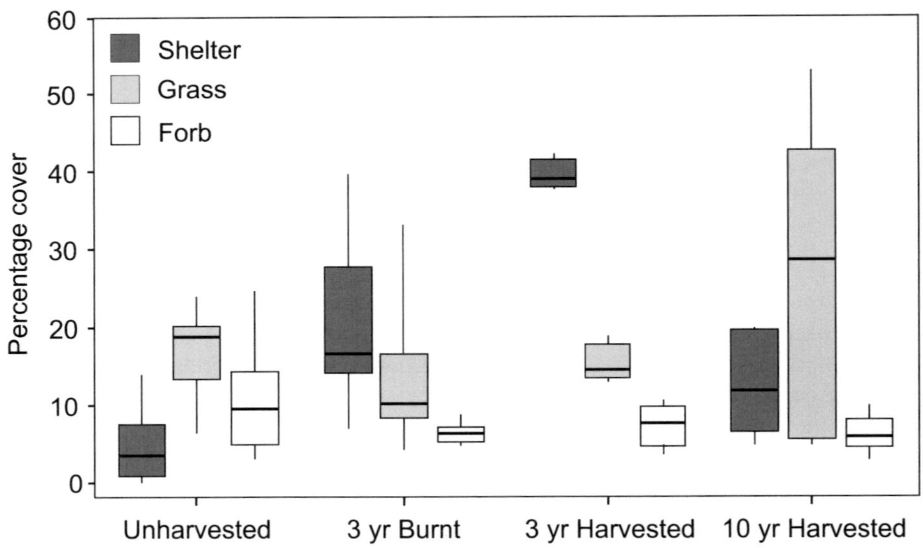

Fig. 2. The percentage cover of shelter vegetation, grass and forbs within the four habitat types. Boxes represent the interquartile range, horizontal lines represent medians and vertical lines represent minimum and maximum values.

The ordination plot from the principle components analysis (Figure 3) shows how the 18 sites differ with respect to the cover of forbs, grass, shelter, tree canopy and litter and woody debris. The X- and Y-axes in the figure (PC1 and PC2) represent 55.6% and 27.5% of the variance in the data respectively. As shown by the vectors, sites separated along PC1 differed predominantly with respect to shelter vegetation and canopy cover while sites separated along PC2 differed with respect to grass and litter and woody debris. The degree to which sites of the same habitat type are clustered in the same section of Figure 3 represents the variance within each of the habitat groupings. By this measure, three year old harvested sites and, to a lesser extent, unharvested forest sites were relatively similar to others in the same group. In contrast, sites within the ten year old harvested and three year old burnt habitats were highly variable with respect to at least some of the measured variables. Analysis of similarities (ANOSIM) indicated that there were strong to moderate differences between all habitat types except for ten year old harvested and three year old burnt sites, which were very similar (Table 1).

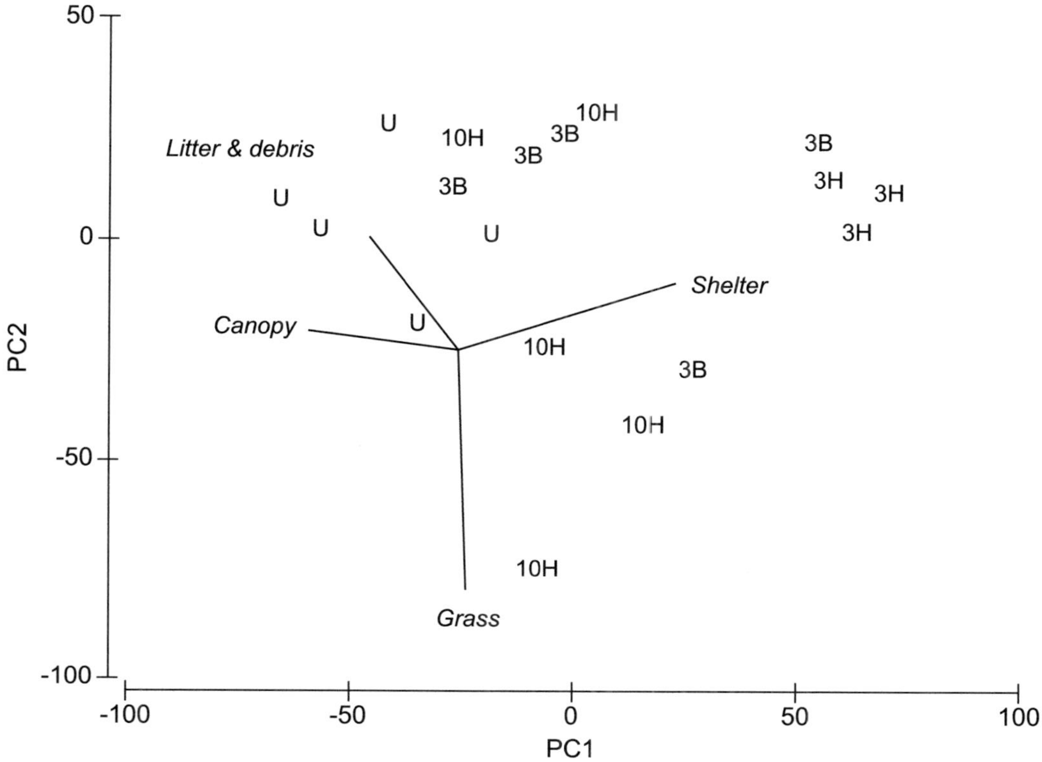

Fig. 3. Ordination plot from the principle components analysis showing how the 18 sites differed with respect to forbs, grass, shelter, canopy and litter and woody debris. The vectors (lines) on the diagram represent the strength and direction of the correlations between the original variables and the first two principle components, PC1 and PC2. U, unharvested forest; 3H, three year old harvested forest; 10H, ten year old harvested forest; 3B, three year old burnt forest.

Contrast	R	P-value
3 yr Harvested vs Unharvested	0.97	0.02
10 yr Harvested vs Unharvested	0.27	0.06
3 yr Burnt vs Unharvested	0.34	0.02
3 yr Harvested vs 10 yr Harvested	0.62	0.02
3 yr Harvested vs 3 yr Burnt	0.42	0.07
3 yr Burnt vs 10 yr Harvested	0	0.43

Table 1. Results of the Analysis of similarities (ANOSIM) procedure testing for multivariate differences between habitat types. R ranges between 0 and 1 and is a measure of effect size.

Contrast	Mean difference (pellets/ha/d)	LCL	UCL	P-value
Swamp wallaby				
3 yr Harv. vs Unharvested	268.7	187.2	350.2	<0.001
10 yr Harv. vs Unharvested	68.7	-1.9	139.3	0.061[1]
3 yr Burnt vs Unharvested	90.9	20.3	161.5	0.02
3 yr Harv. vs 10 yr Harv.	200.0	118.5	281.5	<0.001
3 yr Harv. vs 3 yr Burnt	177.8	96.3	259.3	<0.001
3 yr Burnt vs 10 yr Harv.	22.2	-48.4	92.8	0.51
Eastern grey kangaroo				
3 yr Harv. vs Unharvested	-40.2	-137.6	57.2	0.39
10 yr Harv. vs Unharvested	-43.0	-127.4	41.4	0.29
3 yr Burnt vs Unharvested	-27.3	-108.1	60.7	0.56
3 yr Harv. vs 10 yr Harv.	2.8	-94.6	100.2	0.95
3 yr Harv. vs 3 yr Burnt	-16.5	-113.9	80.9	0.72
3 yr Burnt vs 10 yr Harv.	19.3	-65.1	103.7	0.63

Table 2. Contrasts between habitat types for swamp wallabies and eastern grey kangaroos. A positive mean difference indicates that the value of the first listed habitat is larger than the second. LCL and UCL are lower and upper 95% confidence limits respectively. [1] Interpreted as evidence that 10 yr old harvested sites are used more than unharvested sites.

3.2 Relative density

The relative density of swamp wallabies was influenced by both habitat type and season ($P < 0.001$ in both cases; Figure 4a). The P-value associated with the habitat type by season interaction was also small (0.03), although this effect was primarily driven by very low density at unharvested sites during spring (Figure 4a), with patterns of use during the two seasons similar overall. Consequently, we consider it reasonable to interpret the main effects of habitat type and season. The ANOVA r^2 (SS_{Factor} / SS_{Total}) showed that habitat type explained 52.9% of the variance in the data, relative to 16.2% and 7.9% for the effects of season and the interaction respectively. Contrasts between the four habitat types (Table 2) demonstrate the following habitat ranks: 3H > 3B = 10H > UH, where 3H is three year old harvested, 3B is three year old burnt, 10H is ten year old harvested and UH is unharvested forest. Relative density was greater in winter (mean ± 95% CI: 164.0 ± 28.7) than in spring (66.8 ± 28.7).

In contrast to swamp wallabies, there was no evidence that the relative density of eastern grey kangaroos differed between habitat types ($P = 0.71$; Figure 4b), or of a season by habitat type interaction ($P = 0.18$), although a statistically significant seasonal effect was again detected (mean ± 95% CI: winter 129.0 ± 28.7 compared to spring 75.6 ± 28.7; $P = 0.01$). The ANOVA r^2 indicated that in sum, habitat type, season and their interaction only explained 26.3% of the variance in the data. Contrasts between the four habitat types (Table 2) show substantial overlap in relative density for every habitat contrast.

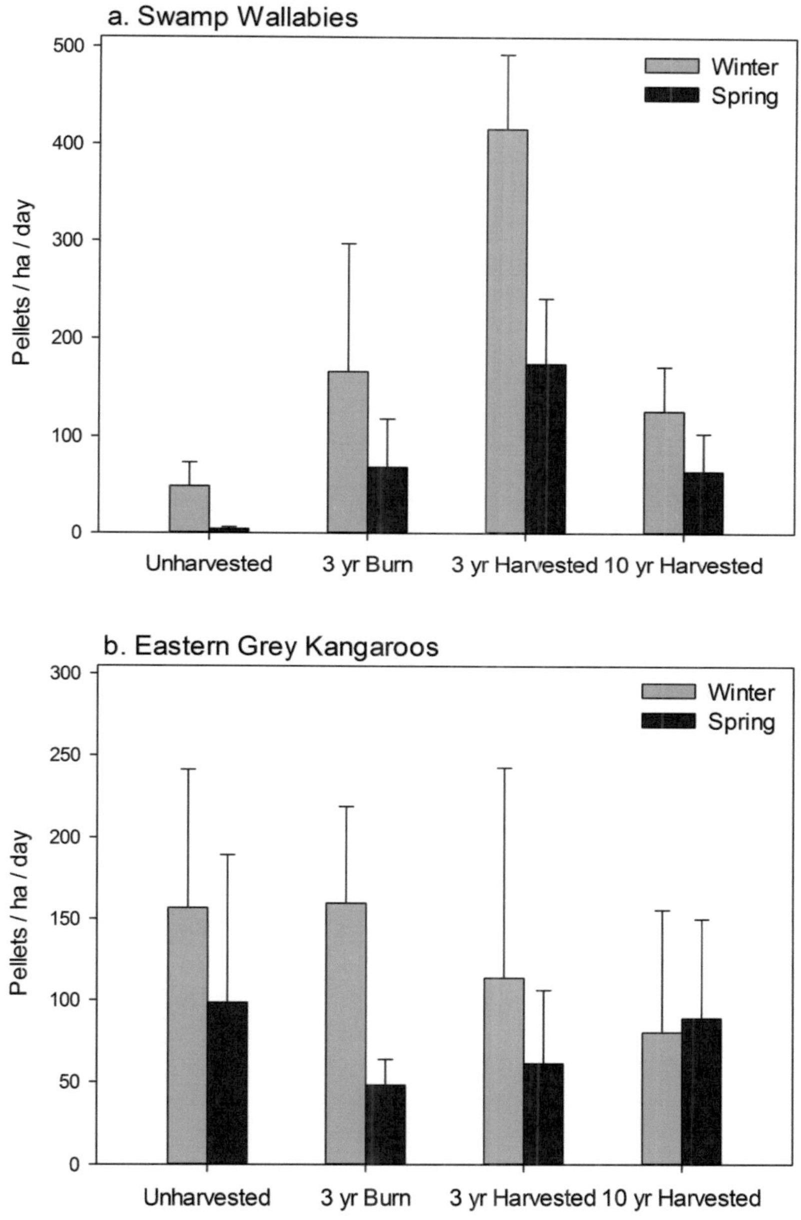

Fig. 4. Relative density of (a) swamp wallabies and (b) eastern grey kangaroos in the four habitat types during winter and spring. Errors bars represent the upper 95% confidence limit.

3.3 Relationships between relative density and resources

The model best describing patterns of swamp wallaby density across the 18 sites included the effects of season, shelter and their interaction, with a probability of 0.72 that it was the best in the set (Table 3, top). The high adjusted r^2 value (80.5%) associated with this model indicates a good fit to the data. A plot of the model (Figure 5a) shows that wallaby density was strongly and positively correlated with shelter during both winter and spring, but with a steeper rate of increase during winter. Predictor weights for season, shelter and food (forbs) were 1, 1 and 0 respectively, indicating the importance of season and shelter as predictors of wallaby density, and the negligible effect of food.

The model best predicting the relative density of eastern grey kangaroos included the effects of season, food (grass), and their interaction with a probability of 0.49 that it was the best in the set (Table 3, bottom). This relatively low Akaike weight indicates substantial model selection uncertainty, and the adjusted r^2 value of 21.9% demonstrates only a poor to moderate fit to the data. A plot of the model (Figure 5b) shows that grass cover was a better predictor of kangaroo density in spring than in winter. Predictor weights for season, food and shelter were 0.81, 0.77 and 0.37 respectively, indicating that season and grass cover were

Models	**Δ AIC**	**Akaike weight**	**Adj. r^2 (%)**
Swamp wallabies			
Season × shelter	0	0.72	80.5
Season × food + season × shelter	3.0	0.16	79.8
Season × food × shelter	3.5	0.12	80.4
Shelter	54.3	0.00	55.1
Food × shelter	21.3	0.00	53.1
Null	54.2	0.00	0
Season × food	55.2	0.00	9.8
Food	55.7	0.00	0
Eastern grey kangaroos			
Season × food	0	0.49	21.9
Season × food + season × shelter	2.6	0.13	19.7
Season × shelter	3.2	0.10	14.1
Season × food × shelter	3.4	0.09	21.2
Null	3.4	0.09	0
Food	4.7	0.05	0
Shelter	5.1	0.04	0
Food × shelter	7.1	0.01	0

Table 3. Model selection results for swamp wallabies (top) and eastern grey kangaroos (bottom). The model with a Δ AIC value of 0 fits the data best and Akaike weights are interpreted approximately as the probability that their associated model is the best in the set. Season: winter or spring; shelter: percentage shrub cover; food: percentage forb cover for wallabies and percentage grass cover for kangaroos. Models with interactions also contain the main effects of the interacting variables.

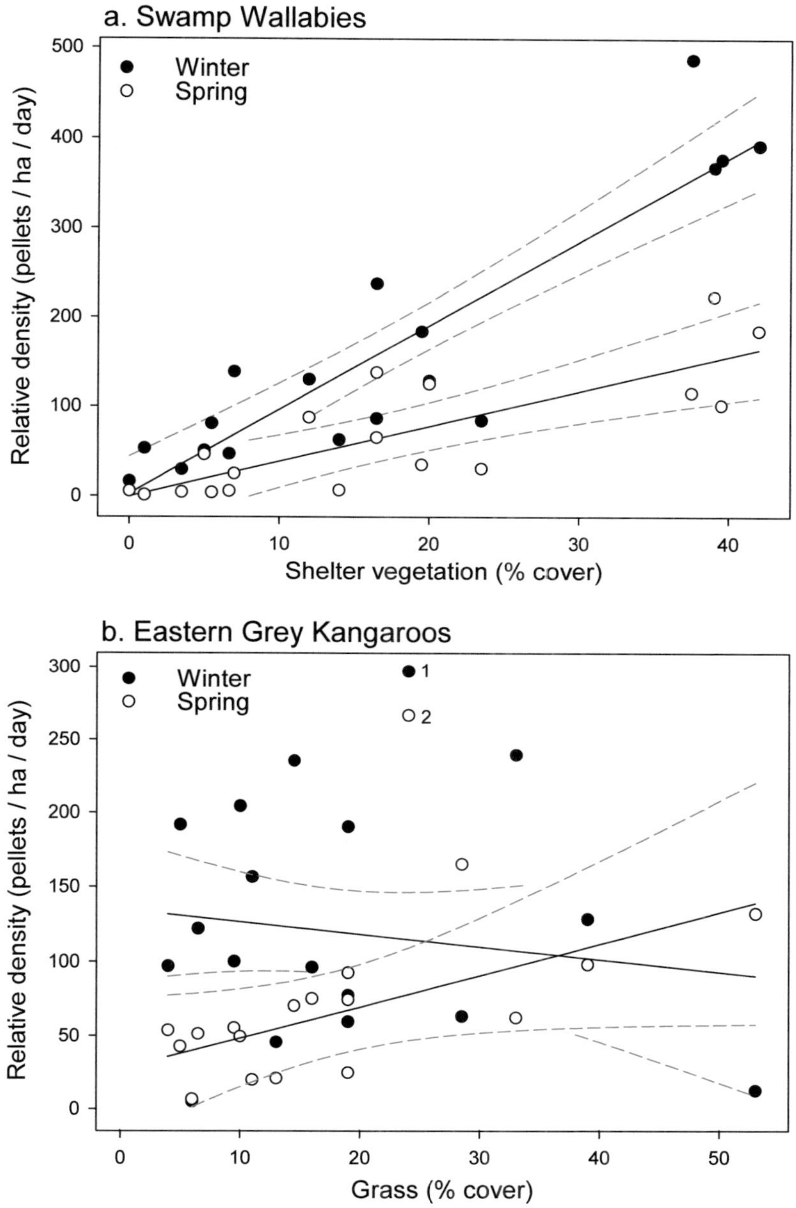

Fig. 5. Predicted relationship between (a) swamp wallaby density and the cover of shelter vegetation and (b) kangaroo density and the cover of grass. Open and closed circles represent the raw data for winter and spring respectively. Dashed lines are 95% confidence limits. The predicted relationship and confidence limits in (b) were calculated after removing the points labelled 1 and 2.

about twice as important as shelter for predicting kangaroo density. All results for eastern grey kangaroos were generated after removing data from an unharvested forest site with very high numbers of faecal pellets in both winter and spring (points labelled 1 and 2 in Figure 5b). Including these data in the analysis did not change the overall result, but further weakened the fit of the model.

4. Discussion

We have quantified the relative density of swamp wallabies and eastern grey kangaroos in a landscape where the distribution and abundance of food and shelter resources had been altered by both timber harvesting and fire. We defined four broad habitat types on the basis of time since disturbance and predicted differential use of at least some of these habitats by both species. We also predicted that density would be correlated with important resources, expecting a positive association between wallaby density and shelter vegetation, and a combined (interactive) effect of food and shelter on kangaroo density. Our finding indicated that the two species responded to the disturbances differently, but not always in the way we expected.

4.1 Resource availability

On the basis of several variables quantifying forest structure and food resources, moderate to large differences were detected between all of the habitat types except for ten year old harvested and three year old burnt sites. Three year old harvested sites had the most distinctive and consistent characteristics, with very high levels of shelter vegetation resulting from densely regenerating eucalyptus seedlings. The similarity between these sites was probably due to their location on the south side of a major ridgeline, and exposure to a consistent harvesting and seedbed preparation treatment during the same season.

In contrast, there were large differences between sites within the three year old burnt and ten year old harvested habitat types, indicating the spatially variable nature of disturbance-habitat relationships. Differences between sites burned by the same fire may be due to spatial variance in fire severity resulting from changes in weather conditions, topography, fuel load or fuel moisture (Bradstock et al., 2010). Although timber harvesting is a more uniform disturbance than fire, post-harvesting tree regeneration may also be influenced by a number of factors (Squire et al., 1991). For example, differences in seedbed conditions are known to affect the germination and establishment of eucalyptus seedlings (Dignan, 2002), and the ten year old sites used in this study did differ in this respect, with harvesting debris burnt on some sites but not on others (Merv Flett, personal communication). More generally, the development of vegetation communities after disturbance, and subsequent differences in resource availability and habitat structure, may be influenced by factors unrelated to the disturbance event, such density dependent competition, post-fire climatic conditions or site specific environmental factors (Bond and van Wilgen, 1996; Keeley et al., 2005). Factors such as these may have contributed to the high variance we observed between sites in the three year old burnt and ten year old harvested habitat classifications.

4.2 Relative density in different habitat types

We detected marked differences in swamp wallaby density among habitat types, with three year old harvested sites containing more than twice the number of faecal pellets then other habitats. Wallaby density was also higher in three year old burnt sites and ten year old harvested sites compared to unharvested forest, but no difference between burnt and ten year old harvested sites could be detected. Our results are consistent with previous work linking high wallaby density to recently harvested areas (Di Stefano et al., 2007; Di Stefano et al., 2009; Lunney and O'Connell, 1988), and more generally with the finding that a range of medium to large herbivorous mammals are attracted to early successional environments (Cederlund and Okarma, 1988; Côté et al., 2004; Fisher and Wilkinson, 2005; Fuller and DeStefano, 2003; Sullivan et al., 2007). For swamp wallabies, density appears to be positively correlated with shelter vegetation which was often high in early successional stands. We discuss this relationship further in section 4.3.

In contrast to swamp wallabies, and contrary to our prediction, we found no differences in the relative density of eastern grey kangaroos among habitat types. This was somewhat surprising given that marked differences in resource availability between some of the habitats, and previous work in a variety of ecosystems showing that eastern grey kangaroos use broadly defined habitat associations selectively (Hill, 1981; Moore et al., 2002; Taylor, 1980). Several studies (e.g. Hill, 1981; Moore et al., 2002; Southwell, 1987) suggest that both food and shelter influence the density of eastern grey kangaroos. In particular, Hill (1981) showed that kangaroo density was high at sites with (a) high levels of both food and shelter, (b) high food and moderate shelter and (c) moderate food and high shelter. Arguably, all four habitat types included in the present study contained at least some sites conforming to one of these three classifications, possibly explaining the similar density profile we observed.

We included season in the statistical model to account for temporal variation in abundance between sampling times, and not because seasonal effects were of particular interest. For both species, density was lower in spring than in winter, and we interpret this change as part of normal temporal fluctuations experienced by many herbivorous mammals (e.g. Ager et al., 2003). Due to the short time span of our study (only one sample in winter and spring, respectively), we are unable to make any reliable inference about seasonal effects.

4.3 Relationships between relative density and resources

Using our simple set of predictor variables we were able to model swamp wallaby density with a high degree of precision. The best model included the effect of season, shelter and their interaction, as was consistent with our original prediction that wallaby density would be positively correlated with the abundance of lateral cover. Although swamp wallabies may prefer sites containing both food and shelter in close proximity (Floyd, 1980; Hill and Phinn, 1993), shelter vegetation alone appears to be a reliable predictor of relative density (Lunney and O'Connell, 1988; Di Stefano, unpublished data).

Swamp wallabies are a cryptic species and probably use sheltered sites to avoid predators such as red foxes (*Vulpes vulpes*) and wedge tailed eagles (*Aquila audax*), both of which were present at our study site. Predation is a process known to influence habitat use by medium to large herbivorous mammals (Ferguson et al., 1988; Rachlow and Bowyer, 1998), and in many cases results in a trade off between the acquisition of shelter and food (Lima and Dill, 1990; Verdolin, 2006). Some species select sites with high shelter and lower food resources when predation risk is high (Ferguson et al., 1988), but others are able to select sites that contain adequate amounts of both shelter and food (Pierce et al., 2004). At our study location, forbs, an important food resource for swamp wallabies (Davis et al., 2008; Di Stefano and Newell, 2008; Hollis et al., 1986), were present in similar abundance regardless of shelter quantity. We have hypothesised elsewhere (Di Stefano et al., 2009) that this enabled swamp wallabies to reduce predation risk by selecting sheltered sites containing adequate food resources, and thus reduce the trade off between shelter and food that often results from predator avoidance behaviour.

We were unable to model the density of eastern grey kangaroos with a high degree of precision. The best model included the effect of season, food (grass), and their interaction, but did not fit the data particularly well, and was associated with substantial model selection uncertainty. The model showed that kangaroo density in spring was positively associated with grass cover, but density in winter was not, an effect that may have been related to increased protein concentrations during the spring sampling period (Taylor, 1984). Overall, these findings were partially consistent with the predicted positive relationship between kangaroo density and grass abundance, but not with our expectation that this effect would be moderated by shelter availability.

Although the density of eastern grey kangaroos is widely believed to be influenced by the co-occurrence of food and shelter (Hill, 1981; Moore et al., 2002; Southwell, 1987), published results are somewhat inconsistent. In some cases forage abundance or quality has been the primary driver of kangaroo density (Taylor, 1980, 1984), while in others the importance of both food and shelter has been clearly shown (Hill, 1981). Further, the relationship between shelter vegetation and kangaroo density remains unclear, with published data identifying positive, negative, and no relationship (Catling and Burt, 1995; Catling et al., 2001; Hill, 1981). In addition, results are probably influenced by the size of the sampling unit used in different studies (Ramp and Coulson, 2004). For example, the multiplicative effect of food and shelter detected by Hill (1981) was derived from estimates of density and resource availability within replicate 25 ha blocks, while we quantified variables within 0.64 ha sampling units. Smaller plots may fail to detect the influence of important resource features occurring outside the plot boundary, while larger plots may miss smaller scale patterns. Unlike swamp wallabies, the density of eastern grey kangaroos appears to be influenced by an interacting array of factors including food abundance, food quality and shelter availability. The influence of these factors are likely to depend on the scale at which they are measured, and to vary throughout the wide range of this species. Additional, multi-scale studies are required to determine how resources influence habitat use across the species' range.

As a caveat, we note that faecal pellet counts have their limitations as an index of relative abundance. One problem with this technique is the inability to detect fine-scale demographic or temporal influences on the relationship between habitat use and resources. Several studies, including some on swamp wallabies and other medium-sized macropods, have found important effects of demography and time of day on patterns of habitat use (Ager et al., 2003; Beyer and Haufler, 1994; Di Stefano et al., 2009; le Mar and McArthur, 2005; Moe et al., 2007; Swan et al., 2008). As such, analyses based on faecal pellet data provide information about broad patterns of habitat use that may hide important demographic or temporal effects.

4.4 Comparison between harvested and burnt sites

Better conservation outcomes may be achieved in managed landscapes if the impacts of human disturbances, such as timber harvesting, are similar to those of natural disturbances, such as fire (Hunter, 1993; Lindenmayer et al., 2006). The existence of three year old harvested and burnt sites in this study enabled us to compare these areas with respect to habitat characteristics, and the density of swamp wallabies and eastern grey kangaroos.

We found that three year old harvested and burnt sites were structurally dissimilar, and that burnt sites had a much more variable array of structural characteristics. This is consistent with previous work identifying increased structural heterogeneity within burnt plots as a major difference between early successional burnt and harvested stands (Lindenmayer and McCarthy, 2002; Schulte and Niemi, 1998).

In relation to macropods we found that three year old harvested and burnt habitats supported very different densities of swamp wallabies, but similar densities of kangaroos. This also reflects other findings, with faunal responses differing between harvested and burnt areas in some instances (Schulte and Niemi, 1998; Simon et al., 2002a) but not in others (Baker et al., 2004; Simon et al., 2002b). Differences between harvesting and wildfire, both in terms of structural characteristics and faunal responses, are likely to be greatest shortly after disturbance and converge over time (Simon et al., 2002b). Findings will also be influenced by forest ecosystem type, characteristics of the disturbance event, disturbance history, focal taxa or species, and spatial context. We did not consider spatial context in our study, but harvested and burnt sites clearly differed with respect to the nature of the surrounding habitat. Quantifying landscape characteristics around each site and assessing their influence on macropod density may be a useful extension of our current work.

An unexpected finding was the similarity between ten year old harvested three year old burnt habitats, both in terms of habitat characteristics and faunal responses. There are two likely reasons for this result. First, the extra regeneration time at ten year old sites enabled the re-development of structural attributes (e.g. tree canopy) that were present in the younger burnt sites but virtually absent from three year old harvested stands. Second, as we have already mentioned, the ten year old sites were subject to variable treatment of post-harvest debris, with burns occurring at some sites but not others. We argue that this variation in operational practice lead to differences in the post-disturbance recovery, resulting in a heterogeneous set of sites closely mimicking the structural heterogeneity found in the burnt landscape.

5. Conclusions

Using faecal pellet counts, we quantified the relative density of swamp wallabies, *Wallabia bicolor*, and eastern grey kangaroos, *Macropus giganteus*, to wildfire and timber harvesting in a native eucalyptus forest in southeastern Australia. We defined four broad habitat types on the basis of time since disturbance (unharvested, three year old harvested, ten year old harvested and three year old burnt) and predicted differential use of these habitats by both species. Multivariate analysis was used to compare sites within habitat types with regard to several habitat attributes including food and shelter resources for our two focal species. We modelled density as a function of these two resources, and predicted a positive correlation between wallabies and shelter vegetation. In contrast, we expected kangaroo density would be influenced by the interaction between food and shelter. In addition, the occurrence of harvesting operations and a fire during the same season provided an opportunity to compare harvested and burnt sites of the same age.

Structural differences were observed between all habitat types except for the three year old burnt and ten year old harvested classifications. Three year old harvested sites differed markedly from most others with respect to shelter vegetation (high) and canopy cover (low).

Swamp wallaby density was highest in three year old harvested sites and lowest in the unharvested forest, and, as predicted, showed a strong positive association with shelter vegetation. In contrast, we could not detect any influence of habitat type on kangaroo density. A statistical model that fit the data relatively poorly indicated that density was positively correlated with food resources (grass) in spring, but not in winter, a finding only partially consistent with our expectations.

Finally, we conclude that fire and timber harvesting at our study location are likely to result in very different early successional landscapes. Three year old harvested sites were structurally simple and spatially homogeneous compared to the more heterogeneous nature of the habitat both within and between similarly aged burnt sites. Kangaroo density did not change in response to these differences, but wallaby density was substantially higher in the harvested areas. The fact that our focal species responded differently demonstrates that the relationship between landscape change and fauna is species specific, depending on both the nature of the change and the requirements of particular species. Nevertheless, it has been shown elsewhere that spatially homogenous and structurally simple patches resulting from timber harvesting will, in general, be less able to support a range of species than more complex and variable landscapes resulting from fire (Lindenmayer and Franklin, 2002; Lindenmayer and McCarthy, 2002; Lindenmayer and Noss, 2006). Consequently, we suggest that forest managers should foster structural and spatial heterogeneity by varying operational harvesting practices, and incorporating fire as a major component of the disturbance regime.

6. Acknowledgements

The University of Melbourne and the Victorian Department of Sustainability and Environment provided funding for this project. Alan York commented on an earlier draft and contributed to the clarity of the manuscript.

7. References

Abrams, M. D. (2003). Where has all the white oak gone? *Bioscience*, Vol. 53, No. 10, pp 927-39.

Ager, A. A., Johnson, B. K., Kern, J. W. & Kie, J. G. (2003). Daily and seasonal movements and habitat use by female rocky mountain elk and mule deer. *J. Mammal.*, Vol. 84, No. 3, pp 1076-88.

Archibald, S. & Bond, W. J. (2004). Grazer movements: spatial and temporal responses to burning in a tall-grass African savanna. *Int. J. Wildland Fire*, Vol. 13, No. 3, pp 377-85.

Attiwill, P. M. (1994). The disturbance of forest ecosystems - the ecological basis for conservative management. *For. Ecol. Manage.*, Vol. 63, No. 2-3, pp 247-300.

Baker, S. C., Richardson, A. M. M., Seeman, O. D. & Barmuta, L. A. (2004). Does clearfell, burn and sow silviculture mimic the effect of wildfire? A field study and review using litter beetles. *For. Ecol. Manage.*, Vol. 199, No. 2-3, pp 433-48.

Bates, D. & Maechler, M. (2009) lme4: Linear mixed-effects models using S4 classes. R package version 0.999375-32. Available at: http://CRAN.R-project.org/package=lme4.

Beyer, D. E. & Haufler, J. B. (1994). Diurnal versus 24-hour sampling of habitat use. *J. Wildl. Manage.*, Vol. 58, No. 1, pp 178-80.

Bond, W. J. & van Wilgen, B. W. (1996). *Fire and plants*, Chapman and Sons, New York.

Bradstock, R. A., Hammill, K. A., Collins, L. & Price, O. (2010). Effects of weather, fuel and terrain on fire severity in topographically diverse landscapes of south-eastern Australia. *Landsc. Ecol.*, Vol. 25, No. 4, pp 607-19.

Burnham, K. P. & Anderson, D. R. (2002). *Model Selection and Multimodel Inference: a Practical Information-Theoretic Approach*, Springer, New York.

Catling, P. C. & Burt, R. J. (1995). Studies of the ground-dwelling mammals of eucalypt forests in south-eastern New South Wales - the effect of habitat variables on distribution and abundance. *Wildl. Res.*, Vol. 22, No. 3, pp 271-88.

Catling, P. C., Coops, N. C. & Burt, R. J. (2001). The distribution and abundance of ground-dwelling mammals in relation to time since wildfire and vegetation structure in south-eastern Australia. *Wildl. Res.*, Vol. 28, No. 6, pp 555-64.

Cederlund, G. N. & Okarma, H. (1988). Home range and habitat use of adult female moose. *J. Wildl. Manage.*, Vol. 52, No. 2, pp 336-43.

Côté, S. D., Rooney, T. P., Tremblay, J. P., Dussault, C. & Waller, D. M. (2004). Ecological impacts of deer overabundance. *Annu. Rev. Ecol. Evol. Syst.*, Vol. 35, No., pp 113-47.

Dale, V. H., Brown, S., Haeuber, R. A., Hobbs, N. T., Huntly, N., Naiman, R. J., Riebsame, W. E., Turner, M. G. & Valone, T. J. (2000). Ecological principles and guidelines for managing the use of land. *Ecol. Appl.*, Vol. 10, No. 3, pp 639-70.

Davis, N. E., Coulson, G. & Forsyth, D. M. (2008). Diets of native and introduced mammalian herbivores in shrub-encroached grassy woodland, south-eastern Australia. Vol. 35, No., pp 684-94.

Di Stefano, J., Anson, J. A., York, A., Greenfield, A., Coulson, G., Berman, A. & Bladen, M. (2007). Interactions between timber harvesting and swamp wallabies (*Wallabia bicolor*): Space use, density and browsing impact. *For. Ecol. Manage.*, Vol. 253, No. 1-3, pp 128-37.

Di Stefano, J. & Newell, G. R. (2008). Diet selection by the swamp wallaby (*Wallabia bicolor*): feeding strategies under conditions of changed food availability. *J. Mammal.*, Vol. 89, No. 6, pp 1540-9.

Di Stefano, J., York, A., Swan, M., Greenfield, A. & Coulson, G. (2009). Habitat selection by the swamp wallaby (*Wallabia bicolor*) in relation to diel period, food and shelter. *Austral Ecol.*, Vol. 34, No. 2, pp 143-55.

Dignan, P. (2002) Effect of time of sowing and seedbed type of *Eucalyptus regnans* germination and establishment on high elevation sites. *Forests Service Research Report No. 381* p. 25. Department of Natural Resources and Environment, Melbourne.

Ferguson, S. H., Bergerud, A. T. & Ferguson, R. (1988). Predation risk and habitat selection in the persistence of a remnant caribou population. *Oecologia*, Vol. 76, No. 2, pp 236-45.

Fisher, J. T. & Wilkinson, L. (2005). The response of mammals to forest fire and timber harvest in the North American boreal forest. *Mammal Rev.*, Vol. 35, No. 1, pp 51-81.

Floyd, R. B. (1980). Density of *Wallabia bicolor* (Desmarest) (Marsupialia: Macropodidae) in eucalypt plantations of different ages. *Aust. Wildl. Res.*, Vol. 7, No. 3, pp 333-7.

Fuller, T. K. & DeStefano, S. (2003). Relative importance of early-successional forests and shrubland habitats to mammals in the northeastern United States. *For. Ecol. Manage.*, Vol. 185, No. 1-2, pp 75-9.

Gasaway, W. C., Dubois, S. D., Boertje, R. D., Reed, D. J. & Simpson, D. T. (1989). Response of radio collard moose to a large burn in central Alaska. *Can. J. Zool.*, Vol. 67, No. 2, pp 325-9.

Gaston, K. J., Blackburn, T. M. & Goldewijk, K. K. (2003). Habitat conversion and global avian biodiversity loss. *Proc. R. Soc. Lond. Ser. B*, Vol. 270, No. 1521, pp 1293-300.

Geffen, E., Hefner, R., Macdonald, D. W. & Ucko, M. (1992). Habitat selection and home range in the Blanford Fox, *Vulpes cana*: Compatibility with the resource dispersion hypothesis. *Oecologia*, Vol. 91, No. 1, pp 75-81.

Hill, G. J. E. (1981). A study of habitat preferences in the grey kangaroo. *Aust. Wildl. Res.* Vol. 8, No. 2, pp 245-54.

Hill, G. J. E. & Phinn, S. R. (1993). Revegetated sand mining areas, swamp wallabies and remote sensing: North Stradbroke Island, Queensland. *Aust. Geog. Stud.*, Vol. 31, No. 1, pp 3-13.

Hollis, C. J., Robertshaw, J. D. & Harden, R. H. (1986). Ecology of the swamp wallaby (*Wallabia bicolor*) in north-eastern NSW. I. Diet. *Aust. Wildl. Res.*, Vol. 13, No., pp 355-61.

Hughes, J. J., Ward, D. & Perrin, M. R. (1994). Predation risk and competition affect habitat selection and activity of Namib desert gerbils. *Ecology*, Vol. 75, No. 5, pp 1397-405.

Hunter, M. L. (1993). Natural fire regimes as spatial models for managing boreal forests. *Biol. Conserv.*, Vol. 65, No. 2, pp 115-20.

Keeley, J. E., Fotheringham, C. J. & Baer-Keeley, M. (2005). Determinants of postfire recovery and succession in Mediterranean-climate shrublands of California. *Ecol. Appl.*, Vol. 15, No. 5, pp 1515-34.

le Mar, K. & McArthur, C. (2005). Comparison of habitat selection by two sympatric macropods, *Thylogale billardierii* and *Macropus rufogriseus rufogriseus*, in a patchy eucalypt-forestry environment. *Austral Ecol.*, Vol. 30, No. 6, pp 674-83.

Lima, S. L. & Dill, L. M. (1990). Behavioral decisions made under the risk of predation - A review and prospectus. *Can. J. Zool.*, Vol. 68, No. 4, pp 619-40.

Lindenmayer, D. B. & Franklin, J. F. (2002). *Conserving forest biodiversity: a comprehensive multiscaled approach*, Island Press, Washington.

Lindenmayer, D. B., Franklin, J. F. & Fischer, J. (2006). General management principles and a checklist of strategies to guide forest biodiversity conservation. *Biol. Conserv.*, Vol. 131, No. 3, pp 433-45.

Lindenmayer, D. B. & McCarthy, M. A. (2002). Congruence between natural and human forest disturbance: a case study from Australian montane ash forests. *For. Ecol. Manage.*, Vol. 155, No., pp 319-35.

Lindenmayer, D. B. & Noss, R. F. (2006). Salvage logging, ecosystem processes, and biodiversity conservation. *Conserv. Biol.*, Vol. 20, No. 4, pp 949-58.

Lugo, A. E. (2000). Effects and outcomes of Caribbean hurricanes in a climate change scenario. *Sci. Total Environ.*, Vol. 262, No. 3, pp 243-51.

Lunney, D. & O'Connell, M. (1988). Habitat selection by the swamp wallaby, *Wallabia bicolor*, the red-necked wallaby, *Macropus rufogriseus*, and the common wombat, *Vombatus ursinus*, in logged, burnt forest near Bega, New South Wales. *Aust. Wildl. Res.*, Vol. 15, No. 6, pp 695-706.

Lutze, M. T., Campbell, R. G. & Fagg, P. C. (1999). Development of silviculture in the native State forests of Victoria. *Aust. For.*, Vol. 62, No. 3, pp 236-44.

Mahoney, S. P. & Virgl, J. A. (2003). Habitat selection and demography of a nonmigratory woodland caribou population in Newfoundland. *Can. J. Zool.*, Vol. 81, No. 2, pp 321-34.

Moe, T. F., Kindberg, J., Jansson, I. & Swenson, J. E. (2007). Importance of diel behaviour when studying habitat selection: examples from female Scandinavian brown bears (*Ursus arctos*). *Can. J. Zool.*, Vol. 85, No. 4, pp 518-25.

Moore, B. D., Coulson, G. & Way, S. (2002). Habitat selection by adult female eastern grey kangaroos. *Wildl. Res.*, Vol. 29, No. 5, pp 439-45.

Murphy, B. & Bowman, D. (2007). The interdependence of fire, grass, kangaroos and Australian Aborigines: a case study from central Arnhem Land, northern Australia. *J. Biogeogr.*, Vol. 34, No. 2, pp 237-50.

Pearson, S. M., Turner, M. G., Wallace, L. L. & Romme, W. H. (1995). Winter habitat use by large ungulates following fire in northern Yellowstone National Park. *Ecol. Appl.*, Vol. 5, No. 3, pp 744-55.

Pierce, B. M., Bowyer, R. T. & Bleich, V. C. (2004). Habitat selection by mule deer: Forage benefits or risk of predation? *J. Wildl. Manage.*, Vol. 68, No. 3, pp 533-41.

Quinn, G. P. & Keough, M. J. (2002). *Experimental Design and Data Analysis for Biologists*, Cambridge University Press, Cambridge.

Rachlow, J. L. & Bowyer, R. T. (1998). Habitat selection by Dall's sheep (*Ovis dalli*): maternal trade-offs. *J. Zool.*, Vol. 245, No., pp 457-65.

Ramp, D. & Coulson, G. (2004). Small-scale patch selection and consumer-resource dynamics of eastern grey kangaroos. *J. Mammal.*, Vol. 85, No. 6, pp 1053-9.

Ryan, K. C. (2002). Dynamic interactions between forest structure and fire behavior in boreal ecosystems. *Silva. Fenn.*, Vol. 36, No. 1, pp 13-39.

Schaefer, J. A. & Pruitt, W. O. (1991). Fire and woodland caribou in southeastern Manitoba. *Wildl. Monogr.*, Vol., No. 116, pp 3-39.

Schulte, L. S. & Niemi, G. J. (1998). Bird communities of early-successional burned and logged forest. *J. Wildl. Manage.*, Vol. 62, No. 4, pp 1418-29.

Simon, N. P. P., Schwab, F. E. & Otto, R. D. (2002a). Songbird abundance in clear-cut and burned stands: a comparison of natural disturbance and forest management. *Can. J. For. Res.*, Vol. 32, No. 8, pp 1343-50.

Simon, N. P. P., Stratton, C. B., Forbes, G. J. & Schwab, F. E. (2002b). Similarity of small mammal abundance in post-fire and clearcut forests. *For. Ecol. Manage.*, Vol. 165, No. 1-3, pp 163-72.

Southwell, C. (1987). Macropod studies at Wallaby Creek II. Density and distribution of macropod species in relation to environmental variables. *Aust. Wildl. Res.*, Vol. 14, No. 1, pp 15-34.

Southwell, C. (1989). Techniques for monitoring the abundance of kangaroo and wallaby populations, In: *Kangaroos, Wallabies and Rat-Kangaroos*, (eds G. Grigg, P. Jarman and I. Hume) pp. 659-93. Surrey Beatty, Sydney.

Southwell, C. J. & Jarman, P. J. (1987). Macropod studies at Wallaby Creek III. The effect of fire on pasture utilization by macropodids and cattle. *Aust. Wildl. Res.* Vol. 14, No. 2, pp 117-24.

Squire, R. O., Dexter, B. D., Eddy, A. R., Fagg, P. C. & Campbell, R. G. (1991) Regeneration Silviculture for Victoria's Eucalypt Forests. SSP Technical Report No. 6. p. 38. Department of Conservation and Environment, Melbourne.

Sullivan, T. P., Sullivan, D. S., Lindgren, P. M. F. & Ransome, D. B. (2007). Long-term responses of ecosystem components to stand thinning in young lodgepole pine forest IV. Relative habitat use by mammalian herbivores. *For. Ecol. Manage.*, Vol. 240, No. 1-3, pp 32-41.

Swan, M., Di Stefano, J., Greenfield, A. & Coulson, G. (2008). Fine-scale habitat selection by adult female swamp wallabies (*Wallabia bicolor*). *Aust. J. Zool.*, Vol. 56, No. 5, pp 305-9.

Taylor, R. J. (1980). Distribution of feeding activity of the eastern grey kangaroo, *Macropus giganteus,* in coastal lowland of southeast Queensland. *Aust. Wildl. Res.* Vol. 7, No. 3, pp 317-25.

Taylor, R. J. (1983). The diet of the eastern grey kangaroo and wallaroo in areas of improved and native pasture in the New England Tablelands. *Aust. Wildl. Res.*, Vol. 10, No. 2, pp 203-11.

Taylor, R. J. (1984). Foraging in the eastern grey kangaroo and wallaroo. *J. Anim. Ecol.*, Vol. 53, No. 1, pp 65-74.

Thompson, I. D., Baker, J. A. & Ter-Mikaelian, M. (2003). A review of the long-term effects of post-harvest silviculture on vertebrate wildlife, and predictive models, with an emphasis on boreal forests in Ontario, Canada. *For. Ecol. Manage.* Vol. 177, No., pp 441-69.

Triggs, B. (2004). *Tracks, Scats and Other Traces: A Field Guide to Australian Mammals* Oxford University Press, Melbourne.

Troy, S., Coulson, G. & Middleton, D. (1992). A comparison of radio-tracking and line transect techniques to determine habitat preferences in the swamp wallaby (*Wallabia bicolor*) in south-eastern Australia, In: *Wildlife Telemetry: Remote Monitoring and Tracking of Animals*, (eds I. G. Priede and S. M. Swift) pp. 651-60. Ellis Horwood, Chichester.

Tufto, J., Andersen, R. & Linnell, J. (1996). Habitat use and ecological correlates of home range size in a small cervid: The roe deer. *J. Anim. Ecol.*, Vol. 65, No. 6, pp 715-24.

Verdolin, J. L. (2006). Meta-analysis of foraging and predation risk trade-offs in terrestrial systems. *Behav. Ecol. Sociobiol.*, Vol. 60, No. 4, pp 457-64.

Wilson, B. A. & Friend, G. R. (1999). Responses of Australian mammals to disturbance: A review. *Aust. Mammal.*, Vol. 21, No., pp 87-105.

8

Forest Life Under Control of Microbial Life

Doaa A. R. Mahmoud
National Research Center, Chemistry of Natural and Microbial Products Department, Cairo Egypt

1. Introduction

Microbial life has been present for at least 3,500 million years, and the earth itself was only formed about 4,600 million years ago. "For much of its history, Earth was a planet of microbes. Even today, microbes dominate the living world in terms of biomass and number". Microbes play numerous key roles in global change, often as silent partners in human activities such as agriculture, mining and water treatment and also microbes are responsible for transforming many of the Earth's most abundant compounds, and thus are central to global changes causing concern (Kenneth Todar 2009).

In the magic world of microorganisms you will wonder how microbes tiny as they are but considered the engines that keep the wheel of life turning. Without microorganisms, the ecosystems of the world would collapse and die. However, it is quite clear that microbe's life can destroy other livings life, one of such livings life is the forest life. Although microorganisms can destroy trees, in contrast, a healthy, balanced community of microorganisms is extremely important for trees health. You will wonder if you think deeply of theses tiny organisms how are world's worst diseases caused by microbes can turned into weapons of mass destruction, on the other hand how microbes can offer a great deal of hope for solving many of the problems our world faces.

The microorganisms are at the bottom of the food chain. Their survival is essential for the survival of all the other species. Loss of forest microbes, however, would have a severe effect throughout the entire forest ecology. The forest is dependent upon the actions of microbes, to sustain the base-level of food supply. Microbes dispose of dead matter by decaying and rotting it. They provide nutrients both from their by-products and from themselves, for other usually larger and more complex life-forms - they are food. Many species of plant could consequently suffer extinction, if microbial populations were destroyed or degraded. Although microbes are numerous and there are many species, they can be sensitive to the smallest of environmental change - changes in water acidity, levels of sunlight, toxins, etc. They can be quite fragile. However, they and their diversity are keys to the survival of the forest. http://www.theguardians.com/Microbiology/gm_mbr10.htm

The microbial world of the forest, though lacking the appeal of the more exotic larger creatures, should be regarded as just as important as any other forest population. Therefore efforts should be directed towards understanding and preserving their habitat and the way they interact with the forest environment.

In this chapter you will recognize the positive face of microorganisms that offer exchangeable benefits with trees and the negative face that exert damage on trees. Microbes are weapon with two faces, therefore we do not need sterile life from microbes but we need to spread beneficial microbes against pathogenic microbes, because the former have magical effects on overcoming the later problems via antagonism activity. To realize how microbial life can control forest life you need a journey in the forest for understanding the relation between trees and microbes, moreover for understanding the aspects of microbial interactions with trees.

But before going in both journeys try to imagine what if we loose forests:

Forests offer watershed protection, timber and non-timber products, and various recreational options. Forests prevent soil erosion help in maintaining the water cycle, and check global warming by using carbon dioxide in photosynthesis. Forest lands are instrumental in the beauty and spiritual impact of our landscape. If the forest disappears, the entire ecosystem begins to fall apart, with dire consequences for all of us.

Fig. 1. Funny photo illustrates the disappearance of trees due to their anger from human activities.

Now let's start our journey in the forest and in the microbial world.

2. Journey in the forest

No exciting journey more than that in the forest. In the forest world you can see an area with a high density of trees. It functions as habitats for organisms, hydrologic flow modulators, and soil conservers, constituting one of the most important aspects of the biosphere. A typical forest is composed of the overstory (or upper tree layer of the canopy) and the understory. The understory is further subdivided into the shrub layer, herb layer, and sometimes also moss layer. In some complex forests, there is also a well-defined lower tree layer. Although a forest is classified primarily by trees a forest ecosystem is defined intrinsically with additional species such as fungi.

Forests can be classified in different ways and to different degrees of specificity. One such way is in terms of the "biome" in which they exist, combined with leaf longevity of the dominant species (whether they are evergreen or deciduous). Another distinction is whether the forests composed predominantly of broadleaf trees, coniferous (needle-leaved) trees, or mixed (Martin et al 2007).

1. Boreal forests occupy the sub arctic zone and are generally evergreen and coniferous.
2. Temperate zones support both broadleaf deciduous forests (*e.g.*, temperate deciduous forest) and evergreen coniferous forests (*e.g.*, Temperate coniferous forests and Temperate rainforests). Warm temperate zones support broadleaf evergreen forests, including laurel forests.
3. Tropical and subtropical forests include tropical and subtropical moist forests, tropical and subtropical dry forests, and tropical and subtropical coniferous forests.

- Physiognomy classifies forests based on their overall physical structure or developmental stage (old growth or second growth).
- Forests can also be classified more specifically based on the climate and the dominant tree species present, resulting in numerous different forest types (e.g., Ponderosa pine/Douglas-fir forest)

In the following figures you can see some examples of forests classification

Old growth forest

Spiny forest

Conifer forest

temperate deciduous forest

Fig. 2. Photos for different types of forest adopted from forest Wikipedia, the free encyclopedia

In temperate deciduous forest, the weather may range from cold with moderate amounts of snow to warm and rainy. Because of their moderate temperatures and other factors, these are ideal places for certain types of bacteria. The tropical rainforests are the richest places in the world in terms of biodiversity. A larger variety of microorganisms live in rainforest than anywhere else. Rainforest organisms have been found to possess amazing properties that are of potential benefit to mankind. Rubber, medical products and treatments, food-stuffs such as potatoes, chocolate and spices all originate from the rainforests. The rain forest has acid soil due to the volume of water moving through the soil.

3. Journey in microbial world

If you have given a microscope you can enter a whole world of "inner space", you will find it really quite fascinating. You will find it fascinating enough to decide to make the study of these tiny forms of life your life's work. You will discover that the basic structure of microbe is unicellular but although that, many species of microbes live together in diverse communities, in the same way that trees, flowers, fish, birds and bugs interact. Earth is a microbial planet-always has been, and always will be. http://www.voanews.com/english/news/a-13-2009-02-13-voa31-68673292.html. But what is the origin of microbes and how it related to the plants?

Kenneth Todar (2009) told us the origin of microbes' story as follows: when life arose on Earth about 4 billion years ago, the first types of cells to evolve were prokaryotic cells. For approximately 2 billion years, prokaryotic-type cells were the only form of life on Earth. The primitive earth's atmospheric gases, such as ammonia (NH_3), hydrogen (H_2) and hydrogen sulfide (H_2S) could be oxidized to produce energy that allowed conversion of CO_2 to cellular (organic) material. As organic material developed, it became the substrate to support the growth and metabolism of other cells that use simple organic compounds as their source of energy. The use of inorganic chemicals as a source of energy is called **chemolithotrophy**; the use of organic chemicals as energy sources is called **chemoheterotrophy**. Thus, chemolithotrophy and chemoheterotrophy were the first two types of metabolism to evolve. An important group of archaea that were involved in this process were the **methanogens**, which grow by using H_2 as an energy source and CO_2 as a carbon source, resulting in the production of the simplest of all organic molecules, methane (CH_4). Archaea and bacteria probably arose from a universal ancestor but are thought have split early during the evolution of cellular life into the two groups of prokaryotes that we recognize today. **Photosynthesis** (metabolism which uses of light as an energy source) developed in bacteria about 3.2 billion years ago. The first type of photosynthesis to appear is called **anoxygenic photosynthesis** because it does not produce O_2. Anoxygenic photosynthesis preceded **oxygenic photosynthesis** (plant-type photosynthesis, which produces atmospheric O_2) by half a billion years. However, oxygenic photosynthesis also arose in prokaryotes, specifically in a group of bacteria called **cyanobacteria**, and existed for millions of years before the evolution of plants. As molecular oxygen (O_2) began to appear in the atmosphere, organisms that could use O_2 for respiration began their evolution, and "aerobic" respiration became a prevalent form of metabolism among bacteria and some archaea.

Even was the above story true or imaginary, at least it gives us outline about the probable origin of microbes and how could plant may originated from them .Also it may explain the underlying mechanisms behind the interactions between plants (trees) and microbes which are still poorly known.

Microorganism is the technical name of microbe. It means an organism of microscopic or submicroscopic size they are grouped or classified in various ways. Most microbes belong to one of four major groups: bacteria, viruses, fungi, or protozoa. The most important groups for forest are fungi, bacteria and viruses. Therefore these three groups will be described here to some extent in details.

3.1 What are fungi?

Fungi are a distinct kingdom of organisms that are neither plants nor animals. The basic fungal vegetative structure is the microscopic hyphae, a thread like-tube that maybe separated into cells by the formation of cross-walls (septa). Unlike plants and animals, fungi absorb all their food from external sources. The hyphae grow through or on their food substrate and sometimes form a visible mycelium. Many fungi are saprophytic or parasitic (callan et al 2001). There is considerable variation in the structure, size, and complexity of various fungal species. For example, fungi include the microscopic yeasts, the molds seen on contaminated bread, and the common mushrooms.

Fungi can be recognized by the following five characteristics:

1. The cells of fungi contain nuclei with chromosomes (like plants and animals, but unlike bacteria).

2. Fungi cannot photosynthesize (they are heterotrophic, like animals)
3. Fungi absorb their food (they are osmotrophic)
4. They mostly develop very diffuse bodies made up of a spreading network of very narrow, tubular, branching filaments called hyphae. These filaments exude enzymes, and absorb food, at their growing tips. Although these filaments are very narrow, they are collectively very long, and can explore and exploit food substrates very efficiently.
5. They usually reproduce by means of spores, which develop on, and are released by, a range of unique structures (such as mushrooms, cup fungi, and many other kinds of microscopically small fruiting bodies).

3.2 What are bacteria?

Bacteria are large domain of single-cell, prokaryote microorganisms. Typically a few micrometers in length, bacteria have a wide range in shape ranging from spheres to rods and spirals. Bacteria are ubiquitous in every habitat on Earth, growing in soil, acidic hot springs, radioactive waste (Fredrickson et al 2004), water, and deep in the Earth's crust, as well as in organic matter and the live bodies of plants and animals. There are typically 40 million bacterial cells in a gram of soil and a million bacterial cells in a milliliter of fresh water; in all, there are approximately five nonillion (5×10^{30}) bacteria on Earth (Whitman et al 1998), forming a biomass on Earth, which exceeds that of all plants and animals (Hogan 2010) . Bacteria are vital in recycling nutrients, with many steps in nutrient cycles depending on these organisms, such as the fixation of nitrogen from the atmosphere and putrefaction. However, most bacteria have not been characterized, and only about half of the phyla of bacteria have species that can be grown in the laboratory (Rappé and Giovannoni 2003). After the discovery that prokaryotes consist of two very different groups of organisms that evolved independently from an ancient common ancestor, these two groups called Bacteria and Archaea http://en.wikipedia.org/wiki/Bacteria

Many bacteria can move using a variety of mechanisms: flagella are used for swimming through water; bacterial gliding and twitching motility move bacteria across surfaces; and changes of buoyancy allow vertical motion (Bardy and Jarrell 2003). Flagella are semi-rigid cylindrical structures that are rotated and function much like the propeller on a ship.

Swimming bacteria frequently move near 10 body lengths per second and a few as fast as 100. This makes them at least as fast as fish, on a relative scale (Dusenbery 2009).

Bacterial species differ in the number and arrangement of flagella on their surface; some have a single flagellum (monotrichous), a flagellum at each end (amphitrichous), clusters of flagella at the poles of the cell (lophotrichous), while others have flagella distributed over the entire surface of the cell (peritrichous). The bacterial flagella is the best-understood motility structure in any organism and is made of about 20 proteins, with approximately another 30 proteins required for its regulation and assembly (Dusenbery 2009).The flagellum is a rotating structure driven by a reversible motor at the base that uses the electrochemical gradient across the membrane for power (Macnab 1999). This motor drives the motion of the filament, which acts as a propeller.

3.3 What are viruses?

A virus is a small infectious agent that can replicate only inside the living cells of organisms. Most viruses are too small to be seen directly with a light microscope. Viruses infect all types of organisms, from animals and plants to bacteria and archaea. Virus particles consist of two or three parts: the genetic material made from either DNA or RNA, long molecules that carry genetic information; a protein coat that protects these genes; and in some cases an envelope of lipids that surrounds the protein coat when they are outside a cell. http://en.wikipedia.org/wiki/Virus. Viruses spread in many ways; viruses in plants are often transmitted from plant to plant by insects that feed on the sap of plants, such as aphids

Viruses are very small (submicroscopic) infectious particles (virions) composed of a protein coat and a nucleic acid core. They carry genetic information encoded in their nucleic acid, which typically specifies two or more proteins. Translation of the genome (to produce proteins) or transcription and replication (to produce more nucleic acid) takes place within the host cell and uses some of the host's biochemical "machinery". Viruses do not capture or store free energy and are not functionally active outside their host. They are therefore parasites (and usually pathogens) but are not usually regarded as genuine microorganisms.

Most viruses are restricted to a particular type of host. Some infect bacteria, and are known as bacteriophages, whereas others are known that infect algae, protozoa, fungi (mycoviruses), invertebrates, vertebrates or vascular plants. However, some viruses that are transmitted between vertebrate or plant hosts by feeding insects (vectors) can replicate within both their host and their vector. This web site is mostly concerned with those viruses that infect plants but we also provide some taxonomic and genome information about viruses of fungi, protozoa, vertebrates and invertebrates where these are related to plant viruses. (Brüssow et al 2004).

4. Relation between forest and microbes

The forest support multitudes of microbial life in many different relationships with the plants that grow there. Life in a forest is self- sustaining and does not require fertilization. In a natural forest setting, leaves, needles and foliage fall and gradually decompose creating a natural fertility process and nutrient source for trees. In an urban or residential environment, the leaf litter that would eventually decompose and provide natural fertilizers for trees is cleared away. Grass is planted under trees which supports a large bacterial component in direct conflict with trees normal association with fungi.

All rainforests (tropical, subtropical and temperate) are under threat from human activity at the present time. They are being destroyed at an alarming rate that could potentially lead to many different types of environmental catastrophe, not only in the local forest zone, but globally. The greatest threats come from deforestation (tree removal by various means and for various purposes) and mining.

Deforestation may be done to create farmland, to build hydro-electric plants, to sell the lumber, or through careless or accidental burning.

Rainforest microbes are extremely efficient at breaking down and recycling waste organic matter - the leaf litter and layers of detritus on the ground. As a result, almost no nutrients reach the forest soil and it is consequently poor. Removal of the trees he soil to dry out and the little humus that allows t exists to deteriorate. This causes the rainforest microbes to die and the soil becomes largely inert, biologically. (http://www.theguardians.com/Microbiology/gm_mbr10.htm)

There is no doubt that microorganisms are the foundation of the cycles that take the essential elements of life from the soil to the tree, they influencing tree growth but they can destroy forests which offer watershed protection, timber and non-timber products, and various recreational options. The forests which prevent soil erosion help in maintaining the water cycle, and check global warming by using carbon dioxide in photosynthesis. The forest lands which are instrumental in the beauty and spiritual impact of our landscape. If the forest disappears, the entire ecosystem begins to fall apart, with dire consequences for all of us.

One of the most important findings so far in the literature indicates that the substitution of a forest area for an area of cultivated plants could reduce the diversity of the bacteria associated with the surface of leaves by more than 99 % (Ekelund et al 2009).

Although microorganisms can destroy tress, in contrast, a healthy, balanced community of microorganisms is extremely important for trees health. Some of these organisms protect trees from disease-causing organisms that would otherwise infect trees. Microbes are weapon with two blades or coin with two faces, one negative face and one positive face.

4.1 Positive face of microbes

The positive face represents the group of microorganisms which exerts influence on health of trees. Those who has positive power; they always engaged with growing or benefit strategy of trees. Their purpose is not to dominate but to get exchangeable benefits. We can call them non-pathogenic microbes. Scientists (ecologists, biologists etc) explain the interactions or the interdependence between the different forest species by means of two terms: Symbiotic relationships and nitrogen cycle.

Symbiotic Relationship between two or more species is called:

Mutualism, if the relationship is beneficial for both of them.
Commensalism, if the relationship is beneficial for one species and neutral to the other, that is, neither beneficial nor harmful for the other species.
Parasitism, if the relationship is beneficial for one species and harmful for the other species.
Neutralism, if there is no relationship between the species.
Amensalism, if the relationship is harmful for one species and neutral to the other.
Synnecrosis, if the relationship is harmful for both the species.

Nitrogen cycle: All living organisms require nitrogen compound for getting their necessary nutrients. Even though air contains nitrogen in abundance-79% by volume, it cannot be used by organisms directly. It is an inert gas and needs to be converted into usable forms of nutrients for the "users"- whether they are plants, trees by processing, either by natural biological process or by man made industrial processing. After the "users" consume their nutrients and eject their excrements to the atmosphere, they must be further acted upon by the natural biological process or the industrial processing to recycle the excrements back into nitrogen again before it can be passed on to the atmosphere. Plants require nitrogen compound in the form of ammonia, urea etc. They get them from the soil through their roots by absorption. "Nitrogen-fixing" bacteria like rhizobia and mycorrhizae fungi enable the roots of the trees and plants to get the required nitrogen compound from the soil (Jenkins and Groombridge 2007)

As mentioned before, the most important groups for forest are fungi, bacteria and viruses; therefore in this chapter you will recognize the positive and the negative face of each microbial group separately.

This is a **mutually beneficial symbiotic relationship between the trees and the fungi,** since the plants get their nutrients from the soil through the fungi and mycorrhizal fungi get their nutrients in the form of sugars, starches, proteins etc from the plant roots.

4.1.1 Positive face of fungi

There are two main types of fungi associated with trees and they are:

- **Decomposers**, which are associated with wood, leaf litter, plant and animal matter.
- **Food gathering, or mycorrhizal fungi**, which form symbiotic relationships with the roots of trees.

Fungi form a key component of the microbial populations influencing tree growth and uptake of nutrients. In addition to increasing the absorptive surface area of their host tree root systems, the hyphae of these symbiotic fungi provide an increased area for interactions with other microorganisms, and an important pathway for the translocation of energy-rich plant assimilates to the soil.

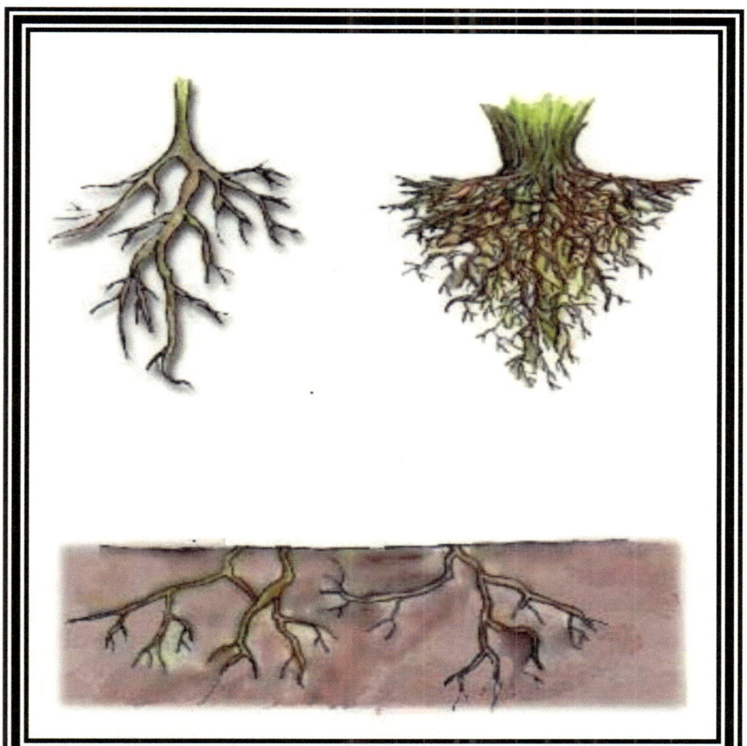

Fig. 3. Rhizosphere is the zone of soil in which interactions between living plant roots and micro-organisms are found.

Traditionally, the influence of plant assimilates on microbial communities has been defined in relation to the rhizosphere, the narrow zone of soil surrounding living roots. The rhizosphere is characterized by increased microbial activity stimulated by the leakage and exudation of organic substances from the root.

Many fungi develop only on dead or dying trees and are not pathogens. These fungi are Saprophytes and are living on dead tree tissues or organic debris as opposed to pathogens which usually gain their sustenance from living trees as Parasites. Other fungi, especially certain mushroom, are actually beneficial to trees. Many of these types of fungi form highly specialized, mutually beneficial associations called Mycorrhizae with the roots of living trees. In these associations the fungi receive sugar and other dietary essentials from the trees, and in return enhance the tree's ability to extract phosphorus and other nutrients from the soil. (Grayston et al 1997).

There is relationship between ancient trees and fungi and it is a very close one. Within the woodland ecosystem, fungi play an important role in recycling nutrients and in individual trees- within and between cells, from the leaves in the canopy down to the root hairs. As a tree ages the relationship between trees and fungi remains the same but the species may change. Fungi can be extremely long lived, perhaps even everlasting, as some species are known to grow continuously. Each tree creates a unique and dynamic support system for fungi and, contrary to previous opinion, it is likely that rather than being detrimental to the tree, fungi actually prolongs its life. http://www.ancient-tree-hunt.org.uk/ancienttrees/treeecology/fungi.htm

4.1.2 Positive face of bacteria

In the forest you can not only see fungi but also bacteria. Certain bacteria are efficient at breaking down inorganic minerals into nutrients. This process, called mineral weathering, is especially important in acidic forest soils where tree growth can be limited by access to these nutrients. Mineral-weathering bacteria can release necessary nutrients such as iron from soil minerals. This gives trees with increased concentrations of mineral-weathering microbes an advantage over other trees.

You also can see cyanobacteria on large long trees. Cyanobacteria are more abundant in mosses high above the ground, and that they "fix" twice much nitrogen as those associated with mosses on the forest floor.

Mycorrhizal fungi are beneficial fungi that act as a secondary root system for trees, plants and shrubs. This secondary root system provides up to a 700% increase in the plant's ability to absorb important nutrients and water. In nature, certain species of beneficial bacteria promote healthy plant growth and soil fertility. These "good" bacteria are called rhizobacteria and displace harmful soil life that may be present. While common in natural settings, their populations are often very low in urban and residential landscapes, nursery potting soils and other man-made landscapes (Grayston et al 1997).

There is an inherent difficulty in determining whether mycorrhizosphere bacteria are specifically associated with roots or mycorrhizal fungi, or they simply form opportunistic associations with a range of other organisms. But as the following figures indicate, presence of bacteria in the soil either in root zoon or other is very important for healthy growth.

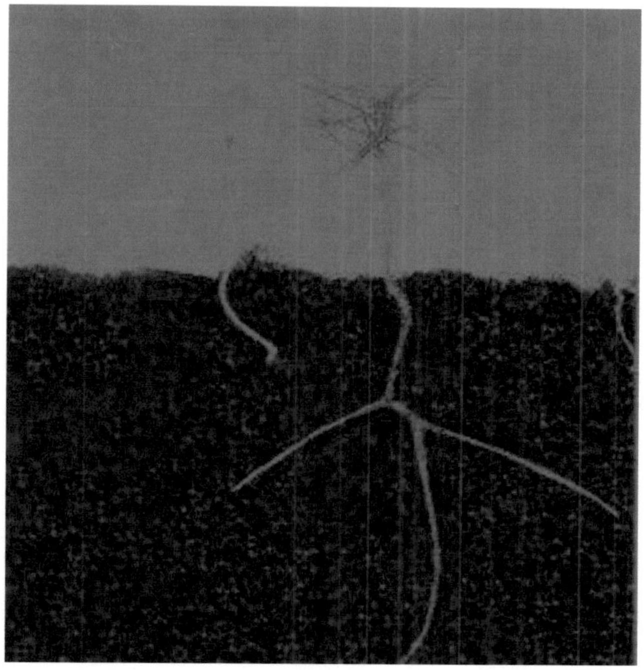

Fig. 4. Plant without Bacteria modified photo from http://www.ganeshtree.com/

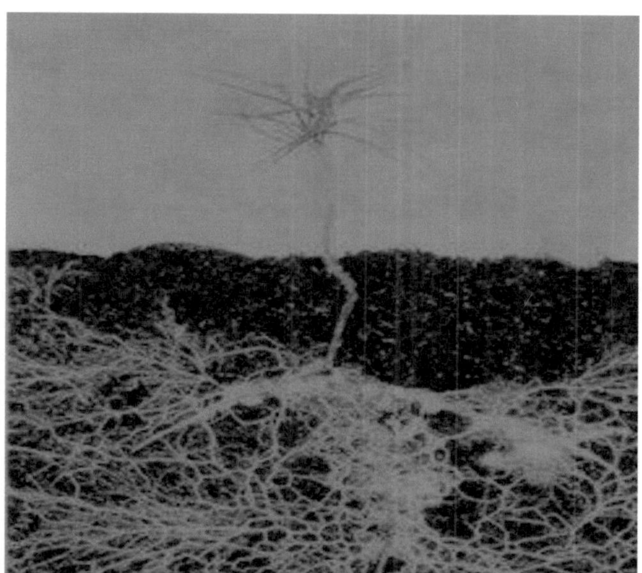

Fig. 5. Plant with Bacteria modified photo from http://www.ganeshtree.com/

Actually it is interesting to mention that there are bacteria live within the plant tissue naturally and they called endophytes, their role inside the plants still unknown but scientists expected that these bacteria improve plant growth, therefore scientists succeeded to isolate different types of bacteria like *Enterobacter*, *Burkholderia*, etc and tested the addition of these type of bacteria to different trees, they observed that microbe- supplemented trees grew faster. Scientists also sequenced the genes of theses bacteria for production of plant growth promoting enzymes, hormones, and other metabolic factors that might help explain how the bacteria improve plant growth. (Safihy et al 2009)

The relation between microbes and trees is very complicated

Each tree species has its own distinct and unique community of hundreds of species of bacteria, the function of these microorganisms in the forest's normal dynamics could be much more important than originally thought. The bacterial community varies not only in relation the various species of trees, but also based on the location of trees in different environments, taking into consideration, for example, the position of trees within a given forest or comparison of the same plant species in forest that are distant from each other.

The surfaces of tree leaves, trunks and roots contain complex biofilms comprised of diverse microorganisms that interact amongst themselves, with the plant, with animals and the atmosphere. Each plant species has a unique set of bacteria associated with it. And each part of the plant has a different community. On the leaf, bark or root, the same bacterial species are not repeated (**Castro** 2011).

Based on information obtained using molecular technology, the scientists believed that cultivated plants would probably have a much lower quantity and different type of associated bacteria than those found on forest species. A comparison has made utilizing soybean, sugarcane and eucalyptus. Surprisingly, the cultivated plants and forest trees are statistically very similar in terms of the estimated diversity and wealth of bacterial species. Nevertheless, the types of bacteria that live on the leaves of cultivated plants are very different from those that live on plants in the forest.

The role of bacterial community in the forest maybe due to that, nitrogen is an essential element for plant growth and is not found in the forest soil. In order to maintain the forest, the nitrogen must come from the environment. Nitrogen is captured from the air by microorganisms that live on tree leaves and bark (**Castro** 2011).

Why do plants in the forest rarely get sick? Probably because they have the natural protection of the microorganisms that are living there. When a plant is domesticated and grown in large monocultures, it rapidly loses the capacity to inhibit growth of pathogenic microorganisms, and with this, the results are outbreaks of disease on plantations. Protecting the plant could be one of the other functions that other bacteria have (**Castro** 2011)

Another explanation for importance of bacteria is, forest litter bacteria are especially important for completion of organic matter mineralization. Bacterial contribution to the decomposition process is complementary to that of other microbiota. Some studies have examined bacterial interrelations with the abundance and/or activity of other biota, e.g. fungi and forest litter composition, fungi and actinomycetes (Alek hina *et al.* 2001), fungi and nematodes (Mikola and Sulkava 2001), fungi and microarthropods , fungi and protozoa, fungi, algae, testate amoebae and microarthropods. However, it seems that field research on forest litter bacteria quantitatively assessing the whole complex of ecological interactions with the abundance and/or activity of fungi, protozoa (flagellates, ciliates, amoebae), nematodes, microarthropods, and forest litter composition is not common (Frouz *et al.* 2001).

4.1.3 Positive face of virus

In nature there is no positive face of virus as the word virus is Latin word referring to poison and other noxious substances.But in laboratory, Scientists can use and manipulate the viral genome for insertion of desirable genes into plants by using the viral lysogenic mechanism.

5. Negative face of microbes

The negative face represents the group of microorganisms which exerts damage on trees. Their purpose is to dominate and damage. We can call them pathogenic microbes.

On contrast to the importance of such microbes to forest life, each year millions of dollars worth of valuable timber and other tree resources are lost because of invasive microbial tree diseases. Diseases are caused by a variety of factors or agents which are divided into two general groups: non-living (abiotic) and living (biotic). Biotic agents are called Pathogens. These pathogens usually are fungi, bacteria and viruses. We have to know some common types of disease problems affecting trees.

Viruses and bacteria are important causes of plant and tree disease in many parts of the world. The diseases that caused by bacteria are, leaf and shoot infections, bacterial canker, fire blight, wetwood, scorches and yellows. The most important bacterial forest diseases are scorches and yellows. Viral diseases in trees are unimportant except in certain cases, for the effects are often subtle. Their Symptoms are often confused with mineral deficiency, ozone damage, or drought. Many fungi associated with tree diseases. One group of fungus-like organisms and three main groups of fungi are associated with diseases of forest trees. Some species are associated with diseases of roots, cankers, leaf blights, needle casts and vascular wilts-, leaf spots and anthracnose.

Microbial diseases	Other diseases
1) Fruiting bodies of a canker fungus - signs of canker infections **2)** Heart rot (internal) and sporophore of a heart rot fungus at a broken branch stub **3)** Sporophore of a butt-rot fungus at base of tree **4)** Sporophores of a root rot fungus arising from a damaged root **5)** Crown gall - a gnarled swelling ("tumor") caused by a bacterium **6)** Severed root resulting from construction damage - site of entry for root and butt-rot fungi	**1)** Leaf spots - a foliage disease **2)** Twig dieback - evidence of cankers and/or stress and decline **3)** Mistletoe - a parasitic seed plant **4)** Wilt - evidence of moisture deficiency, vascular wilt disease or root rot **5)**Vascular streaking (internal) - evidence of vascular wilt disease **6)** Branch canker at a branch stub **7)** Nematode damage to small tree roots lesions (upper) and galls (lower)

Table 1. Some common types of disease problems affecting trees

5.1 Negative face of fungi

Fungal diseases can quickly destroy a healthy tree. Deciduous, fruit bearing, ornamental and evergreen trees are all susceptible to fungal diseases. Some varieties of trees are hardier than others, but still can become infected with a fungal disease. A fungus invades the tree when a spore germinates on a tree and eventually spreads and feeds on the host tree. Fungi can cause damage for trees or other plants by different ways, each fungus has its own method for invading tree but some trees have their own tricks for defense and here are some examples for such bad invasion:

*Some fungi like *Verticillium dahliae* produces microsclerotia (thick-walled viable fungal cells) that can survive in soil for many years and invade the tree directly through the root system or through root wounds (Hiemstra and Harris 1998; Tjamos et al. 2000).The fungus spreads in the vascular tissue via spores, produces toxins that kill cells, and blocks the trees ability to transport nutrients and water. The tree in turn produces defense compounds that attempt to isolate infected cells to limit fungal movement in the tree. The isolation of infected vascular tissue reduces the flow of water from the roots upward. Symptoms of Verticillium wilt can develop any time during the growing season. Leaves may wilt in portions of the canopy or on scattered branches and may become chlorotic and necrotic (faded green, yellow, or brown); branch dieback may occur, and a general decline of the tree ensues.

Photo of *Verticillium dahliae* adopted from en.wikipedia.org/wiki/Verticillium_dahliae

*Several species of fungi like *Aureobasidium apocryptum* (synonym *Kabatiella apocrypta*) and *Colletotrichum gleosporioides* .The fungus over winters on diseased leaf or stem tissue, and produces infectious spores in the spring. During cool, rainy periods of spring, the fungal spores are carried by wind and rain to newly emerging leaves of trees. Symptoms on infected leaves appear as scattered dead areas (red, black, brown, or tan) developing along and between leaf veins. These lesions may enlarge rapidly and kill large areas of leaf tissue. Leaves may become distorted, and significant leaf drop can occur in late spring. Severe infection can result in extensive defoliation. The fungus can infect beyond the leaf surface into twigs causing cankers, which may eventually kill small branches (Berry 1985).

Photo of anthracnose: (*Aureobasidium apocryptum*) adopted from www.ipmimages.org

*The fungus *Neonectria galligena* causes a disease known as Nectria canker, this fungus is slowly growing pathogen usually does not kill the tree, but reduces the value of the tree for veneer and weakens the tree causing breakage at the point of the canker on the main stem. The fungus overwinters in the callus (wound-closing) tissue produced by the tree in response to infection. Infections occur on twigs, branches, and trunks in spring or early summer during moist periods, and the spores are dispersed by wind or rain. Infection occurs naturally through leaf scars or through wounds from improper pruning, storm damage, frost cracks, and other mechanical damage. Symptoms of Nectria canker appear as dark, water-soaked, depressed areas of the bark on stems or branches. Infected small twigs may become girdled, wilt, and die suddenly. The tree responds to infection by trying to isolate the fungus by the production of a ridge of callus tissue. As the fungus continues to grow from these perennial cankers into healthy wood, the tree produces another ridge of callus tissue. Concentric ridges of callus tissue then develop over time producing a round or elongated target like canker on the tree. This canker disease grows slowly when the tree is dormant or under stress, and kills the bark, cambium, and outer sapwood (Sinclair and Lyon 2005).

Photo of *Neonectria galligena* adopted from http://donsgarden.co.uk/homepages

**Eutypella parasitica* is the fungus that causes a disease known as Eutypella canker. This disease causes mortality by girdling trees and the elliptical perennial cankers are entry points for decay fungi. The resulting decay and malformation of the trunk of the tree results in wood useless for veneer and makes the tree susceptible to wind breakage. Symptoms of Eutypella canker large, tar-like spots surrounded by yellow-orange zones appear on the upper surface of the leaves.

Photo of *Eutypella parasitica* adopted from www.forestyimages.org

*Other types of fungi known as decay fungi cause stem trunk, root, and butt rots. The fungi develop slowly in trees and their presence may only be evident once trees start to die or produce conks. *Inonotus glomeratus* causes a white to light brown, spongy canker rot. Branch stubs or wounds are the primary entry point for infection by this canker rot fungus. *Phellinus igniarius* causes trunk rot. *Climacodon septentrionalis* infects through wounds and causes a spongy white rot. Decay is extensive by the time conks appear. *Ganoderma lucidum* causes a butt and trunk decay, and *Laetiporus sulphureus* more commonly causes rot in butt and stem.

Photo of Inonotus glomeratus adopted from http://www.herbarium.iastate.edu/fungi/index.html

*Sapstreak disease is a disease caused by the fungus *Ceratocystis virescens* (Houston 1993). The fungus produces a distinctive banana oil odor; noticeable if profuse sporulation is found. Infection occurs through roots, buttress roots, or lower stem wounds. Symptoms of sapstreak disease are a sparse crown with unusually small leaves. Branch dieback occurs, trees may fail to leaf-out the next year, or the tree may die suddenly. The diseased wood has a yellow-green stain with red flecks that darkens after exposure to air, and eventually turns to a light brown.

*Armillaria root disease is a destructive disease of the roots and butts of trees and the fungus most often associated with this disease is *Armillaria mellea* (Shaw and Kile 1991), although other species (*A. calvescens, A. gallica, A. ostoyae,* and *A. sinapina*) have been reported to be associated with this disease .Trees infected with *Armillaria* can have reduced growth, rapid mortality, wood decay (white rot), and susceptibility to wind throw. *Armillaria* can infect non-wounded root systems of trees, and the fungus spreads from tree to tree through roots, and via rhizomorphs which infect intact bark. These fungi produce a honey-colored mushroom around the base of the tree in late summer or fall during moist periods .Disease caused by *Armillaria* species contribute to tree decline by the interaction with other factors such as site, tree age, drought, insects, etc.

*Pecan scab is the most serious disease of pecan and is caused by the fungus *Cladosporium caryigenum*. The fungus overwinters on infected shucks, leaves, leaf petioles, and stems. In the spring, during periods of warm, rainy weather, spores germinate and are dispersed long distances by rain and wind. Disease symptoms first appear on immature leaves, leaf petioles, and nut shucks as small, olive-green spots that turn black. Leaf infections do not cause serious defoliation (Kromroy 2004).

*Anthracnose is a fungal leaf spot of pecan caused by *Glomerella singulata*. The spores are spread in spring and early summer during rainy periods. Symptoms start as brown to black sunken lesions on the leaves and shucks. Cream- to salmon-colored spores in concentric rings may also appear on the shucks. *Gnomonia* leaf spot caused by the fungus *Gnomonia caryae* is characterized by reddish-brown (liver colored) circular spots.

*Botryosphaeria canker (branch dieback) results from infection by the fungus *Botryosphaeria dothidea* and is most often associated with stressed trees. Trees affected with this canker disease exhibit leaf wilting on twigs and branches. The infected branch turns black, cankers enlarge, and the pith of the branch is black or dark brown. This fungal disease infects through natural openings in the bark or through pruning wounds. The fungus is dormant in the winter, and spores are spread by wind in the spring. Cankers will girdle and kill twigs and branches. Twig death is usually generalized within a tree, and stress may be caused by defoliation from other pathogens, drought, or shading (Sinclair and Griffiths 1994).

*Phytophthora root rot incited by the fungus-like organism *Phytophthora cinnamomi* is a soil-borne pathogen that causes root and collar rot. Symptoms appear as root lesions that develop upward in the trunk. The pathogen spreads by spores that can be carried over large distances by water and in soils. Infection occurs through the roots, causes root cell death (rotting), and therefore prevents transport of water from the roots. Moderate levels of soil compaction and moisture contribute to the susceptibility of root rot on American chestnut (*Castanea dentata*) seedlings (Rhoades et al. 2003). Root rot reduces tree vigor and causes mortality.

*Chestnut blight is a devastating disease caused by the fungus *Cryphonectria parasitica* (Anagnostakis 2000). The fungus enters through cracks or wounds in the bark and eventually kills the cambium. The fungus is spread by wind, rain, insects, and birds Airborne spores carried by birds and insects that visit colonized bark "account" for overland spread of the fungus.

*Heart rot caused by the sulfur fungus, *Laetiporus sulfureus*, is a brown heart rot of living trees that will also decay dead trees. The fungus can invade through bark wounds and dead branch stubs. Symptoms include a slightly depressed or cracked bark in areas of the tree with dying branches. Massive clusters of bright, sulfur-yellow to orange, shelf-like conks are produced annually (usually in fall), and become brittle and white wit age (Sinclair and Lyon 2005).

5.2 Negative face of bacteria

Not only fungi have the ability to invade trees but so do bacteria. But diseases caused by bacteria are not much as those of fungi and here are some of bacterial diseases.

*Bunch disease is caused by a *phytoplasma* [related to gram-positive bacteria] (Davis and Sinclair 1998) that may be transmitted from tree to tree by insects (leafhoppers). The pathogen can also be transmitted through grafts. Symptoms of bunch disease are branches with excessive lateral stem growth and compacted growth of leaves on these stems (witches'-broom). Upright shoots form on the trunks and main branches. Bunch disease is very obvious in the spring and early summer as the diseased shoots leaf out earlier than non-infected shoots. Symptoms may occur throughout the tree or be limited to individual branches. Terminal branches infected with bunch disease do not produce a normal crop of nuts. Branches may fail to go dormant in the fall and be killed by frost or winter injury.

*Bacterial leaf scorch affects some trees *Xylella fastidiosa* (Wells et al. 1987) is a gram-negative bacterium that lives in an infected tree's water conducting tissue (xylem) and is transmitted to other healthy trees by insects that feed on xylem fluid, such as leafhoppers, treehoppers, and possibly spittlebugs .The bacterium multiplies in the xylem tissue and, together with the overproduction of defense compounds produced by the tree in response to infection physically, blocks the xylem. Water transport then becomes limited to leaves, branches, and roots. Symptoms of bacterial leaf scorch first appear in late summer to early fall and can be identified by a characteristic marginal leaf scorch (Gould and Lashomb 2005). Leaves develop an irregular pattern of light and reddish brown tissue separated from green tissue by a yellow halo. As the disease progresses, leaves curl and drop prematurely, branches die, and the tree declines. Symptoms of other physiological or cultural problems can be mistaken for bacterial leaf scorch.

*Ash yellow is a disease caused by a *phytoplasma* (bacteria-like organism lacking a cell wall) that inhabits the phloem (Griffiths et al. 1999) and substantially reduces growth, induces premature decline, and death of *Fraxinus* spp. Both green (*F. pennsylvanica*) and white (*F. americana*) ash are susceptible to ash yellows, in addition to other ash species. The *phytoplasma* is spread by phloem-feeding insects such as leafhoppers (Hill and Sinclair 2000). The *phytoplasma* is introduced into the phloem tissue, via the insect saliva, and spreads throughout the tree killing the tissue. Symptoms of ash yellows include small, chlorotic leaves growing in tufts at the end of branches, branch dieback, bark cracks, thin chlorotic crowns, epicormic sprouting (along branches or at ground level), early fall coloration, loss of dominant growth habit, general decline, and witches' brooms (cluster of spindly shoots) on the trunk of the tree.

*Bacterial blight caused by the bacterium *Xanthomonas arboricola*. The bacterium overwinters in twig cankers, and dormant buds and catkins. The bacteria invade new young shoots, leaves, catkins, and fruit through natural openings, wounds, or insect damaged areas. During periods of spring rainfall or wet weather, conditions are suitable for bacterial infection. Symptoms of bacterial blight on young walnut shoots appear as black spots or lesions. The lesions may girdle the shoot and extend into the pith to form cankers. The blight causes small, irregular-shaped brown to black spots on all tissues of the leaves (midrib, veins, rachis, and petiole). Unless infection is severe, defoliation does not usually occur and most infected leaves remain on the tree. Infected catkins turn black and become shriveled and distorted (Bentz and Sherald 2001).

5.3 Negative face of virus

Viruses cause many important plant diseases and are responsible for huge losses in crop production and quality in all parts of the world. Infected plants may show a range of symptoms depending on the disease but often there is leaf yellowing (either of the whole leaf or in a pattern of stripes or blotches), leaf distortion (e.g. curling) and/or other growth distortions (e.g. stunting of the whole plant, abnormalities in flower or fruit formation).

Some important animal and human viruses can be spread through aerosols. The viruses have the "machinery" to enter the animal cells directly by fusing with the cell membrane (e.g. in the nasal lining or gut).

By contrast, plant cells have a robust cell wall and viruses cannot penetrate them unaided. Most plant viruses are therefore transmitted by a vector organism that feeds on the plant or (in some diseases) are introduced through wounds made, for example, during cultural operations (e.g. pruning). A small number of viruses can be transmitted through pollen to the seed while many that cause systemic infections accumulate in vegetative-propagated crops. The major vectors of plant viruses are insects. This forms the largest and most significant vector group and particularly includes, **aphids** which transmit viruses from many different genera, including *Potyvirus, Cucumovirus* and *Luteovirus. Potato virus* http://www.dpvweb.net/intro/index.php

Viruses differ from fungi and bacteria in that they do not produce spores or other structures capable of penetrating plant parts. Since viruses have no active methods of entering plant cells, they must rely upon mechanically caused wounds, vegetative propagation of plants, grafting, seed, pollen, and being carried on the mouth parts of chewing insects. Some viruses are introduced into plants through small wounds caused by handling and by insects chewing on plant parts. Once the virus enters the host, it begins to multiply by inducing host cells to form more virus (Viruses do not cause disease by consuming or killing cells but rather by taking over the metabolic cell processes, resulting in abnormal cell functioning. Abnormal metabolic functions of infected cells are expressed. Infected plants serve as reservoirs for the virus and the virus can be transmitted easily (either mechanically or by insects) to healthy plants (Pfleger and Zeyen 2008).

Symptoms and effects of viral diseases:

As mentioned before symptoms are often confused with mineral deficiency, ozone damage, or drought. Many say that viral diseases in trees are unimportant, for the effects are often subtle.

- leaves are mottled with necrotic and chlorotic lesions,
- ring spots, and yellowing
- Effects:
    - stunted growth
    - decreased photosynthesis and increased respiration
    - reduction in cold tolerance
    - rarely, death results

http://www.forestpathology.org/bacteria.html

Viruses are widespread pathogens in agriculture crops and weeds; the latter are considered important reservoir of viruses .Therefore, it can be expected that they are also present in forest ecological system. Viruses can be the cause of severe disease of no woody and woody plants. Thousands of viruses' diseases have been described worldwide. A considerable number of viruses have caused epidemic in agriculture crops when environmental conditions, host range and other factors were favorable. Information of viruses' diseases of forest tree is rare .One reason may be that, the interest of research in forest pathology was traditionally focused on insect damage and fungal diseases .A second reason is the technical difficulty of demonstrating, isolating and transmitting viruses in woody plants. And last but not least, the economic importance of loss of a relatively small number of forest trees has been considered as negligible compared with agricultural crops such as fruit trees (Nienhaus 1985).

It can be stated that a number of viruses are inciting factors for decline of forest trees. Viruses are predisposing factors leading to early senescence reduces the regeneration capacity of the host plant and the juvenile metabolic vigor is lost. Under abiotic stress conditions the infected trees have less potential for recovery from inciting factors than non infected trees.

After reading the negative face of microbes for trees, the question arises here is to which degree the potential for recovery of infected trees, the answer might be in the following section.

6. Control of plant disease

Diagnosis of plant problems can be a difficult and at times impossible task but it is important. Therefore, the goal is a correct diagnosis so that management procedures can be implemented successfully.

As mentioned above, plant diseases are caused mainly by fungi, bacteria and viruses. Therefore plant diseases need to be controlled.

Different approaches may be used to prevent, mitigate or control plant diseases. One of such approach is using chemicals. Today, there are strict regulations on using chemicals;

consequently, some researchers have focused their efforts on developing alternative inputs to synthetic chemicals for controlling diseases. Among these alternatives are those referred to as biological control? The terms "biological control" equal "biocontrol" have been used in plant pathology.

But what is biological control? It is the suppression of damaging activities of one organism by one or more other organisms, often referred to as natural enemies. In plant pathology, the term applies to the use of microbial antagonists to suppress diseases.

Biological control offers an environmentally friendly approach to the management of plant disease. The microbiologists and plant pathologists try to gain a better knowledge of biocontrol agents, to understand their mechanisms of control and to explore new biotechnological approaches. A screening approach was developed to assess the potential of plant-associated bacteria to control plant pathogens. The study of plant-associated bacteria and their antagonistic potential is important not only for understanding their ecological role and the interaction with plants and plant pathogens but also for any biotechnological application. Plant-associated bacteria can be used directly for biological control of soil borne plant pathogens or indirectly for the productions of active substances (Doaa et al 2008).

As biocontrol of plant disease involves the use of an organism or organisms to reduce disease, several microorganisms have shown potentialities as biological control agents against important plant diseases caused by soil-borne pathogens. They are *Pseudomonas fluorescens* against the take-all disease of wheat caused by *Gaeumannomyces graminis, Erwinia caratovora* infection of wheat, *Thielaviopsis basicola* infection of tobacco and damping off caused by *Pythium* in cotton; *Bacillus subtilis* against *Fusarium roseum* wilt of com: *Trichoderma harzianum* against *Alternaria spp.* infection of radish; *Penicillium oxalicum* against root rot of peas; non-pathogenic *Fusarium oxysporum* against *Fusarium* wilt of cucumber; *Trichoderma viride* against *Verticillium* wilt of tomatoes; *Cytophaga* sp. against damping off of conifer seedlings; *Bacillus sp., Penicillium sp.,* and *Alcaligenes sp.* against crown gall disease of cherry seedlings caused by *Agrobacterium tumefaciens; Chaetomium globosum* against damping off of sugar beets; *Pseudomonas putida* against *Fusarium solani* wilt of beans; *Bacillus sp.* and *Pseudomonas sp.* against *Fusarium oxysporum* wilt of carnations. http://www.microbiologyprocedure.com/microbial-products-influencing-plant-growth/biological-control-of-plant-diseases.html

Here are photos for how can one microorganism kill other microorganism to assess the potential of microbes to control plant pathogens, adopted from previous work for the author.Over the past one hundred years, research has repeatedly demonstrated that phylogenetically diverse microorganisms can act as natural antagonists of various plant pathogens (Cook 2000).The interactions between microorganisms and plant hosts can be complex. Interactions that lead to biocontrol can include antibiosis, competition, induction of host resistance, and predation (Cook and Baker 1983).However, fewer isolates can suppress plant diseases under diverse growing conditions and fewer still have broad-spectrum activity against multiple pathogenic taxa. Nonetheless, intensive screens have yielded numerous candidate organisms for commercial development. Some of the microbial taxa that have been successfully commercialized and are currently marketed as EPA-registered biopesticides in the United States include bacteria belonging to the genera *Agrobacterium, Bacillus, Pseudomonas,* and *Streptomyces* and fungi belonging to the genera *Ampelomyces, Candida, Coniothyrium,* and *Trichoderma* http://www.plantmanagementnetwork.org/pub/php/review/biocontrol/

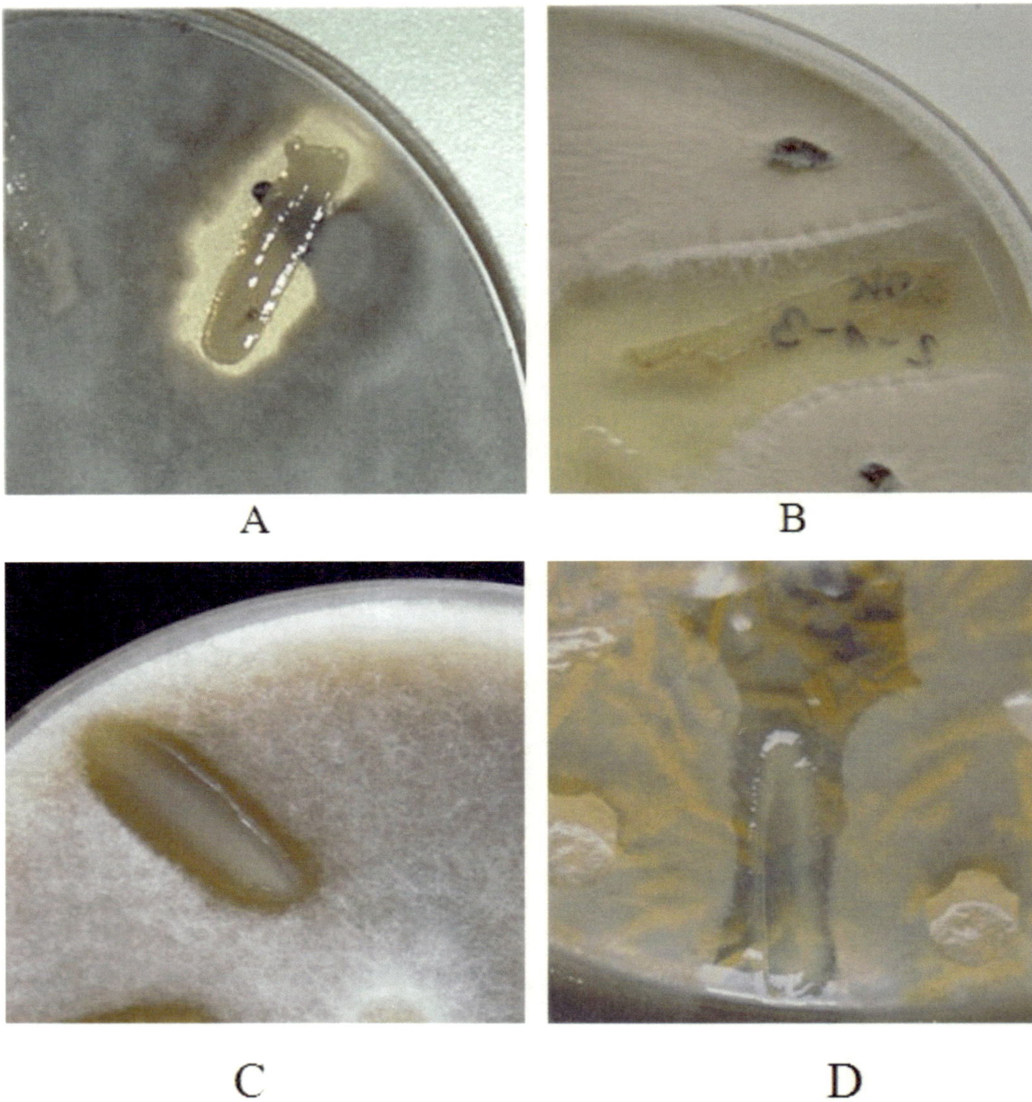

Fig. 6. Antagonistic activity of Paenibacillus *polymyxa* against plant pathogens, A (*Phoma betae*), B (*Rhizoctonia solani*), C (*Pythium ultimum*), D (*Aphanomyces cochlioides*). (Adopted from Doaa et al 2008)

Screening is a critical step in the development of biocontrol agents. The success of all subsequent stages depends on the ability of a screening procedure to identify an appropriate candidate. Many useful bacterial biocontrol agents have been found by observing zones of inhibition in Petri plates .However, this method does not identify biocontrol agents with other modes of action such as parasitism, induced plant resistance, or some forms of competition.

Screening methods for parasitism include burying and retrieving propagules of the pathogen to isolate parasites. For competition, methods include looking for microbes that quickly colonize sterilized soil and have the ability to exclude other organisms that attempt to invade the space, and looking for microbes that colonize the infection court. Primary screens for new biocontrol microbes are still undertaken(Larkin and Fravel 1998) and it seems likely that continued prospecting will be required to diversify the potential applications of biocontrol as well as replace more widely used biocontrol products should resistance develop.

7. Probable mechanisms of biocontrol

The phenomenon of disease suppressive soils has fascinated plant pathologists for decades. Observed in many locations around the world, suppressive soils are those in which a specific pathogen does not persist despite favorable environmental conditions, the pathogen establishes but doesn't cause disease, or disease occurs but diminishes with continuous monoculture of the same crop species. The phenomenon is believed to be biological in nature because fumigation or heat-sterilization of the soil eliminates the suppressive effect, and disease is severe if the pathogen is re-introduced.

There are three main mechanisms by which one microorganism may limit the growth of another microorganism: antibiosis, mycoparasitism, and competition for resources. Antibiosis is defined as inhibition of the growth of one microorganism by another as a result of diffusion of an antibiotic. Antibiotic production is very common among soil-dwelling bacteria and fungi, and in fact many of our most widely used medical antibiotics (e.g., streptomycin) are made by soil microorganisms. Antibiotic production appears to be important to the survival of microorganisms through elimination of microbial competition for food sources, which are usually very limited in soil

Another mechanism of biocontrol is destructive mycoparasitism. This is parasitism of a pathogenic fungus by another fungus. It involves direct contact between the fungi resulting in death of the plant pathogen, and nutrient absorption by the parasite.

Microorganisms compete with each other for carbon, nitrogen, oxygen, iron and other micronutrients. **Nutrient competition** is likely to be the most common way by which one organism limits the growth of another, but demonstrating that this is actually responsible for biological control is quite challenging.

In most terrestrial habitats, microbial competition for the soluble form of iron, Fe3+, is keen. Some fungi and bacteria produce very large molecules called siderophores which are efficient at chelating Fe3+. Individual strains can have their own particular siderophores and receptors which can bind Fe3+ in such a way that the iron becomes inaccessible to other microorganisms, including pathogens. In some cases, siderophore production and competitive success in acquiring Fe3+ is the mechanism by which biocontrol agents control plant diseases. Siderophores produced by certain strains of *Pseudomonas* have been implicated in disease suppression of several fungal diseases, but none of these biocontrol organisms have yet been developed commercially. http/www. Understanding the mechanisms of biocontrol.

8. Recommendation

It is critical that effective safe means of controlling and eliminating many of these invasive tree diseases must found and applied. Major challenges include reducing pathogen without affect biodiversity, as the microbial communities in the forest systems will be modified due to tillage, herbicides and pesticides. Microbial diversity in these systems will be reduced and the functional consequences of this loss of diversity are still uninvestigated but may alter plants life style. Increased environmental awareness has progressively led to a shift from conventional intensive management to the opposite direction.

We do not need sterile life from microbes but we need to spread beneficial microbes against pathogenic microbes, because the former have magical affects on overcoming the later problems via antagonism activity.

9. References

Alek hina L. K., Polyanskaya L. M., Dobrovol'skaya T. G. 2001 - Populationdynamics of microorganisms in the soils of the Central Forest Reserve (model experiments)– Eurasian Soil Science, 34: 88–91.

Anagnostakis, S.L. 2000. Revitalization of the majestic chestnut: chestnut blight disease. Online (www.apsnet.org). *APSnet Feature*, The American Phytopathological Society, St. Paul, MN. 9p.

Bardy S, Ng S, Jarrell K (2003). "Prokaryotic motility structures". *Microbiology* 149 (Pt 2): 295–304. doi:10.1099/mic.0.25948-0. PMID 12624192.

Bentz, J. and J. Sherald. 2001. Transmission of the xylem-limited bacterium *Xylella fastidiosa* to shade trees by insect vectors. In:Ash, C.L. (ed.), *Shade tree wilt diseases*, The American Phytopathological Society, St. Paul, MN: 203-208.

Berry, F.H. 1985. Anthracnose diseases of eastern hardwoods. Forest insect and disease leaflet No. 133. USDA Forest Service. Online (www.na.fs.fed.us/spfo/pubs/fidls/anthracnose_east/fidl-ae.htm).

Brüssow H, Canchaya C, Hardt WD (2004). "Phages and the evolution of bacterial pathogens: from genomic rearrangements to lysogenic conversion". *Microbiology and Molecular Biology Reviews* 68 (3): 560–602. doi:10.1128/MMBR.68.3.560-602.2004. PMC 515249. PMID 15353570.

Callan B. (2001). Introduction to forest disease. Common Tree Diseases of B.C. http://www.pfc.cfs.nrcan.gc.ca/diseases/CTD/index_e.html

Castro F.(2011). The microbes of forests. Agência FAPESP. Free news letter(Brasil)

Cook, R. J. 2000. Advances in plant health management in the 20th century. Ann. Rev. Phytopathol. 38:95-116.

Cook R. J. and Baker K. F. 1983. The Nature and Practice of Biological Control of Plant Pathogens. American Phytopathological Society, St. Paul, MN.

Davis, R.E. and W.A. Sinclair, 1998. Phytoplasma identity and disease etiology. Phytopathology 88: 1372-1376.

Doaa A.R. Mahmoud, Abeer A. Mahmoud and A.M. Gomaa (2008). Antagonistic Activities of Potato Associated Bacteria viaTheir Production of Hydrolytic Enzymes with Special Reference to Pectinases Research Journal of Agriculture and Biological Sciences, 4(5): 575-584

Dusenbery, David B. (2009). *Living at Micro Scale*, p. 136. Harvard University Press, Cambridge, Mass. ISBN 978-0-674-03116-6

Ekelund, F., Saj, S., Vestergard, M., Bertaux, J. & Mikola, J. (2009).The "soil microbial loop" is not always needed to explain protozoan stimulation of plants. SOIL BIOLOGY & BIOCHEMISTRY 41, 2336-2342

Fredrickson JK, Zachara JM, Balkwill DL. (2004). "Geomicrobiology of high-level nuclear waste-contaminated vadose sediments at the Hanford site, Washington state". *Applied and Environmental Microbiology* 70 (7): 4230–41.

Frouz J., Keplin B., Pizl V., Tajovsky K., Star y J., Lukes ova A., Novakova A., B alik V., Hanel L., Materna J., Duker C., Chalupsky J., Rus ek J., Heinkele T. 2001 - Soil biota and upper soil layer development in two contrasting post-mining chronosequences - Ecological Engineering, 17: 275–284.

Gould, A.B. and J.H. Lashomb. 2005. Bacterial leaf scorch of shade trees. Online (www.apsnet.org). APSnet Feature, The American Phytopathological Society, St. Paul, MN. 18p.

Grayston, S.J., Vaughan, D. and Jones, D. (1997) Rhizosphere carbon flow in trees, in comparison with annual plants: the importance of root exudation and its impact on microbial activity and nutrient availability. Applied Soil Ecology 5, 29–56

Griffiths, H.M., W.A. Sinclair, C.D. Smart, and R.E. Davis. 1999. The phytoplasma associated with ash yellows and lilac witches'-broom: *'Candiatus Phytoplasma fraxini'*. *International Journal of Systematic Bacteriology* 49:1605-1614.

Hiemstra, J.A. and D.C. Harris. 1998. *A compendium of* Verticillium *wilt in tree species.* Ponsen and Looijen, Wageningen, the Netherlands. 80p.

Hill, G.T. and W.A. Sinclair. 2000. Taxa of leafhoppers carrying phytoplasmas at sites of ash yellows occurrence in New York State. *Plant Disease* 84:134-138.

Hogan C.M. 2010. *Bacteria.* Encyclopedia of Earth. eds. Sidney Draggan and C.J.Cleveland, National Council for Science and the Environment, Washington DC

Houston, D.R. 1993. *Recognizing and managing sapstreak disease of sugar maple.* USDA Forest Service, Northeastern Forest Experiment Station, Research Paper NE-675. 11 p.

Kenneth Todar 2009 http://textbookofbacteriology.net/themicrobialworld/origins.html

Kromroy, K.W. 2004. *Identification of* Armillaria *species in the Chequamegon-Nicolet national forest.* USDA Forest Service, North Central Research Station, Research Note NC-388, St. Paul, MN. 10 p.

Jenkins M. D. , Groombridge B. (2007) World Atlas of Biodiversity: Earth's Living Resources in the 21st Century, World Conservation Monitoring Centre, United Nations Environment Programme, retrieved 20 March 2007

Larkin, R. P. and Fravel, D. R. 1998. Efficacy of various fungal and bacterial biocontrol organisms for control of Fusarium wilt of tomato. Plant Dis. 82: 1022-1029.

Macnab RM (1 December 1999). "The bacterial flagellum: reversible rotary propellor and type III export apparatus". *J. Bacteriol.* 181 (23): 7149–53.

PMC

Mikola J., Sulkava P. 2001 - Responses of microbial-feeding nematodes to organic matter distribution and predation in experimental soil habitat - Soil Biology and Biochemistry, 33: 811–817.

Nienhaus F. (1985). Infectious disease in forest trees caused by viruses, mycoplasma-like organisms and primitive bacteria. Experientia 41p 597-603

Pfleger F.L. and Zeyen R. J. (2008) Tomato-Tobacco Mosaic Virus Disease http://www.extension.umn.edu/distribution/horticulture/dg1168.htm

Rappé and Giovannoni (2003)Rappé MS, Giovannoni SJ (2003). "The uncultured microbial majority". *Annual Review of Microbiology* 57: 369–94.

Rhoades, C.C., S.L. Brosi, A.J. Dattilo, and P. Vincelli. 2003. Effect of soil compaction and moisture on incidence of phytophthora root rot on American chestnut (*Castanea dentata*) seedlings. *Forest Ecology and Management* 184:47-54.

Safiyh T., Craig G., Sébastien M., Lee Newman, Adam H., Nele W., Tanja B., Jaco V., and Daniel v. (2009) Genome Survey and Characterization of Endophytic Bacteria Exhibiting a Beneficial Effect on Growth and Development of Poplar Trees. *Applied and Environmental Microbiology*, Vol. 75, No. 3, p. 748-757 doi: 10.1128/AEM.02239-08

Shaw, C.G. and G.A. Kile. 1991. *Armillaria* root disease. USDA Forest Service, *Agriculture Handbook* 691, Washington, DC. 233 p.

Sinclair, W.A. and H.M. Griffiths. 1994. Ash yellows and its relationship to dieback and decline of ash. *Annual Review of Phytopathology* 32:49-60.

Sinclair, W.A. and H.H. Lyon. 2005. *Diseases of trees and shrubs*. 2nd edition. Cornell University Press, Ithaca, NY. 680p.

Tjamos, E.C., Rowe R.C., Heale J.B., and Fravel D.R. 2000. *Advances in Verticillium research and disease management*. The American Phytopathological Society, St. Paul, MN. 376p.

Whitman WB, Coleman DC, Wiebe WJ (1998). "Prokaryotes: the unseen majority". *Proceedings of the National Academy of Sciences of the United States of America* 95 (12): 6578–83.

Wells, J.M., B.C. Raju, H.Y. Hung, W.G. Weisburg, L. Mandelco-Paul, and D.J. Brenner. 1987. *Xylella fastidiosa* gen. nov., sp. nov.: Gram-negative, xylem-limited, fastidious plant bacteria related to *Xanthomonas* spp. *International Journal of Systematic Bacteriology* 37(2):136-143.

http://www.theguardians.com/Microbiology/gm_mbr10.htm

http://www.voanews.com/english/news/a-13-2009-02-13-voa31-68673292.htmlhttp://en.wikipedia.org/wiki/Bacteria

http://en.wikipedia.org/wiki/Virus

http://www.theguardians.com/Microbiology/gm_mbr10.htmhttp://www.ancient-tree-hunt.org.uk/ancienttrees/treeecology/fungi.htm

http://www.dpvweb.net/intro/index.php

http://www.forestpathology.org/bacteria.html

http://www.microbiologyprocedure.com/microbial-products-influencing-plant-growth/biological-control-of-plant-diseases.html

http://www.plantmanagementnetwork.org/pub/php/review/biocontrol/

9

Assessment of *Lepthosphaeria polylepidis* Decline in *Polylepis tarapacana* Phil. Trees in District 3 of the Sajama National Park, Bolivia

Mario Coca-Morante
Departamento de Fitotecnia y Producción Vegetal
Facultad de Ciencias Agrícolas
Pecuarias
Forestales y Veterinarias "Dr. Martin Cárdenas"
Universidad Mayor de San Simón
Cochabamba
Bolivia

1. Introduction

The Sajama National Park (SNP) was the first protected area (1939) in Bolivia (Fig. 1). Nowadays it is a National Park and Natural Management Area (Daza von Boeck 2005). The SNP contains a forest of the native Andean tree known as *queñua* or *quehuiña* (*Polylepis tarapacana* Phil). Forest of this type is found only in the Bolivian Andes (Argollo *et al.* 2006), where it suffers from human disturbance, including tree felling, man-made fires, the grazing of domestic animals (Toivonen *et al.* 2011) and firewood and coal extraction (Fjeldså & Kessler 2004). Indeed, its continued existence is threatened (Rivera 1998; mentioned by Daza von Boeck 2005).

The SNP occupies some 100,000 ha, with its *queñua* forest forming a belt around it (Daza von Boeck 2005). The work reported in this chapter focuses on an area on the southwestern side of District 3 of the SNP's *queñua* forest (to the northeast of the Oruro Department in the Sajama Province, on the western *Altiplano*, covering part of the Curahuara de Carangas and Turpo jurisdiction) (Fig. 1). Located between 68°40S-69°10W and 17°55S-18°15W, the SNP lies at an altitude of 4200-6600 m in the mountains below the Sajama Peak (6524 m) (Fig. 2) (Daza von Boeck 2005). The mean annual temperature in the area of the *queñua* forest is 10°C; the maximum temperature reached is 22°C, but winter temperatures can be as low as -30°C. The mean annual rainfall is 280 mm, with a range of 90-400 mm. The study area in District 3 lies at an altitude of 4200-4300 m.

2. Conservation status of the *queñua* forest

The SNP's *queñua* forest has become fragmented over centuries of human activity, leaving its animal and plant biodiversity seriously threatened (Argollo *et al.* 2006). According to

IUCN criteria, 10 of the 13 species of *Polylepis* in the Bolivian Andean region are threatened or almost threatened, the latter category being that into which *P. tarapacana* currently falls (Gareca *et al.* 2010).

Fig. 1. Land sat satellite image TM (19/07/2011) showing the peak Sajama (light blue colour).

3. Biological factors affecting the survival of the SPN *queñua* forest

The SNP's *queñua* forest is also at risk from disease. During systematic studies of *Polylepis* in Bolivia, Kessler (pers. comm.) observed malformations of the branches - black knots similar to those formed on cherry (*Prunus* sp.) and plum trees (*Prunus domestica*). The latter author proposed that the problem might be caused by *Apiosporina morbosa* (Schwein). However, morphological and molecular analyses performed by Macía *et al.* (2005) showed *Leptosphaeria polylepidis* M.J. Macía, M. and. Palm & M.P. Martin **sp. nov**. to be the causal agent.

Leptosphaeria Ces. & Of causes different diseases in annual species. Its anamorph states are known as *Camarosporium, Hendersonia, Phoma, Rhabdospora* and *Stagonospora* (Hawksworth *et al.* 1995). *Leptosphaeria* species are known to affect different members of the family Rosaceae

Fig. 2. South face of the snow-covered Sajama Peak (6542 m), view from the Patacamaya-Tambo Quemado international road. A: Geographical location of the Sajama National Park; B: Partial view of District 3 containing the studied *queñua* forest (white segmented line).

(Huhndorf 1992, mentioned by Macía *et al.* 2005). According to Stoykov (2004), the fungal families Phaeosporaceae and Leptosphaeriaceae (Pleosporales) are mainly saprotrophs (very rarely hemibiotrophs) that affect herbaceous stems, leaves and the floral parts of different plants.

Using transects in the southwestern sector of the SNP, beginning at a place known as Huito, Pinto Alzérreca (2007) found that of 377 trees some 35% had black knots on their branches. However, no direct relationship between the abundance of these knots and plant health was reported, nor was tree mortality found to be related to the presence of the fungus. However, in the same year, a technical report published by the SNP recorded a great increase in the mortality of *queñua* trees, particularly in District 3, naming *L. polylepidis* as the likely cause.

4. The aims of the present study were

i. to describe the decline symptoms produced in *queñua* trees caused by *L. polylepidis*
ii. to estimate the incidence and distribution of *L. polylepidis* decline in District 3 of the SNP's *queñua* forest.

5. Methodology

5.1 Examination of a permanent plot

A permanent plot of dimensions 100x100 m (correcting for ground undulations) was marked out in an area with representative tree density and with a structure and slope typical of District 3 of the SNP's *queñua* forest (Fig. 2). The boundaries of the plot were set using a device providing geographic positioning system (GPS) readings, a tape and compass; these boundaries ran S-N and W-E. The plot was divided into 25 segments of equal size. All *queñua* trees within the plot were labelled at a height of 1.3 m and their coordinates recorded.

5.2 Disease assessment and spatial distribution

The health of the *queñua* trees in the plot was assessed by recording the number of: *i)* apparently healthy trees (S), *ii)* diseased trees (E), *iii)* dead trees (M) and *iv)* burnt trees (Q). Trees with wilted leaves and branches and with black knots on the latter were considered diseased. The spatial distribution of the trees in each health category was determined according to Madden *et al.* (2007), with a regular pattern defined as $\sigma^2 < \mu$, a randomised pattern defined as $\sigma^2 = \mu$, and an aggregate or clustered pattern defined as $\sigma^2 > \mu$, where σ^2 = the variance of the size of the subpopulations, and μ = the mean size of the subpopulations.

5.3 Identification of the disease-causing agent

Sample pieces of branches (approximately 10-20 cm in length) were collected from: *i)* 10 trees with branches showing symptoms of decline, *ii)* 15 trees with black knots on the branches and, *iii)* 15 apparently healthy trees (Table 1). Attempts to isolate the causal agent of disease involved placing 1 cm-long branch samples in a moisture chamber at 24°C for 72 h, and culturing other 1 cm-long branch samples on two media *i) Queñua* Dextrose Agar (QDA) (*queñua*=250 g extract of leaves and branches), and, *ii)* Potato Dextrose Agar (PDA), according to the method of French and Hebert (1989). Fungi were identified using semi-permanent slides with lactophenol according to Macía *et al.* (2005). The anamorph state was characterised according to Sutton (1980) and Câmara *et al.* (2002).

6. Results

6.1 Symptoms and the causal agent of decline

Table 1 shows the proportion of apparently healthy trees, trees with black knots and trees with clear symptoms of decline that were positive for *L. polylepidis*.

Sample	Number of samples	Potato dextrose agar (PDA)	Queñua dextrose agar (QDA)
Apparently healthy trees	15	0%+	6%+
Trees with black knots	15	100%+	100%+
Trees with symptoms of decline	10	100%+	100%+

+ = positive for *L. polylepidis*

Table 1. Isolation of *L. polylepidis* from *P. tarapacana* samples with decline symptoms, black knots or no disease symptoms.

The decline caused by *L. polylepidis* is characterized the yellowing of the apical leaves, followed by progressive die-back of the branches from the tip downwards, by gradual defoliation of the branches and death within a few years (Fig. 3A, B, C). Slicing the bark

Fig. 3. Signs and symptoms of decline in *P. tarapacana* trees. A: partial wilting of branches; B: totally wilted dead tree (left) and apparently healthy tree (right); C: close-up showing partial wilting; D: discoloration of the vessels; E: internal view of an apparently healthy branch (upper) and of one with wilting symptoms (lower); F: stromatic bodies under the bark; G: stromatic bodies under the bark and inside the wood; H: bitunicate asci containing eight ascospores of *L. polylepidis* extracted from the inner stromatic bodies.

from partially or completely dry (dead) branches revealed a dark coloration (Fig. 3D, E), with abundant stromatic bodies visible under the ritidoma (Fig. 3F). Figure 3 G, shows black, spherical bodies incrusted in and below the bark (Fig. 3G). These spherical bodies, formed by the causal agent, contain a gelatinous mass composed of asci, ascospores and pseudo-paraphysae (Fig. 3H). These asci are bitunicate, cylindrical-clavate and contain eight ascospores with three transverse septa. The spores are brown when mature (Fig. 3H).

Most of the samples placed in the moisture chamber showed randomised ostiolate pycnial bodies distributed over the bark (Fig. 4F). These were partially immersed in the bark and

Fig. 4. Anamorphic state of *L. polylepidis* with samples showing symptoms of decline in the moisture chamber experiment. A: pieces of a branch of *P. tarapacana* showing decline symptoms; B: vascular discoloration symptoms in branches with decline; C: stromatic bodies under the ritidoma; D: samples in the moisture chamber; E: samples in the moisture chamber after 72 h of incubation at 24°C; F: close-up of stromatic bodies (red circle) and spore cloud leaving the pycnidial structure of the anamorphic fungus (*Phoma* spp.) (Red colour arrows); G: Spore cloud leaving an ostiole (green arrow); the blue arrow shows the pycnidial body; H: pycnidial squash showing the exiting spores: I: small spores and elliptical conidia; note the almost hyaline nature of the anamporphic fungus characteristic of *Phoma* spp.

liberated their spores in a creamy-coloured cloud. The spores were small and hyaline (Fig. 4G). According to Sutton (1980), these are characteristic of *Phoma* spp. growing on QDA and PDA.

6.2 Black knots on branches and their cause

Black knots were found on the branches of both declining and apparently healthy trees (Fig. 5A). The stromatic bodies were spherical and compact (Fig. 5B, C). In cross section a thick, dark brown pseudo-parenchymatic wall was seen, with an ascus containing ascospores at

Fig. 5. A: black knots (stromatic bodies) on branches of *P. tarapacana*; B and C: close-up to the stromatic bodies; D: vertical section of conidioma showing layers with brown-melanised cells of scleroplectenchyma tissue containing mature asci within; E: ascus and ascospores of *L. polylepidis*; F: ascospores of *L. polylepidis*; G: isolation of *L. polylepidis* from apparently healthy branches (left) on QDA, and from a sample with stromatic bodies on PDA; H: stromatic bodies formed on PDA after 12 days; I: ascospores extracted from these stromatic bodies.

the centre (Fig. 5D, E). Once again, the asci were typically bitunicate, cylindrical-clavate and contain eight ascospores with three clear brown septa (Fig. 5E, F). Fifteen samples of branches with black knots were all positive for *L. polylepidis* on DQA and DPA (Table 1). After three weeks on DPA, isolates from the black knots formed stromatic bodies (Fig. 5G, H). Inside these bodies gelatinous masses, formed by asci and ascospores were seen (Fig. 5I).

Only one of 15 apparently healthy branch samples returned a positive result for *L. polylepidis* (Table 1) (Fig. 5G). Black stromatic bodies were seen after 15 days of incubation on QDA (Fig. 5G); these contained an ascus mass and ascospores characteristic of *L. polylepidis* (Fig. 4I).

6.3 Incidence and spatial distribution

Ninety eight *queñua* trees were recorded in the experimental plot. Fifty three (54%) showed symptoms of decline (sometimes with and sometimes without black knots on the branches), twenty-one (21%) trees were dead (due to diseases and other, non-established causes), and the remaining 23 (23%) were apparently healthy (Fig. 6A). The spatial distribution of plant disease (E) in the plot showed an aggregated or clustered pattern ($\sigma^2 > \mu$ 4.3 > 2.1) (Fig. 6B). The diseased trees were distributed in 18 quadrants (72% of the total 25) and the apparently healthy trees (S) in seven (28% of the total 25) (Fig. 6C). Eleven quadrants (1, 7, 8, 9, 12, 13, 14, 16, 18, 19 and 21) had 1-2 diseased trees, and seven quadrants (3, 10, 11, 15, 17, 23 and 25) from 2-7 diseased trees (Fig. 6C).

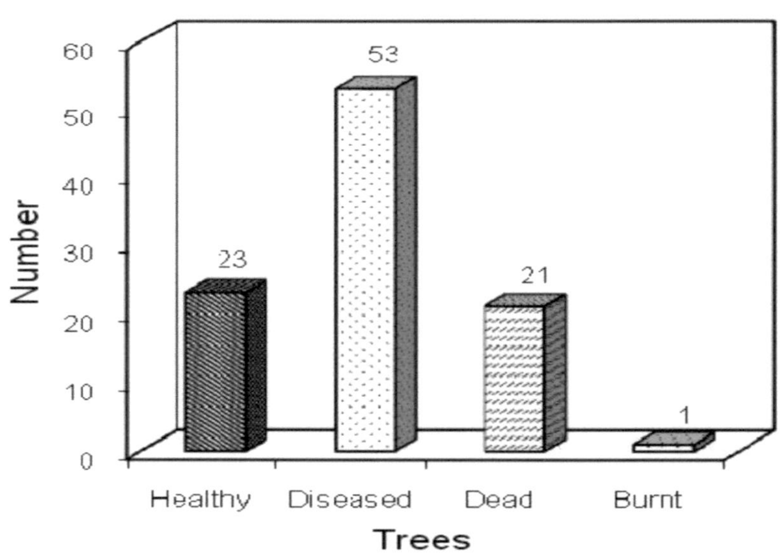

A

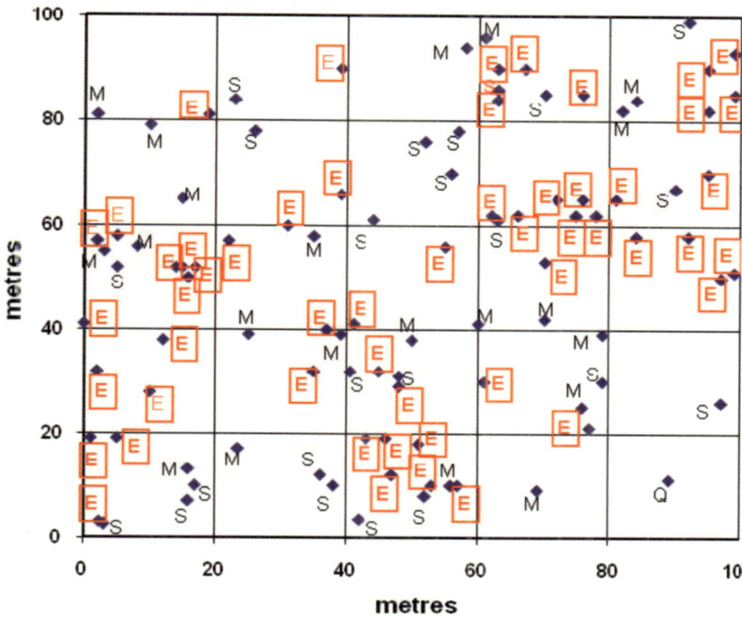

B

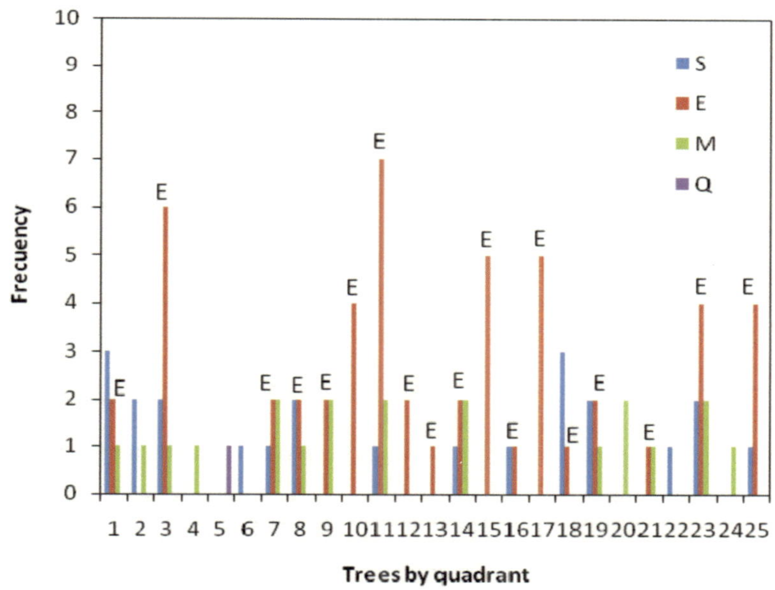

C

Fig. 6. Incidence, spatial distribution and frequency distribution of diseased, apparently healthy, dead and burnt trees in the experimental plot. A: Incidence of disease; B: Distribution of *queñua* trees in the plot (E=diseased tree; M=dead tree; S= apparently healthy tree and Q=burnt tree; B: Health status of the *queñua* trees in the plot; C: Frequency of healthy trees (S, blue), diseased trees (E, red), dead (M, green) and burned (Q, black) by plot quadrant.

7. Discussion

L. polylepidis is regarded as a specific pathogen of *Polylepis* spp., and was recorded on *P. tarapacana* by Macia *et al.* (2005). Coca-Morante (2008) has also recorded decline symptoms and black knots caused by *L. polylepidis* in some *P. besseri* trees growing in the Sach'aloma forest (Cochabamba, Bolivia). The climatic conditions at Sach'aloma (3800 m) are, however, totally different to those of District 3.

Decline among the trees in the studied plot was shown by wilting and/or black knots on the branches, though apparently healthy trees may also have black knots. Wilting begins apically, becoming evermore extended and intense, until the tree suffers complete defoliation and death. The formation of black knots on the branches is the only sign of *L. polylepidis* infection on living *P. tarapacana* (Macía *et al.* 2005; Pinto Alzérreca and Robledo 2006; Pinto Alzérreca 2007).

Pinto Alzérreca (2007) indicates a lack of any direct relationship between the abundance of black knots (*galls* in her terminology) and the health of the plant. This author also indicated tree mortality not to be related to the presence of fungi. However, the present results indicate that the black knots on the branches plus decline symptoms are associated with the death of *queñua* trees.

Many of the samples with symptoms of decline that were cultured in the moisture chamber showed structures of *Phoma* spp., the anamorphic state of *L. polylepidis* (Sutton 1980; Hawksworth *et al.* 1990; Câmara *et al.* 2002). The telemorphic state of *L. polylepidis* would appear to cause moncyclic disease, while the anamorphic *Phoma* spp. state appears to be associated with polycyclic epidemics (Madden *et al.* 2007).

Black knots and symptoms of decline are usually seen in young branches. This is probably due to the ease with which the pathogen can gain access to and develop in their tissues. It is likely that the stromatic bodies formed by the pathogen under the ritidoma are related to the discoloration of the vasculature, a gradual consequence of the xylem and phloem becoming obstructed. Infection is therefore associated with the wilting seen in affected trees. According to Guest & Brown (1997), wilting results from the physical blockage of xylem vessels caused by the pathogen and, to some extent, the host response to the presence of the pathogen. Symptoms vary from yellowing, vascular browning, tylosis formation and the gumming of the vascular system, through to the general wilting of the plant. In other tree species, the vascular tissues and surrounding cortical tissues are also colonized by wilt pathogens such as the Dutch elm pathogen (*Ophiostoma ulmi*), the oak wilt pathogen (*Ceratocystis fagacearum*) and the persimmon wilt pathogen (*C. diospyri*). These pathogens are not true vascular wilt fungi, however, since their presence is not restricted to the vascular tissue. The damage caused by these agents depends largely on the extent of their cortical invasion.

The spatial distribution of disease in the plot showed an aggregated pattern. According to Madden *et al.* (2007), in aggregated patterns the points on a surface do not have an equal probability of being occupied by an individual i.e., the trees in the present plot do not have an equal probability of being infected. The dead trees (n=21) in the present work could have been killed by the studied disease but, of course, may have died of other causes. However, the detection of a single apparently healthy tree (7%) infected with *L. polylepidis* is indicative that some healthy trees are probably in the initial phases of infection. Several years may pass before symptoms become visible (generally this type of disease is associated with polyetic epidemics) (Madden *et al.* 2007). According to the distribution of health category frequency by quadrants, it can be seen that the disease is very common in the plot area. If 7% of apparently healthy trees are infected, the disease may be having an important impact on the decline of *queñua* trees in District 3.

Decline in conjunction with black knots, at different levels of severity, was seen in 54% of the trees in the plot. However, Pinto Alzérreca (2007), who used transects in the same sector, reported 35% of 377 examined trees to show black knots. The disease therefore appears to have extended since that time. However, it is difficult to determine whether the disease is truly becoming more or less important in the SNP since no more historical data on decline symptoms or black knots are available. According to Garrett *et al.* (2009), if a disease becomes important in an area in which it was not important in the past, this may be due to changes in the climate favourable to the pathogen.

Climate change may indeed be having some effect on the *Polylepis/Lepthosphaeria* (plant/pathogen) pathosystem, and in the future may modify the spatial distribution of diseased trees. The Andean nations are likely to be among the most affected by climate change (Marengo *et al.* 2008). According to Vuille *et al.* (2003), western Bolivia can expect to experience slightly drier conditions, while Nuñez *et al.* (2008, mentioned by Marengo *et al.* 2009) suggest that northwestern Argentina and the Bolivian *Altiplano* will experience higher temperatures during the summer months and a 40% reduction in rainfall by the year 2100, leading to increasing aridity in the region. The changes experienced to date may be associated with the greater incidence of this disease.

8. Conclusion

These results strongly suggest that *L. polylepidis* is affecting *P. tarapacana* in District 3 of the SNP, causing decline and black knots on branches. Disease incidence appears to be high and to show several levels of severity. A worsening situation may be developing.

9. Acknowledgment

The author are very grateful to Dr. Edgar Gareca León of the *Universidad Mayor de San Simon*, for his valuable comments and suggestions, to the authorities and forest rangers of the Sajama National Park (*Servicio Nacional de Parques* SERNAP) for their help in fieldwork, and to Profesor Freddy Espinoza Colque and the students of the *Escuela de Ciencias Forestales* (ESFOR), *Facultad de Ciencias Agrícolas, Pecuarias, Forestales y Veterinarias "Dr. Martin Cárdenas"* (FCAPFyV), *Universidad Mayor de San Simón* (UMSS), Cochabamba, Bolivia, for their help in establishing the experimental plot and the collection of data; to Ing. Javier Burgos from *Centro de Investigaciones y Servicios en Teledeteccion* (CISTEL) *of the FCAPFyV, UMSS* for provided the satellite image; to Adrian Burton for version English review.

10. References

Argollo J., Claudia Solís and Villalba R. 2006. History of High Mountain Forests of *Polylepis Tarapacana* at the Bolivian Central Andes. 8th International Symposium on High Mountain Remote Sensing Cartography. Pp:-9-16.

Câmara M.P.S., Palm M.E., van Berkum P. & O'neill N. R. 2002. Molecular phylogeni of Leptosphaeria and Phaeosphaeria. Mycologia, 94(4), pp. 630-640.

Coca Morante M. 2008. Informe de evaluación fitosanitaria preliminar del bosque de *Polylepis besseri* en las alturas de Sach'aloma, Vacas, Cochabamba, Bolivia. Facultad de Ciencias Agrícolas, Pecuarias, Forestales y Veterinarias "Dr. Martin Cárdenas", Universidad Mayor de San Simón. Cochabamba, Bolivia. 2pp.

Daza von Boeck, R. 2005. Programa de Monitoreo Parque Nacional Sajama. Servicio Nacional de Parques (SERNAP) - BIAP.

French E. R. y T. T. Hebert. 1980. Métodos de investigación fitopatología. Instituto Interamericano de Ciencias Agrícolas. San José, Costa Rica. 289 pp.

Fjeldså J., & Michael Kessler. 2004. Conservación de la diversidad de los bosques de Polylepis de las tierras Altas de Bolivia: Una contribución al manejo sustentable en los Andes. Editorial FAN. DIVA Technical Report 11. Centro para la Investigación de la Diversidad Cultural y Biológica de los Bosques Pluviales Andinos (DIVA). 214 pp.

Guest D. I, & J. F. Brown. 1997. Infection processes. pp: 245-262. In: Plant pathogens and plant diseases (Brown J.F. & H. J. 0gle). University of New England, Armidale NSW 2351 Australia.

Garrett K. A, M. Nita, E.D. De Wolf, L. Gomez and A.H. Sparks. 2009. Plant Pathogens as Indicators of Climate Change. pp: 425-437. In: Climate Change: Observed Impacts on Planet Earth. Published by Elsevier B.V.

Gareca E., Martin Hermy, Jon Fjeldså, Olivier Honnay. 2010. Polylepis woodland remnants as biodiversity islands in the Bolivian high Andes. Biodivers Conserv (2010) 19:3327–3346.

Hawksworth D. L., P. M. Kirk, B.C. Sutton and D. N. Pegler. 1995. Ainsworth & Bisby's Dictionary of the fungi. International Mycological Institute. CAB INTERNATIONAL.

Macía, M.J., Mary E. Palm & Maria P. Martin. 2005. A new species of *Leptosphaeria* (Ascomycotina, Pleosporales) on Rosaceae from Bolivia. Mycotaxon Volume 93, pp. 401-406.

Madden L.V., Gareth Hughes, and Frank van den Bosch. 2007. The Study of Plant Disease Epidemics. The American Phytopathological Society. St. Paul, Minnesota. 421 pp.

Marengo J. A., J. D. Pabón, Amelia Díaz, Gabriela Rosas, Grinia Ávalos, E. Montealegre, M. Villacis, Silvina Solman, and Maisa Rojas. 2008. Climate Change: Evidence and Future Scenarios for the Andean Region.

Pinto Alzérreca K. y G. Robledo. 2006. Estudio de las interacciones observadas entre la infección de *Leptosphaeria polylepidis* y la salud de *Polylepis tarapacana* (Rosaceae) en bosques del Parque Nacional Sajama (Dpto. de Oruro, Bolivia). II Congreso de Ecología y Conservación de Bosques de Polylepis - Cuzco, Perú. pp: 121.

Pinto Alzérreca, K. A. 2007. Estudio de las interacciones observadas entre la infección de *Leptosphaeria polylepidis* y la salud de *Polylepis tarapacana* (Rosaceae) en bosques de la ladera suroeste del nevado Sajama (Parque Nacional Sajama, Oruro, Bolivia). Tesis de grado. Facultad de Ciencias Puras y Naturales, Carrera de Biología, Universidad Mayor de San Andrés. 108 pp.

Sutton B.C. 1980.The Coelomycetes: Fungi Imperfecti with Pycnidia, Acervuli and Stromata. Commonwealth Mycological Institute. Kew, Surrey, England. pp: 378-391.

Stoykov D. Y. 2004. A contribution to the study of Leptosphaeriaceae and Phaeosphaeriaceae (Pleosporales) in Bulgaria, I. Mycologia Balcanica 1:123-128.

Toivonen J. M., M. Kessler, K. Ruokolainen and D. Hertel. 2011. Accessibility predicts structural variation of Andean Polylepis forests. Biodivers Conserv 20:1789–1802.

Vuille M., R. S. Bradley, M. Werner and F. Keimig. 2003. 20TH Century climate change in the tropical Andes: observations and model results. *Climatic Change* 59: 75–99.

Section 3

Amelioration of Dwindling Forest Resource Through Plantation Development

10

Research Development and Utilization Status on *Jatropha curcas* in China

Li Kun[1,2], Liu Fang-Yan[1,2] and Sun Yong-Yu[1,2]
[1]*Research Institute of Resources Insects, CAF, Kunming*
[2]*Desert Ecosystem Station in Yuanmou County, SFA*
China

1. General situation of *J. curcas* resources in China

1.1 Growth habit

J. curcas has a wide range of adaptation, and is able to endure drought and barren; as its root system is well developed, it can grow on the barren wasteland (cobbly soil, coarse soil, and limestone open ground etc.); it is photophilous, and loves warm and thermal climate. Wild *J. curcas* are mainly distributed in dry and hot subtropical zone and humid tropical rain forest. They usually grow on the flat ground, hills and river valley barren mountain slopes, and their height above sea level shall be 700m-1600m. *J. curcas* has certain cold resistance, and is able to endure a short-time low temperature of -5°C. It is not demanding of the soil fertility of its growing place, as long as the pH value is 5-6 and the soil drainage is good. They are usually planted as hedge around the land, and there are also semi-wild *J. curcas*, which often grow among the shrubs beside the flat-ground road, and are scattered (Li et al., 2007; Gou & Hua, 2007).

1.2 Resources distribution

J. curcas in China are mainly distributed in the south subtropical dry and hot river valley areas. They are planted or wild in provinces (areas) of Guangdong, Guangxi, Yunnan, Sichuan, Guizhou, Taiwan, Fujian, and Hainan etc, and are planted in the west, southwest and middle part of Yunnan, as well as in the drainage basins of Yuan River, Jinsha River, and Lancang River (Liu et al., 2008); In Guangxi, they mainly grow in Qinzhou, Bobai, Rong County, Cangwu, Nanning, Yining, Longzhou, Ningming, Baise, Lingyun, and Du'an etc. In, Sichuan Province, they are wild or planted in Panzhihua, Yanbian, Miyi, Ningnan, Dechang, Huili of Xichang, Jinyang, and Yanyuan etc. In Guizhou, they are mainly distributed in the south and southwest part; the south subtropical dry and hot river valley areas at the drainage basins of Nanpan River, Beipan River and Hongshui River are the cultivation centers; and they are sparsely distributed in cities or counties of Luodian, Wangmo, Ceheng, Zhenfeng, and Xingyi etc. In recent years, the new *J. curcas* forestation area in Zhenfeng, Wangmo, Ceheng, Luodian etc has exceeded 1500hm, and *J. curcas* experimental forest demonstration base and the seedlings base have been preliminarily established in Zhenfeng County, Xingyi City, and Luodian County (Yu et al., 2006; Yang et al., 2006; Yuan et al., 2007).

1.3 Research history

In the past 3O years, the Chinese researches on *J. curcas* mainly focus on the following aspects: the research on *J. curcas* plant resources, the research on the isolation and expression of antifungal proteins and relevant gene of *J. curcas* seed, the research on the separation and purification of curcin and on the gene cloning action mechanism, the rapid propagation of *J. curcas* and the establishing of Medicinal-plant Network Database. After the national "Eighth Five-Year Plan", "Ninth Five-Year Plan" and "Tenth Five-Year Plan", the research and development of biomass energy in China has made considerable achievements. In recent years, as energy shortage in China has become increasingly serious, the development of *J. curcas* has aroused wide concern. In 2004, the Ministry of Science and Technology listed the "biomass fuel oil technological development" into the "Tenth Five-Year Plan" National Key Programs for Tackling Key Problems. The Resource Entomology Research Institute of Chinese Academy of Forestry and the Sichuan University etc have conducted relevant researches for many years, and have yielded substantial results. Xiangtan University and Hunan Academy of Forestry Sciences have also carried out the research on the extractive technique of *J. curcas* oil. In recent years, the development and exploitation of forestry biomass energy have been highly valued by the National Development and Reform Committee and the State Forestry Bureau. In 2006, the State Forestry Bureau established the forestry biomass energy office, adopting *J. curcas* as the preferred tree species for forestry biomass energy development. The local government and scientific research departments made great efforts in the research and development, and not a few enterprises and self-employed entrepreneurs also took part in the planting and development. Nowadays, China has not only set up the Southwest Region Energy-plant Database, screening out 8-9 fine varieties of *J. curcas*, established a 133.3hm seedlings base in Panzhihua and the high-quality and high-yield *J. curcas* planting technology system, but also has established the productive technology and enterprise standard for the new-type motor biodiesel made from *J. curcas*. The extraction of biodiesel from *J. curcas* has gone through experimental stage, and the products have also passed the trial of Chengdu Bus Company, and can absolutely be used for automobile engines. However, the production and exploitation of oil plants in China has just started, and the extraction and processing of some energy crop oil is still in the preliminary test stage (Li et al., 2007).

2. Basic research progress of *J. curcas*

2.1 Biologic characteristics

J. curcas, alternate names are *J. curcas*, fake peanut tree, Qingtongmu, Huangzhong Tree, smelly tung tree, Liangtong, Shuiqi, tung oil tree etc, this plant is a small arbor, 2-5m high, bark is smooth, sap is milk white; leaves are alternate, circular, and have long petiole, clustering tree tops; entire margin or lobed into 3-5 margins, 10-15cm long. Oblong and light green petals, there are nodes on the upper pedicels of staminate flowers, and no node on the upper pedicels of pistillate flowers, the flower is tiny, sepals are quinquepartite; 8-12 pistils, 2-4 cells in an ovary. Fruit is yellow, spherical, like a loquat; mature seed is black, episperm is gray black, smooth, when the hull is removed, there will be three peanut-like seeds inside, the seed is oblong and 18-20mm long. There are about 200 kinds of J. plants all over the world. And mainly five kinds of J. plants are planted in China, that is: *J.curcas, J.podagrica, J.multifida, J.gossypifolia*, and *J.integerrima* (Liu et al., 2008).

2.2 Reproductive ecology

J. curcas is monoecious and diclinous, cymose. Its inflorescence formation and flowering time is in April, *J. curcas* in China are mostly budding in early March; with the rising of air temperature, the inflorescence grows and develops, and the inflorescence goes through a growth stage of 15-20d. Its bloom stage is mainly from the last ten-day of April to the first ten-day of May, during which period 80% of the flowers will come out (Liu et al., 2010a). The flowers of *J. curcas* mostly aggregate at the stem or the top of lateral branch, and some of them are formed at the leaf axil. The rachis top of indefinite inflorescence does not immediately develop into a flower, and still keeps the ability of germinating and producing a lateral flower. The rachis top of definite inflorescence will soon develop into a terminal flower (pistillate flower, usually grows at the top or leaf axil of the inflorescence), and therefore loses the growing ability. 2-4 lateral rachises will grow under the terminal flower, and the top of lateral rachis will soon develop into a terminal flower (which is also mostly pistillate flower), and 2-4 lateral rachises will grow under this terminal flower; the blossoming order is that the uppermost and innermost flowers mature and come out, while for the downmost and outmost flowers, some of them mature and come out and some fail to mature and therefore dry up and fall off; as a result, the staminate flowers that actually pollinate the terminal flower (pistillate flower) are far less than the staminate flowers that have ever grown (Li et al., 2007; Yang et al., 2007; Liu et al., 2010b). In short, pistillate flowers are mostly terminal flowers, and staminate flowers are distributed around the terminal flower. In the inflorescences that grow on top of the branch, except for the terminal flowers on the definite rachis and a few terminal flowers on the lateral rachis, there are all staminate flowers; in the inflorescences that grow nearer the top, there are appreciably fewer and fewer pistillate flowers, and some inflorescences solely have staminate flowers, without any pistillate flower (Luo et al., 2008). In one and the same inflorescence, there are far more staminate flowers than pistillate flowers. The ratio of pistillate flowers to staminate flowers is approximately 1: 10. Therefore, pistillate flowers have a high probability of being pollinated, and the rate of fruit setting is 100% (Guo et al., 2007).

2.3 Seed biology

The seed of *J. curcas* contains various ingredients, mainly including fatty substance, protein and polypeptide, terpenoids and some micromolecule substance, yet the substance contents vary according to different producing places (Table 1 & 2). The fatty substance of the *J. curcas* is mainly distributed in the kernel, and contains many chemical constituents such as alcohol, acid, ketone, and anthracene (Deng et al., 2005). Its fat content is high, and oil content is 40%-60%, exceeding the common oil crops such as rapes and soybeans, and its oil is of good fluidity. The seed oil of *J. curcas* can blend well with diesel, gasoline, and ethyl alcohol, and after blending, there is no separation for a long period of time; by chemical or biological conversion, the biodiesel that is superior to the existing 0# diesel can be obtained. The protein content of *J. curcas* seed is 18.2%, and its main constituent is curcin, whose toxicity is similar to the ricin of the seed of Ricinus communis and the crotin of the seed of Croton tiglium (Lin et al., 2004). Curcin is capable of inhibiting the gastric carcinoma cells, mouse myeloma cells and human hepatoma from proliferation in vitro. At present, the full-length cDNA sequence and gene sequence of curcin has been cloned, and have expressed the active curcin mature protein and the conserved domain protein in the escherichia coli (Lin et al., 2003; Chen et al., 2003).

Compounds	*Retention time (min.)*	*Percentage (%)*
2-propyldecan-1-ol	11.73	0.95
2-octyldodecan-1-ol	12.53	0.59
1,2-benzenedicarboxylic acid	14.61	16.65
Hexadecanoic acid	15.16	18.02
2-hydroxy-cyclopentadecanone	15.23	1.88
9-hexadecenoic acid	16.02	1.48
9-octadecenoic acid	16.3	44.04
9,12-octadecadienoic acid[Z, Z]	16.94	0.52
Znthracene 9,10-dimethyl	17.47	0.46

(Cited from Li et al., 2000)

Table 1. Chemical composition and its content within mixed seed samples of *J. curcas*

Fatty acid compostion	*Cities or Counties*					
	panzhihua	binchuan	ningnan	yongsheng	shuangbai	luodian
Myristic Acid (C14:0)	0.09	0.04	0.07	0.06	0.05	0.11
Palmitic Acid (C16:0)	17.25	16.84	16.64	13.47	14.88	16.41
palmitoleic acid (C16:1)	1.08	0.53	0.9	0.74	0.87	1.13
Seventeen carbonic acid (C17:0)	0.13	0.08	0.14	0.11	0.13	0.12
Seventeen nonadecenoic acid(C17:1)	0.06	0.03	0.06	0.07	0.05	0.07
Stearic acid (C18:0)	7.42	7.83	7.96	6.43	9.81	5.58
Oleic Acid (C18:1)	40.31	42.44	44.91	40.26	46.03	37.91
Linoleic Acid (C18:2)	32.69	32.67	28.02	38.04	27.78	37.02
Linolenic acid (C18:3)	0.4	0.21	1.06	0.38	0.4	0.64
Arachidic Acid (C20:0)	0.22	0.11	0.25	0.25	—	0.66
Twentieth nonadecenoic acid(C20:1)	0.23	—	—	0.1	—	0.15
Behenic acid (C20:2)	0.13	—	—	0.1	—	0.18
Total	100	100	100	100	100	100
Saturated fatty acid	25.11	24.54	25.06	20.32	24.87	22.88
Unsaturated Fatty Acid	74.89	75.46	79.94	79.68	75.13	77.12

"—" indicates Not test
(Cited from Li et al., 2006)

Table 2. Fatty acid composition of *J. curcas* seeds in different areas (%)

2.4 Good-seed breeding basis

As *J. curcas* is equally fit for seedling propagation and asexual reproduction, therefore, "sexual propagation, asexual utilization" is also applicable to the good-seed breeding of such tree species. However, in the process of new-variety breeding, clonal breeding shall be adopted to meet the requirements for a new variety and to ensure the consistency and stability. In a manner of speaking, the new-variety breeding for *J. curcas* oil-plant energy forest is to breed the clonal varieties that are full-seed and high-quality, and have high and stable yields. As is mentioned above, some of the genesiological characteristics of *J. curcas* need to be further researched to enable the experiments such as mating seed production and hybrid seed production go smoothly. At present, the good-seed breeding research of *J. curcas* adopts "a combination of sexual and asexual"; the selected good individual plants of *J. curcas* are priorly used for building the cutting orchard, and the clonal test plantation is built by cuttage propagation, so as to select the good new varieties for production application; furthermore, provenance/family determination is carried out to select the good provenance or family, and the good family within the good provenance will go through further asexual reproduction; and then clonal breeding is carried out to appraise and select the good new varieties for building the oil-plant energy forest an enlarged scale and for cultivation and utilization.

Fig. 1. The good-seed breeding experiments of *J.curcas*

2.5 Cultivation physiological ecology

The planting and reproduction modes of *J. curcas* usually are direct seeding forestation, seedling transplanting, hardwood cutting method, hedge planting method and tissue culture method etc. The direct seeding forestation is fit for rainy season, the rate of emergence for large-area seeding is usually 50%-70%, and the emergence time varies, generally 10-40d; this method is easy and convenient. Seedling transplanting can cultivate strong seedlings, beneficial to resisting plant diseases and insect pests and improving survival rate. Hedge planting method is a preferable planting method for planting beside the fertile agricultural arable land. Hardwood cutting method is a simple and convenient planting method of *J. curcas*, as long as the tree pond is of certain depth, which is usually around 30cm, and reasonable time is chosen, normally it can survive. In the conducted cuttage experiments of *J. curcas*, the cuttings' survival rate is above 80% (Li & Liu, 2006). The tissue culture technique of *J. curcas* is approaching to maturity, and usually the explants such as embryonal axis, petiole and cotyledon are used for induction. *J. curcas* can be used for dibble-seeding afforestation, its budding ability is good, and it is also fit for cuttage propagation. Those through seminal propagation can germinate the typical taproot and 4 lateral roots, while those through cuttage propagation can not germinate the taproot. Normally, a plant of about 3m can be grown into in 3 years. Those through seminal propagation can bear fruits in 3-4 years, and those through cuttage propagation can bear fruits in 1 year or so (Li et al., 2007).

3. Research on the cultivation techniques of *J. curcas*

3.1 Division of ideal habitats for *J. curcas*

China is extremely abundant in *J. curcas* resources. The distribution of *J. curcas* resources across the country and their ecological habits in different provenances have been generally found out after investigation The adaptability evaluation and the division of ideal habitats for *J. curcas* have been carried out, and meanwhile, introduction breeding is practiced and the *J. curcas* seedlings breeding base is established.

In Yunnan Province and Sichuan Province, according to the analysis of *J. curcas*'s demand for ecoclimate conditions, the *J. curcas*'s ideal and less ideal habitats in the provinces are proposed, and meanwhile the ideal habitats for *J. curcas* are determined after considering the conditions of forest site, such as Lijiang City, Chuxiong City and Panzhihua City. For provinces such as Zhejiang and Fujian where there is no resources distribution, the growth performance of different provenances is compared, to determine the suitability of planting the *J. curcas* there, or to screen out the provenances and varieties that suit the local ecological conditions.

3.2 Good-seedling breeding for *J. curcas*

In order to meet the large-scale forestation's requirements for good seedlings, the researches on seed development, budding characteristics, pre-sowing treatment etc become particularly important, and seedlings breeding research has become the hot topic in cultivation research. According to research findings, *J. curcas* seed reaches its physiological maturity 58d after flowering; the seeds in different geographical provenances have no much difference in apparent characters, while their thousand seed weight is of marked difference, therefore the thousand seed weight can act as the selection index for good material. The optimum temperature for seed germination of *J. curcas* is 25-35°C; seed coat and seeding depth have something to do with seed germination, while sunlight makes no difference to germination; the seed of *J. curcas* belongs to the orthodox seed of light neutrality. Research on the pre-sowing treatment of *J. curcas* has shown that dressing the seeds with 98% concentrated sulfuric acid for 30min is the best treatment, and after seeding, the treated seeds can sprout

1-2d earlier than the control group. Hardwood cutting is the main method for *J. curcas* afforestation, and the cutting seedlings have the characteristics of branching quickly, growing fast, and bearing fruits early; the one-year or two-year branches that are 25cm long are preferable; ABT1 rhizogenic powder and KMnO4 can both enhance the *J. curcas* cuttings to take root, and temperature, sunlight, and soil etc also affect the root-taking of cutting branches.

3.3 Afforestation technique

According to the ecological habits of *J. curcas*, the reproducing area shall preferably have a mean annual temperature of 18-28.5°C, a extreme minimum temperature of -4°C, long frost-free season, long duration of sunshine, and a high yearly accumulative temperature (Yu & Ding, 2009). Preparation of soil shall adopt zonal clearing or blocky soil preparation; density of plantation is normally 840-1110 plant/hm, while for areas with high management level, the afforestation can go by 1665 plant/hm. When transplanting the nursery stocks, decomposed farmyard manure shall be applied to the bottom of the holes, and after revival from transplanting, the lack shall be timely supplied and fertilizer for seed bed shall be additionally applied. As the *J. curcas* forest land is in good ecological environment and hydrothermal conditions, weeds multiply fast, therefore timely nurturance is necessary so as to increase the survival rate and preservation rate of afforestation, and enhance the afforestation effect.

Fig. 2. Artificial forest of *J. curcas* in dry-hot valley of JinSha River

3.4 Main plant diseases and insect pests

The main plant diseases of *J. curcas* are 11 types: *J. curcas* leaves brown spot, cuttings blight, cuttings canker, damping off, powdery mildew, sootymould, anthracnose, seed mildew and rot, rot, gray mold, and leaf spot. And the insect pests are Maladera sp, wireworm, Tetrangchus urticae Koch, Aphis nerii Boyer de Fonscolombe, Nipaecoccus vastator Maskell, aphid, Poecilocoris rufigenis Dallas, white ant, cutworm, leaf miner, and inchworm. Termite, cold damage, rats, and crickets can also endanger the *J. curcas*, and prevention and control measures have been formulated according to local situations.

Fig. 3. The main diseases within *J. curcas* grown in dry-hot valley of Red River

4. Existing problems and development tendency

4.1 Problems and suggestions

At present, there are a lot of problems existing in the development and exploitation of *J. curcas* in China: (1) Raw material resources of *J. curcas* are deficient. For a long time, the importance attached to the research and development of energy plants is far from enough, therefore, insufficient resources have been cultivated, biodiesel raw-material forest falls short, and productivity is low. The bottleneck problem of development is the source of feed, thus can not stand comparison with petroleum diesel. (2) The intrinsic drawback of *J. curcas* oil, which restricts the extent of its application. *J. curcas* oil molecule is big, and viscosity is high, prone to cause incomplete combustion. (3) Improvement of productive technology and degradation of higher aliphatic acid, the cultivation techniques and biotechnology of raw materials, oil processing technology system assembly and cost reduction etc. At present, biodiesel is prepared through the agency of ester exchange reaction catalyzed by homogeneous base catalyst such as sodium (potassium) hydroxide or sodium methoxide, which will make the reaction mixture saponified into the colloidal state, making it hard to separate; during after-treatment, it is likely to lead to wastage of catalyst and discharge of wastewater. (4) A lot of issues related to commercial application of *J. curcas* fuel oil have yet to be solved, such as the preferential policies of tax abatement or tax exemption (Liu et al., 2006).

4.2 Development tendency and prospects

J. curcas has great potential for market development, further research and exploitation will bring considerable economic benefit to mankind, and its exploitation prospect is broad. As the raw material for production of biodiesel, *J. curcas* is being exploited as one of the important energy plants, and is the tree species most likely to substitute for fossil energy in the future, having immense exploitation potential. The arboreal and shrubby oil plants that are the raw materials for biomass fuel oil can be planted in the inferior land in China, including the extensive mountainous areas, sand areas, and arid valley areas, to develop the biomass fuel. Planting biomass fuel plants in these areas not only can provide abundant regenerative raw materials for the biomass fuel oil industry of China, but also will help improve ecological environment, increase farmers' income, and provide a new approach to building a well-to-do society in villages.

The seed of *J. curcas* contains many active constituents, has important farm chemical and medical value, and is a bioenergy plant material with high comprehensive values. As to the research results of molecular biology in recent years, DNA and RNA extraction method has been established, and several kinds of functional genes have been successfully cloned. Furthermore, the plant regeneration and transgenic system has been established, therefore, *J. curcas* has s great prospect as the model plant for energy plant research. However, *J. curcas* is notoriously terrifying because of its high toxicity. Many researchers only researched into poison prevention or making toxicants out of its toxicity etc, while few have considered making use of its medicinal value. And it is believed by some scholars that the possibility of deriving the medicine-origin lead compound from poisonous plants is high. There have been reports on successfully developing new drugs out of poisonous plants, and aroused the close attention from extensive medical workers; therefore, phytotoxin as the medicine origin of lead compound has a promising application prospect (Gou & Hua, 2007). *J. curcas* contains many toxic ingredients, especially the curcin contained in its seeds, an in-depth research of which is expected to bring good economic benefit.

5. Acknowledgements

This paper funded by National Science and Technology Support Program(2011BAD38B04; 2007BAD50B04) .We thank the resident of Yang Jun-Hua, our colleagues of Luo Chang-Wei, Cui Yong-Zhong and Zhang Chun-Hua, and graduate students of Wang Xiao-Qin and Peng Hui for their participation and friendship.

6. References

Chen Yu, Wei Qin, Tang Lin, Chen Fang. Proteins in vegetative organs and seeds of *Jatropha curcas* L.and those induced by water and temperature stress. Chinese Journal of Oil Crop Sciences, 2003,4:98-104

Deng Zhi-Jun, Cheng Hong-Yan, Song Song-Quan. Studies on *Jatropha curcas* seed. Acta Botanica Yunnanica, 2005,27(6):605-612

Gou Yuan, Hua Jian. Discussion on Utilization,Latest Development and Prospects of *Jatropha Curcas* Resources. Resource Development & Market, 2007,23(6):519-522

Guo Cheng-Gang, Wang Chao-Wen, Li Jian-Fu, Gou Ping. Phenology and dynamic observation of flower development within *Jatropha curcas*.Modern Agricuture Science and Technology, 2007,1:12-13

Li Hua, Chen Li, Tang Lin, Chen Fang. Physicochemical characteristics and fatty-acid composition of seed oil of *Jatropha curcas* from Southwest China. Chinese Journal of Applied Environmental Biology, 2006, 12(5): 643-646

Li Kun, Yin Wei-Lun, Luo Chang-Wei. Breeding System and Pollination Ecology in *Jatropha curcas*. Forest Research, 2007,20(6):775-781

Li Wei-Li, Yang Hui, Lin Nan-Ying, Xu Yi-Li, Xie Quan-Lun. Study on seed soil chemical composition of renewable energy plant-*Jatropha curcas*. Journal of Yunnan University (Natural Sciences), 2000, 22 (5):32

Li Xiang-Yong, Liu Fan-Zhi. Cutting propagation and transplantation within *Jatropha curcas*. Guangxi Tropical Agricuture, 2006,105(4):32

Li Zhen-Hua, Guo Yu-Qi, Ma De-Ping, Wang Song, Tian Zheng-Yuan. Exploitation and utilization situation of energy plants-*Jatropha curcas* and its prospect. Henan Agricuture Science, 2007,7:10-12

Lin Juan, Chen Yu, Xu Ying, Yan Fang, Tang Lin. Antitumor effects of curcin from seeds of *Jatropha curcas*. Acta Pharmacologica Sinica, 2003,3:241-246

Lin Juan, Zhou Xuan-Wei, Tang Ke-Xuan, Chen Fang. A Survey of the Studies on the Resources of *Jatropha curcas*. Journal of Tropical and Subtropical Botany, 2004,12(3):285-290

Liu Fang-Yan, Li Kun, Wang Xiao-Qin, Zhang Chun-Hua, Luo Chang-Wei, Cui Yong-Zhong, Peng Hui. Sex-differentiation and number of female flowers for polygamous inflorescences of *Jatropha curcas* in two drainage basins. Journal of ZheJiang Foresty College, 2010a,27(5):684-690

Liu Fang-Yan, Li Kun, Zhang Chun-Hua, Luo Chang-Wei, Wang Xiao-Qin, Cui Yong-Zhong. Study on the floral quantity size and ecological adaptability of *Jatropha curcas*. Journal of Nanjing Forestry University(Natural Sciences Edition), 2010b,34(6):1-5

Liu Huan-Fang, Deng Yun-Fei, Liao Jing-Ping. Floral organogenesis of three species of Jatropha (Euphorbiaceae). Journal of Systematics and Evolution, 2008,46(1):53-61

Liu Jie, Li Qian-Zhu, Yin Hang, Yang Song, Song Bao-An. Advance in the Studies and Developments on the Resources of *Jatropha Curcas*. Journal of Guizhou University(Natural Science Edition), 2006,23(2):105-110

Liu Ze-Ming, Su Guang-Rong, Yang Qin. Investigation and Analysis on Resources and Development Strategies of *Jatropha curcas* L.in Yunnan Province. China Forestry Science and Technology, 2008,22(1):37-40

Luo Chang-Wei, Li Kun, Chen You, Liu Fang-Yan, Sun Yong-Yu. Biological Characteristics of Flowering and Fruiting of *Jatropha curcas* in Yuanjiang Savanna Valley. Journal of Northeast Forestry University, 2008,36(5):7-10

Yang Qin, Peng Dai-Ping, Duan Zhu-Biao. Study on Pollination Biology of *Jatropha curcas* (Euphorbiaceae). Journal of South China Agricultural University, 2007,28(3):62-66

Yang Shun-Lin, Fan Yue-Qing, Sha Yu-Cang, Zhu Hong-Ye, Duan Yue-Tang. Distribution and integration exploitation utilize foreground of *Jatropha curcas* L. resource. Southwest China Journal of Agricutural Sciences, 2006,19:447-452

Yu Shuai-Yong, Ding Gui-Jie. Research Progress of *Jatropha curcas* L. Guizhou Forestry Science and Technology, 2009,37(1):49-54

Yu Shu-Ming, Sun Jian-Chang, Chen Bo-Tao. Exploration and Utilization of *Jatropha curcas* Resources in Guizhou Province. Journal of West China Forestry Science, 2006,35(3):14-17

Yuan Li-Chun, Zhao Qi, Kang Ping-De. Investigation of geographical distribution and evaluation of *Jatropha curcas* in Yunnan province. Southwest China Journal of Agricutural Sciences, 2007,20(6):1283-128

11

Impact of Sustainable Management of Natural Even-Aged Beech Stands on Assortment Structure of Beech in Croatia

Marinko Prka
State company „Hrvatske šume" ltd
Zagreb
Croatia

1. Introduction

When considering assortment structure of the main forest products and the compilation of assortment tables, continuous patterns of biological growth and development of stands (trees) come into conflict with the provisions of the standards for the classification of forest wood products and common practices in wood trade, prescribed by man and changeable with time. Actually the Croatian forestry faces additional uncertainties brought by the application of new standards on our beech stands.

The quantity and quality of broadleaved wood assortments are highly affected by the diversity of habitus and occurrence of faults on and in the tree. Proper development of individual trees and stands is provided by harmonized effects of biotic and abiotic factors. Disturbance of this harmony as a rule causes the occurrence of faults that highly affect the quantity and quality of wood assortments. The occurrence of faults, their size and number, is of random character and it cannot be correlated with measurable tree parameters. Excluding heritage properties, habitus and faults are the result of conditions in which the tree grew until the time of felling. In an organized forestry, the time of felling is different for individual trees. From the standpoint of forest silviculture and harvesting, the determination of the felling time for individual trees (selection, marking by which a specific type of cut is performed) represents the most significant influence of man on stand relationships, and also on the structure of wood assortments.

The tables showing the share of forest wood assortments (assortment tables) are a significant tool necessary for the forestry staff, and common beech is the most represented species in the Croatian forests. When planning fellings and annual allowable cuts, it is necessary to know the quantity and quality of wood assortments, determined in accordance with the applicable standards for the products of forest harvesting. Reliable and usable tables of wood assortments are necessary for the assessment of efficiency of the process of forest harvesting of a certain area, and also for the comparison of work activities performed by specific parts of enterprise. Due to the diversity of their phenotype, broadleaved species are more demanding than conifers with respect to the investigation of assortment structure, i.e. the possibility of achieving the quality of forest wood assortments.

According to Benić (1987) forest assortments are standardized products determined by standards, common practices and trading habits, and they can also be determined by agreement between the producer and purchaser. The products of wood processing are determined by wood species, form of cross-cut, dimensions, method of processing, quality and possible quality deviations. Assortment structure of the stand is determined by assortment shares of individual trees. The selection of trees for felling during rotation is not the result of random decisions, but rather a procedure based on rules and principles arising out of forest management. These rules and principles are the result of a comprehensive scientific development of forestry profession. If the operations of forest workers are based on science, clear (measurable) results of such operating procedure must be obtained. In science, it is considered that the most objective research is the one whose results may be expressed numerically (Kelvin, 1953). For these reasons, with even-aged beech stands, it is more convenient to investigate the assortment structure of individual types of cut (cutting sites) than the assortment structure of the stand. Only with clean cuts and in stands before the final felling, these two expressions have the same meaning.

In forest management, the forestry profession applies the principle of sustainability or the principle of sustainable development. In this way, by operations of forest tending, the existing forests are developed, and by operations of regeneration and reforestation new generations of forests are grown. Natural management is based on operations of forest tending and regeneration whose basic principles can be found in virgin forest. Forests managed in this way represent a strong ecological and economic foothold in all, even unfavorable, life conditions. If from its origin, the forest has been developed under the influence of man who carries out the operations of tending and regeneration based on natural principles, then we talk about natural forest management. (Matić, 2009).

By forest management, man affects directly the assortment structure of stands. According to Matić & Skenderović (1992) in carrying out silvicultural operations of forest tending, difference should be made between positive and negative selection. Tending of stands and trees is based on the fact that the tree phenotype is the result of genotype and impact of the environment or stand conditions. By tending, the spontaneous selection of trees in a stand is replaced by selection based on silvicultural principles. Negative selection is aimed at observing and removing from the stand all unwanted plants until the age when the trees of future can be identified (stand carriers), and after that positive selection is applied by which everything that prevents the development of the identified trees of future is eliminated from the stand. Negative selection involves providing more light to seedlings and cleaning of saplings and young growth. Positive selection involves thinning in mature young-growth, young, medium-aged, older and old stands, and then follow the procedures of natural regeneration (reforestation) of stands by preparatory, seeding and finishing felling. The selection criteria of marking trees for felling is also applied in preparatory felling where trees that decrease or prevent ripening of the best trees and trees whose self seeding is not wanted (less wanted species and similar) are removed. After that, in seeding, shelterwood and final fellings, the decisive role in making decision on marking trees for felling is given to seeding of reforestation area and gradual providing of light to seedlings and saplings, which is regulated by space distribution of the remaining trees in the regeneration area.

The impact of management on the structure of beech stands and production of wood can be clearly seen through almost two centuries of organized forestry in the research area, i.e. the fact that we no longer manage virgin forest stands that developed with minimum or no human assistance. On the other hand, it is a fact that forestry has always known how to implement natural management in Croatian forests. This is best proved by the preservation of their natural structure and diversity that are especially conspicuous in forests where forest management has been applied continuously for almost two centuries and a half. Natural approach to forest management in Croatia can be can be seen in the Zagreb School of Forest Silviculture applied and developed by the Faculty of Forestry of the University of Zagreb (Matić & Anić, 2009).

Fig. 1. Young (5 years) and old (100 years) natural beech stands

Common beech is a broadleaved species and with aging it almost always develops false heartwood, which complicates additionally the investigation of its assortment structure. Roundwood faults are related to irregularities in structure, texture, color and consistency. They reduce technical properties, make processing difficult and decrease the degree of wood utilization. Some wood faults are created as the effect of the phenomenon of growth and development of the tree so that the term "wood fault" is relative. European beech is the tree species of one color, whose aging generates optionally or always colored heartwood of irregular shape. Such colored beech heartwood is called false heartwood, red heartwood, brown heartwood or corewood (*Figure* 2). In Croatia, local population also use the term „kern" (from the German noun der Falschkern - false heartcore).

Fig. 2. Different shape of beech false heartwood

Beech false heartwood, as the phenomenon that affects considerably the quality of industrial wood, has been the object of professional and scientific interest for more than 100 years (Krpan et al., 2006). Numerous theories and interpretations of the formation of false heartwood are hypothetical even today. As stated by Glavaš, Tusson (1905) supposes that false heartwood of beech is formed as the reaction of wood cells to the attack of fungi. It was established later that the primary cause of formation of false heartwood in beech trees is not a biotic but rather an abiotic factor. The formation of false heartwood of beech is the effect of reaction of living wood cells to the penetration of air or oxygen into the tree trunk. Oxygen acts as poison on living cells, and they defend themselves by anatomical and chemical changes (tyloses, oxidation, formation of colored substances) trying to prevent further penetration of air. Substances formed as the result of such cell reactions are not introduced into cell walls, but deposited on them, and this is the basic difference between true and false heartwood (Glavaš, 1999).

The factors of formation of all types of heartwood are divided into obligatory and optional. For the formation of false heartwood, a certain quantity of air must penetrate into the inside of the tree. Optional factors are as follows: natural aging of parenchyma cells, excessive penetration of air into the tree, low temperatures (along with a serious draught in the previous summer), presence of fungi that destroy wood and fungi that change wood color, genetic predispositions and forest silvicultural measures and other human impact. The occurrence and development of individual types of false heartwood is not always caused by influence of one factor only, but rather by a combination of several factors. It has been confirmed that there is a correlation between the formation and degree of development of false heartwood and tree physiology. In the last twenty years, the then knowledge has been considerably updated by Torelli's researches (1984, 1994). According to this author, false heartwood of beech is caused by environmental impact, and all factors that cause the reduction of water content in the inside of the tree trunk are responsible for its formation. At the age between 80 and 90 of beech trees (depending on growth conditions), a certain disruption of physiology balance occurs. At that age, the leaf area and root system of beech trees are not enlarged anymore, despite the diameter increment. This leads to the disruption of balance of the water regime inside the tree and the central part of the tree trunk dehydrates. The process of dehydration of the central part of the tree trunk is, from the physiological aspect, similar to the genetically conditioned formation of false heartwood. The following factors are crucial for the development of false heartwood of beech: the size of tree top and tree diameter, i.e. quick growth of the tree and intensity of tree top reduction. The development of false heartwood is also affected by the soil condition, position, social status of the tree and tree top height. The process of false heartwood formation starts much later on soils of poor quality. Any mechanical damages of trees with dehydrated central part result in the penetration of oxygen into the tree by which the enzymatic process of tyloses formation is initiated.

The share of false heartwood in beech industrial roundwood has a considerable impact on the quality of beech logs. The differences in the quality of discoloured and optionally colored heartwood (false heartwood), under assumption that wood with optionally colored heartwood is sound, are almost the same as differences in characteristics of discoloured and obligatorily colored heartwood. The boundary line of false heartwood does not correspond to the boundary line of the growth ring. At the cross section the boundary line of false heartwood may be radial, star-like and completely irregular. False heartwood may be differently nuanced and it is not always symmetrical with respect to the longitudinal axis of the tree trunk, and drying causes considerable change of color. The largest diameter of false heartwood (red heart) appears between the 1st and 4th meter from the stump, and thereafter it decreases towards the stump and tree top. There is another increase of the diameter of read heart, although lower, between the 6th and 8th meter of tree trunk. In beech tree trunk false heartwood has the form of two cones connected by their bases, and however this shape is not necessarily regular (*Figure 3*). It achieves the greatest width at the point where the formation started (Tomaševski, 1958).

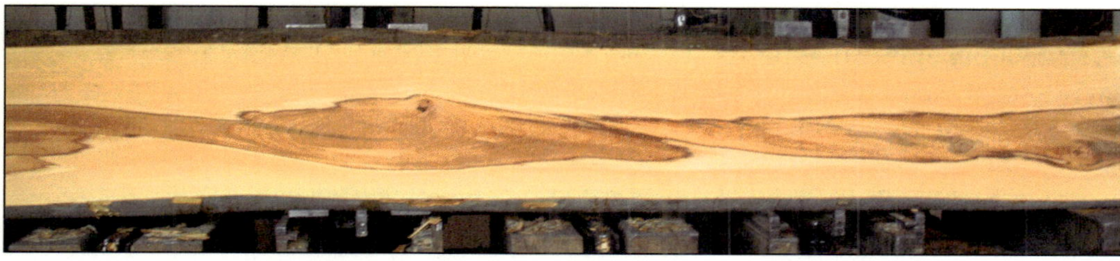

Fig. 3. Distribution of beech false heartwood in the trunk

Conductive elements in parts of wood attacked by false heartwood are blocked by tyloses, and consequently the impregnation can hardly penetrate into the wood (Govorčin et al., 2003). The impregnated beech wood with false heartwood is, therefore, highly susceptible to decay. Right due to susceptibility to decay of beech wood with false heartwood, this phenomenon is extremely important in practice and in wood trading (Glavaš, 1999, 2003). A specific kind of decay in beech trees with false heartwood is the phenomenon of specific white rot. It occurs when rot fungi penetrate into wood through wounds, broken branches, front end of logs, etc., and then develops inside the wood. White rot of beech trees is caused by different types of specific fungi, and most commonly they are as follows: *Schizophyllum commune* Fr., *Hypoxylon coccineum* (Pers.) Wind., *H. fragiforme* (Person ex Fries) Kicky, *Tremella faginea* Britz., *Stereum purpureum* Pers., *Biospora monilioides* Corda and others. White rot in wood with false heartwood is not spread evenly but rather in the form of a leaf, tongue or similar. The reason lies in the fact that different parts of beech wood are differently affected by false heartwood and tyloses and hence they offer different resistance to rot fungi. However, false heartwood cannot prevent the rotting process. It only makes the rotting process slower and more uneven. (Glavaš, 1999, 2003).

The share of false heartwood is prescribed for industrial roundwood by Croatian standards (for wood assortments and quality classes except those of the lowest value), and it is assessed or measured on the front end of industrial roundwood by measuring the diameter of the part where false heartwood is formed and by expressing it in centimeters or as the percentage of the diameter (area) at the place of measurement.

The presence and share of false heartwood in beech trees is unknown until the tree is felled and industrial roundwood processed. From the standpoint of forest harvesting and knowledge of quality or assortment structure of beech stands, the interest in the origin and development of false heartwood of beech trees is quite understandable. Although it has been known for some time that beech coming from Bilogora is highly appreciated in the market due to low presence of false heartwood. Right because it is unknown until felling and processing of trees, false heartwood represents an additional problem in planning revenues and in considering issues related to assortment structure of beech. For these reasons, it is necessary to expand the current knowledge with the information on frequency of occurrence of false heartwood and its impact on the quality of wood assortments by diameter classes based on the age of our even-aged stands and type of cut. It can be concluded that with this respect common beech is the most demanding autochthonous species in Croatia.

Compilation, precision and application in practice of tables showing shares of forest assortments in the annual allowable cut (assortment tables) are connected with serious and numerous difficulties caused by the influence of biotic and abiotic factors on stand development. Reliable and applicable assortment tables should take into account as many factors as possible. Among biotic factors, one of the most important is the impact of man through management, i.e. through implementation of criteria for marking trees by which a certain type of cut is performed. Consequently, it can be concluded that our primary interest, in terms of operating (production) activities, is the assortment structure, which can be achieved by implementing certain types of felling at a certain age of the stand, and not the assortment structure of all trees in the stand of a specific age. The efforts to achieve the best possible quality of all trees in the stand of a specific age are the basis of all management procedures of even-aged beech stands, carried out by work of many generations of forestry experts.

2. Research place and methods

2.1 Place of research

Regarding the ownership title, the economic forests in the Republic of Croatia can be generally divided into state-owned forests and privately owned forests. The aim of „Hrvatske šume" ltd and private owners is to manage forests by preserving and upgrading their biological and environmental diversity and to protect the forest ecosystem. In Croatia forests and forest land owned by the Republic of Croatia are managed by „Hrvatske šume" d.o.o., and for the purpose of performing a part of activities in public domain and improving forest management of forests and forest land of private forest owners a Forest Extension Service has been established.

In the widest sense, the state forests of Forest Administration Branch Office (FABO) Bjelovar involve the area of Northwest (Central) Croatia, they cover a total area of 131.820 hectares and they are located in seven counties: Bjelovar and Bilogora, Brod and Posavina, Koprivnica and Križevci, Požega and Slavonia, Sisak and Moslavina, Virovitica and Podravina and Zagreb county. These forests owned by the Republic of Croatia are managed by „Hrvatske šume" ltd Zagreb through Forest Administration Branch Office Bjelovar. The whole area is divided into 15 forest offices and 34 management units.

In the total growing stock of these forests of almost 33.000.000 m^3, common beech is the most represented species with the growing stock of almost 12.000.000 m^3 or 36 %. The total current annual increment is around 910.000 m^3 with the share of beech of approximately 328.000 m^3 or 37 %. Beech in forests of this area is vital and healthy. This is highly supported by the information on the average 3,3 % of occasional beech felling compared to the annual allowable cut of beech in the area of FABO Bjelovar in the period

2001 to 2009. Within a limited research area (Forest Office Bjelovar) for the same period the average annual share of accidental felling in the annual allowable cut of beech was only 2,3 %.

In the period between 2007 to 2010 forests owned by private forest owners in the Republic of Croatia was under the control of the Forest Extension Service. Since 2010. care for the forests owned by private forest owners in Croatia is returned to the State company „Hrvatske šume" and the Forest Extension Service has been called off. As well known, the Republic of Croatia has a relatively high percentage share of areas covered with forests and forest land (approximately 44 %) compared to the total land area. On the other hand, the share of private forests in the total forested areas of approximately 22 % is significant, and still not so high compared to some neighboring countries that have a considerably higher share of private forests (Slovenia 80 %, Hungary 40 %, etc.).

The surface area of private forests, or forests owned by forest owners in the area of Bjelovar and Bilogora County has not been precisely determined yet, and it is estimated to approximately 20.000 ha. Two developed management programs that are related to a restricted area of Bjelovar (for management units Bjelovarske šume and Trojstvo) cover the total surface area of private forests of approximately 3.000 hectares with the growing stock of almost 450.000 m^3 and 10-year allowable cut of approximately 100.000 m^3. The average growing stock in these two management units is 163 m^3/ha (for MU Bjelovarske šume) and 199 m^3/ha (for MU Trojstvo), respectively, which is more than the estimates used before the development of the management programs. The fact that private forests definitely need professional assistance is also supported by the mixture ratio of tree species where the most represented species are allochthonous locust with more than 30 %, hornbeam with approximately 20 % and soft broadleaved species (lime, alder, etc.) with approximately 15 to 20 % share in the total growing stock. The share of the most significant autochthonous species is small. The share of beech in the growing stock ranges between 6 % (MU Bjelovarske šume) and 10 % (MU Trojstvo). The share of pedunculate oak is up to 4 %, and sessile oak up to 7 %. The key problems of management in this area, too, are caused by fragmentation of forest holdings so that almost 12.000 owners have an average property of 0,23 hectares (MU Bjelovarske šume) to 0,28 hectares (MU Trojstvo). All this speaks of significant differences between private and state forests in this area, and also of considerable, but insufficiently utilized, resources of private forests.

The research was carried out in the management unit „Bjelovarska Bilogora" of Forest Office Bjelovar (FABO Bjelovar). All research compartments belong to the ecological-management type II-D-11 and management class BEECH with a 100-year rotation, whose share in the surface area of the management unit is 76,1 %, and in the growing stock 80,6 %. The management unit „Bjelovarska Bilogora" is located on Southwest and South slopes of Bilogora, at the altitude ranging between 115 m and 307 m above sea level. Its total surface area is 7.632,62 ha, of which 7.444,17 ha is stocked. The management unit is divided into 180 compartments and 533 sub-compartments. In 2003 the total growing stock was 2.317.147 m^3. In the growing stock, the beech as the most represented species accounts for 1.036.386 m^3 or 44,73 %. The total 10-year allowable cut for I/1 management semi-period is 586.231 m^3, of which 443.752 m^3 is main felling, and 142.479 m^3 is thinning. The share of beech in the 10-

year allowable cut is 297.753 m³ (67,2 %) in the main felling and 45.939 m³ (32,2%) in thinning, or a total of 343.692 m³ (58,6 %).

The total allowable cut of the main felling of the Forest Administration Branch Office Bjelovar is approximately 400.000 m³ with the share of beech considerably higher than 50 %. The total felling in the period 2001 to 2009 in the area of Forest Administration Branch Office Bjelovar can be seen in Table 1.

Year	Beech						Allowable cut				
	Main felling		Thinning		Total		Main felling		Thinning		Total
	m³	%	m³	%	m³	%	m³	%	m³	%	m³
2001.	122.384	60,4%	80.340	39,6%	202.724	**33,7%**	325.923	54,1%	276.139	45,9%	602.062
2002.	129.999	59,0%	90.525	41,0%	220.524	**35,7%**	289.554	46,9%	327.778	53,1%	617.332
2003.	171.831	69,6%	74.934	30,4%	246.765	**40,0%**	338.814	54,9%	278.530	45,1%	617.344
2004.	184.957	72,1%	71.729	27,9%	256.686	**39,6%**	366.680	56,5%	282.180	43,5%	648.860
2005.	192.115	70,6%	79.925	29,4%	272.040	**41,2%**	339.610	51,4%	320.912	48,6%	660.522
2006.	205.372	71,2%	82.895	28,8%	288.267	**43,6%**	383.765	58,0%	277.367	42,0%	661.132
2007.	199.731	70,9%	81.923	29,1%	281.654	**41,7%**	388.054	57,4%	288.013	42,6%	676.067
2008.	226.122	72,4%	86.313	27,6%	312.435	**45,8%**	375.802	55,1%	306.020	44,9%	681.822
2009.	227.579	74,6%	77.307	25,4%	304.886	**43,9%**	413.830	59,6%	280.435	40,4%	694.265
Total	1.660.090	69,6%	725.891	30,4%	2.385.981	**40,7%**	3.222.032	55,0%	2.637.374	45,0%	5.859.406

Table 1. Felling from 2001 to 2009 – Forest Administration Branch Office Bjelovar

Year	Beech						Allowable cut				
	Main felling		Thinning		Total		Main felling		Thinning		Total
	m³	%	m³	%	m³	%	m³	%	m³	%	m³
2001.	13.570	55,1%	11.048	44,9%	24.618	**39,7%**	30.991	50,0%	31.011	50,0%	62.002
2002.	6.447	61,8%	3.992	38,2%	10.439	**21,7%**	23.789	49,5%	24.243	50,5%	48.032
2003.	27.702	84,7%	5.018	15,3%	32.720	**47,9%**	43.746	64,0%	24.603	36,0%	68.349
2004.	30.567	87,7%	4.281	12,3%	34.848	**48,4%**	49.513	68,8%	22.426	31,2%	71.939
2005.	36.268	92,1%	3.110	7,9%	39.378	**51,3%**	49.766	64,8%	27.049	35,2%	76.815
2006.	33.606	90,6%	3.489	9,4%	37.095	**49,2%**	51.794	68,7%	23.635	31,3%	75.429
2007.	27.084	81,9%	5.979	18,1%	33.063	**38,1%**	58.030	66,9%	28.760	33,1%	86.790
2008.	34.263	81,2%	7.938	18,8%	42.201	**49,9%**	55.132	65,2%	29.450	34,8%	84.582
2009.	28.379	82,1%	6.194	17,9%	34.573	**41,9%**	53.970	65,4%	28.517	34,6%	82.487
Total	237.886	82,3%	51.049	17,7%	288.935	**44,0%**	416.731	63,5%	239.694	36,5%	656.425

Table 2. Felling from 2001 to 2009 – Forest Office Bjelovar

Along with a continuous growth of the allowable cut in this period (from approximately 600.000 m³ to approximately 700.000 m³), an increasing share of beech trees can also be seen in the average wood volume (from approximately 34 % to approximately 45 %). Similarly, it can be said that the share of the main felling has increased considerably (from approximately 60 % to almost 75 %) in an average wood volume of beech in the area of FABO Bjelovar. The survey of fellings performed in a limited research area (Forest Office Bjelovar, *Table 2*) show the same increasing trend of allowable cuts (from approximately 50.000 m³ to approximately 85.000 m³), increase of beech share in the total wood volume (from approximately 40 % to 50 %) and considerable increase of the main felling (even more than 90 %) in the average allowable cut of beech in the area of Forest Office Bjelovar (*Figure 4*).

These trends are the effect of the growing stock structure of the research area, disproportion of age classes, as well as some inconsistencies in determining the rotation of beech stands, and however they should be taken into account when planning the development of both forestry and wood-processing activities in the area of Bjelovar.

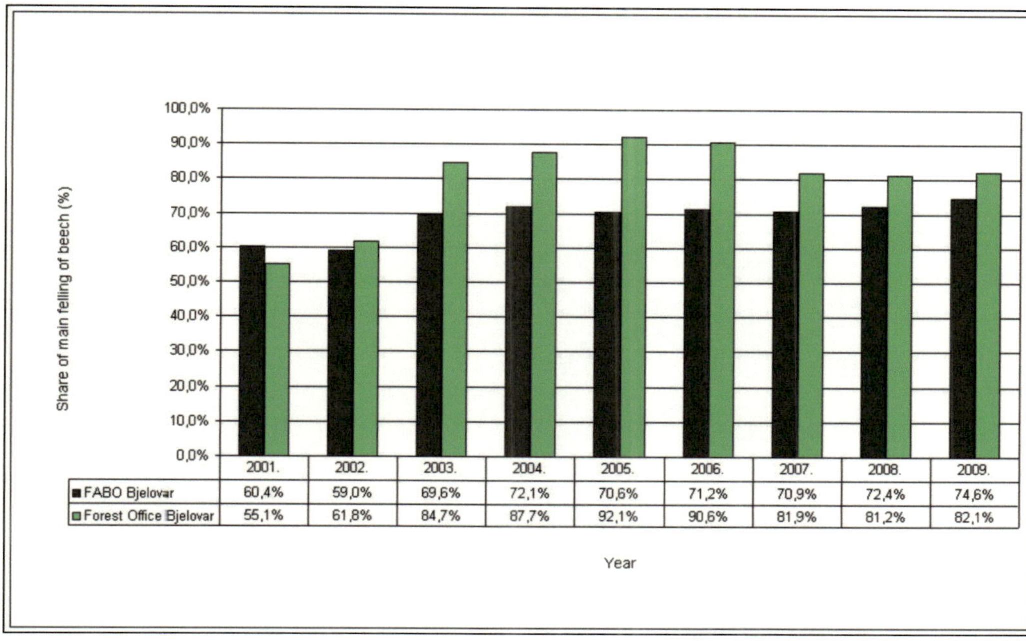

Fig. 4. Share of main felling in felled volume of beech from 2001 to 2009

2.2 Method of work

The age of felling sites ranged between 59 and 91 with thinning operations, between 94 and 110 with preparatory felling, between 100 and 112 with seeding felling and between 98 and 114 with final felling.

Type of felling															
Thinning				*Preparatory fellings*				*Seeding fellings*				*Final fellings*			
Forest block	Number of marked trees	Number of trees in the sample	%	Forest block	Number of marked trees	Number of trees in the sample	%	Forest block	Number of marked trees	Number of trees in the sample	%	Forest block	Number of marked trees	Number of trees in the sample	%
7c	292	60	20,5%	9a	1198	102	8,5%	11a	1667	177	10,6%	20d	394	46	11,7%
13a	665	65	9,8%	11a	683	78	11,4%	21a	2112	108	5,1%	21a	1201	74	6,2%
13b	285	51	17,9%	17a	865	91	10,5%	38a	1308	109	8,3%	42a	1239	118	9,5%
20e	569	66	11,6%	19b	490	58	11,8%	59c	409	41	10,0%	42c	876	104	11,9%
29a	368	46	12,5%	21a	1166	132	11,3%	83a	166	31	18,7%	59c	438	44	10,0%
29b	229	34	14,8%	38a	1164	102	8,8%	94b	650	76	11,7%	75a	547	55	10,1%
37a	631	83	13,2%	42a	456	63	13,8%	95b	439	64	14,6%	83a	445	42	9,4%
37c	335	48	14,3%	42c	394	42	10,7%	-	-	-	-	89b	145	23	15,9%
39b	368	56	15,2%	60a	862	97	11,3%	-	-	-	-	155f	953	57	6,0%
65b	164	24	14,6%	66a	577	64	11,1%	-	-	-	-	166c	135	20	14,8%
66b	163	31	19,0%	73a	888	100	11,3%	-	-	-	-	-	-	-	-
69b	515	67	13,0%	94b	343	54	15,7%	-	-	-	-	-	-	-	-
80b	46	17	37,0%	95b	306	42	13,7%	-	-	-	-	-	-	-	-
82a	159	49	30,8%	-	-	-	-	-	-	-	-	-	-	-	-
162a	371	45	12,1%	-	-	-	-	-	-	-	-	-	-	-	-
162c	282	45	16,0%	-	-	-	-	-	-	-	-	-	-	-	-
Total	5442	787	14,5%	Total	9392	1025	10,9%	Total	6751	606	9,0%	Total	6373	583	9,1%

Table 3. Distribution of model trees according to the standard HRN (1995)

The sample of model trees was formed by random selection of approximately 10 % of marked trees. Moving around the stand at predetermined azimuths, all marked beech trees found in the travel direction were included in the sample.

In the period 1997 to 2007, the field research involved a total of 3.776 model trees. Table 3 and 4 show the number of model trees by research compartment according to the type of cut and applied standard.

Vrsta sjeka - *Type of felling*															
Thinning				*Preparatory fellings*				*Seeding fellings*				*Final fellings*			
Forest block	Number of marked trees	Number of trees in the sample	%	Forest block	Number of marked trees	Number of trees in the sample	%	Forest block	Number of marked trees	Number of trees in the sample	%	Forest block	Number of marked trees	Number of trees in the sample	%
7c	292	59	20,2%	9a	1198	102	8,5%	11a	1667	174	10,4%	11a	721	76	10,5%
13a	665	65	9,8%	11a	683	78	11,4%	38a	1308	109	8,3%	38a	879	102	11,6%
13b	285	51	17,9%	17a	865	91	10,5%	59c	409	41	10,0%	42a	1239	118	9,5%
20e	569	66	11,6%	19b	490	58	11,8%	66a	953	138	14,5%	42c	876	104	11,9%
29a	368	46	12,5%	38a	1164	102	8,8%	73a	1077	155	14,4%	59c	438	44	10,0%
29b	229	34	14,8%	60a	862	97	11,3%	94b	650	76	11,7%	94b	711	133	18,7%
37a	631	83	13,2%	66a	577	64	11,1%	95b	439	64	14,6%	95b	378	68	18,0%
37c	335	48	14,3%	73a	888	100	11,3%	124a	1134	105	9,3%	-	-	-	-
39b	368	56	15,2%	94b	343	54	15,7%	-	-	-	-	-	-	-	-
65b	164	24	14,6%	95b	306	42	13,7%	-	-	-	-	-	-	-	-
66b	163	31	19,0%	-	-	-	-	-	-	-	-	-	-	-	-
69b	515	67	13,0%	-	-	-	-	-	-	-	-	-	-	-	-
80b	46	17	37,0%	-	-	-	-	-	-	-	-	-	-	-	-
82a	159	50	31,4%	-	-	-	-	-	-	-	-	-	-	-	-
162a	371	45	12,1%	-	-	-	-	-	-	-	-	-	-	-	-
162c	282	45	16,0%	-	-	-	-	-	-	-	-	-	-	-	-
Total	5442	787	14,5%	Total	7376	788	10,7%	Total	7637	862	11,3%	Total	5242	645	12,3%

Table 4. Distribution of model trees according to the standard HRN EN

Model trees were processed in accordance with the requirements of the *Croatian standards for forest harvesting products* (former JUS – standards of ex Yugoslavia) of 1995 (HRN D.B4.020, HRN D.B4.022, HRN D.B4.027, HRN D.B4.028, HRN D.B5.023), and «bucking simulation» was made on the same trees in accordance with the Croatian standard *Hardwood Round Timber – Qualitative classification, Part 1: Oak and beech HRN EN 1316-1:1999.* The faults of wood and processed round timber were measured in accordance with the terms of the standards HRN D.A0.101, HRN D.B0.022, and HRN EN 1309-2, HRN EN 1310, HRN EN 1311 and HRN EN 1315. Many other characteristics were also measured or assessed on model trees: diameter at breast height, tree height, trunk height, length of logs, trunk diameter, length of cut logs (1-m and longer), diameters of cut logs (1-m and longer), lengths and diameters of fuel wood up to 4 m in processing large fuel wood, length of waste wood, diameters of waste wood, bark thickness, false heartwood and described tree markings.

Out of the total number of model trees, 693 of them were processed and measured only in accordance with the requirements of the *Croatian standards for forest harvesting products* of 1995, on 2.308 trees the measurements and classification of technical roundwood were carried out in accordance with the requirements of both standards, while 775 model trees were measured and then technical roundwood was classified in accordance with the requirements of the Croatian standard *Hardwood Round Timber – Qualitative classification, Part 1: Oak and beech HRN EN 1316-1:1999.*

In this way the sample for the preparation of assortment tables in accordance with the requirements of Croatian standards for forest harvesting products of 1995 covered 3.001

model trees (*Table 3*). On the other hand, in accordance with the requirements of the Croatian standard HRN EN 1316-1:1999, assortment tables were prepared on the basis of the sample made of 3.082 model trees (*Table 4*).

By processing model trees of the sample in accordance with the requirements of the *Croatian standards for forest harvesting products of 1995*, 10.098 pieces of technical roundwood were produced, whose total volume without bark was 4.337 m^3. The total processed and used net volume of all sample trees according to the requirements of this standard was 7.469 m^3. By bucking the model trees according to the requirements of the Croatian standard *Hardwood Round Timber – Qualitative classification, Part 1: Oak and beech HRN EN 1316-1:1999* 13.507 pieces of technical roundwood were produced, whose total volume without bark was 6.010 m^3, and the total net volume of all sample trees, according to the requirements of the standard, was 8.931 m^3.

The occurrence and characteristics of false heartwood of beech trees were investigated with respect to diameter class of trees and type of cut. A sample of 787 trees was made in thinnings. In preparatory cuts 788 trees were processed, in seeding cuts 467 trees and in final cuts 266 trees, which makes a total of 2.308 trees (*Table 5*).

Beech false heartwood was measured on the front ends of the processed industrial roundwood of the pertaining tree in accordance with the procedure prescribed by the standard, i.e. minimum and maximum diameter of false heartwood is measured on the front ends, and the mean value is taken rounded to the nearest lower centimeter. The mean diameter of the relative front of the log is measured and determined in the same way. Measurements are carried out on both ends (fronts) of industrial roundwood. If false heartwood is only present on one front end of the log, at the other front end only the mean diameter is measured and determined. Absolute and percentage shares of false heartwood are expressed in wood volume of industrial roundwood in accordance with the standard, which prescribes Huber's formula for the calculation of log volume. Smalian's formula is used for the calculation of false heartwood volume in a log:

$$V_k = (g_1 + g_2)/2 * l \qquad (1)$$

Where:
V_k = false heartwood volume,
g_1 = crosscut area of false heartwood at the thicker end of the log,
g_2 = crosscut area of false heartwood at the thinner end of the log,
l = log length.

This formula is known as the formula of two crosscuts, and it is used for accurate determination of the volume of an imperfect paraboloid (Pranjić & Lukić, 1997).

The volume of false heartwood was determined with some simplifications, which are conditioned by the procedure of felling and processing industrial roundwood. For a more accurate determination of false heartwood volume, it would be necessary to make more cross cuts or longitudinal cut in each piece of industrial roundwood, and this cannot be done for obvious reasons. Since a similar way of measuring false heartwood is applied in classifying wood assortments into quality classes and in trading with wood assortments, this way of measurement of false heartwood is acceptable. In the same way, the research of the share of false heartwood is restricted only to industrial roundwood, as in other wood assortments (parts of tree) it has almost no significance. Mathematical and statistical processing of data was carried out by use of computer program *Microsoft Excel 97*.

3. Results and discussion

Total percentage share of beech trees with false heartwood in individual types of cut increases from thinning, where it is 11,7 %, to final cut, where it is 84,6 % (*Table 5*).

Diameter class cm	Thinning		Preparatory felling		Seeding felling		Final felling	
	Number of trees							
	in sample	with false heartwood	in sample	with false heartwood	in sample	with false heartwood	in sample	with false heartwood
17,5	25	0	1	0	-	-	3	0
22,5	104	1	18	1	-	-	3	0
27,5	154	8	61	19	7	2	4	1
32,5	257	27	123	45	25	9	3	2
37,5	133	25	161	79	44	20	8	7
42,5	59	12	177	104	87	52	27	21
47,5	38	9	131	95	114	87	45	39
52,5	8	4	62	42	81	69	65	57
57,5	6	4	35	28	51	41	46	40
62,5	1	1	8	8	38	33	27	24
67,5	-	-	6	6	12	12	17	16
72,5	2	1	4	3	7	7	9	9
77,5	-	-	1	1	1	1	7	7
82,5	-	-	-	-	-	-	2	2
Total	787	92	788	431	467	333	266	225
Percentage	11,7%		54,7%		71,3%		84,6%	

Table 5. Sample size and share of trees with false heartwood

The increase of total share of the number of trees with false heartwood by type of cut may be explained by the increase of the mean diameter at breast height (in terms of the increase of number of thicker trees in the sample) and stand age from thinning to final cut. The average age of the sample by type of cut is 76 years for thinnings, 104 years for preparatory cut, and for seeding and final cut the average age of the sample trees is 106 years.

As the distribution of diameters at breast height are different for individual types of cut, the number of trees with false heartwood was determined by diameter classes for each type of cut. Percentage share of beech trees with false heartwood by diameter classes and type of cut is shown in Table 5 and Figure 5. Diameter class with the highest number of trees with false heartwood increases from thinning to final cut, and diameter class in which more than half of trees with false heartwood may be expected decreases from thinning to final cut.

These results confirm past researches (Torelli 1984, 1994; Prka, 2003, 2005; Krpan at al., 2006) dealing with the origin of the process of formation of false heartwood at the age approximately ranging between 60 and 75. It can be concluded from these researches that the phenomenon of false heartwood does not considerably affect the structure of assortments of thinning stands, and however its impact on the structure of assortments of a shelterwood system (preparatory cut, seeding cut and final cut) cannot be absolutely excluded.

The increase of the number of trees with false heartwood within individual diameter classes from thinning to final cuts implies that the formation of false heartwood depends less on diameter at breast height, and more on stand age. Such distribution of trees in the sample with false heartwood, by type of cut and diameter class, fits fully into the latest researches of the cause of formation of false heartwood in beech trees (Torelli, 1984, 1994). Therefore, these results should be interpreted as the disruption of balance of water regime within the tree and dehydration of the central part of the tree trunk due to the disturbance of physiological balance in older trees.

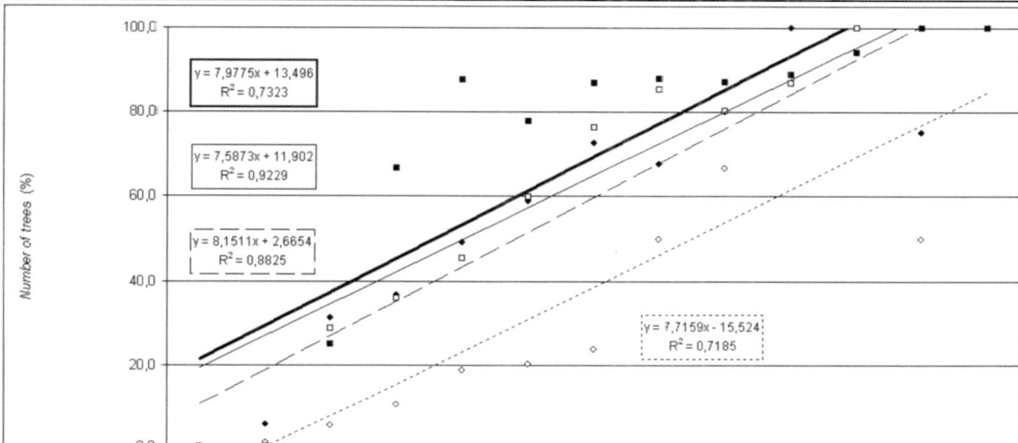

	17,5	22,5	27,5	32,5	37,5	42,5	47,5	52,5	57,5	62,5	67,5	72,5	77,5
◇ Thinning	0,0	1,0	5,2	10,5	18,8	20,3	23,7	50,0	66,7	100,0		50,0	
◆ Preparatory felling	0,0	5,6	31,1	36,6	49,1	58,8	72,5	67,7	80,0	100,0	100,0	75,0	100,0
□ Seeding felling			28,6	36,0	45,5	59,8	76,3	85,2	80,4	86,8	100,0	100,0	100,0
■ Final felling	0,0	0,0	25,0	66,7	87,5	77,8	86,7	87,7	87,0	88,9	94,1	100,0	100,0

Diammeter class (cm)

◇ Thinning ◆ Preparatory felling □ Seeding felling ■ Final felling ------ Thinning — — Preparatory felling ——— Seeding felling ━━━ Final felling

Fig. 5. Share of number of trees with false heartwood

The length of industrial roundwood made from the tree affected by the process of formation of false heartwood was measured during the measurement of industrial roundwood. It represents total length of industrial roundwood made from the tree where false heartwood appears at least at one front end (cross cut) of the log during bucking and cross-cutting of industrial roundwood in accordance with Croatian Standards. According to the way of measurement, it can be seen that the actual longitudinal presence of false heartwood in the tree, i.e. in industrial roundwood, remains partly unknown for understandable reasons. Total length of false heartwood in industrial roundwood is surely somewhat smaller than the length measured in this way (*Table 6, Figure 6*). On the other hand, false heartwood may remain hidden with some logs, i.e. it must not necessarily appear at the front ends of logs. For these reasons, data obtained in this way have only an approximate value.

Table 6 and Figure 6 present the data on mean (average) value of length of industrial roundwood affected by false heartwood by type of cut and diameter class. As the decision on the place of the trunk cut in the production of logs is made based on external characteristics, data on the length of industrial roundwood affected by the process of formation of false heartwood collected in this way have a certain operating value.

Diameter class cm	Thinning	Preparatory felling	Seeding felling	Final felling
	Mean lenght of roundwood with false heartwood m			
22,5	2,35	5,32	-	-
27,5	5,96	5,60	3,92	3,73
32,5	7,23	8,97	7,88	8,65
37,5	10,02	10,11	11,17	14,93
42,5	10,50	12,16	12,14	13,33
47,5	7,16	11,84	13,35	15,61
52,5	5,44	13,36	13,55	16,70
57,5	6,51	12,92	13,91	15,83
62,5	4,16	12,11	14,04	16,61
67,5	-	14,69	13,64	15,31
72,5	3,71	7,20	16,84	14,62
77,5	-	22,47	17,30	16,01
82,5	-	-	-	13,78

Table 6. Mean length of false heartwood in industrial roundwood of beech trees

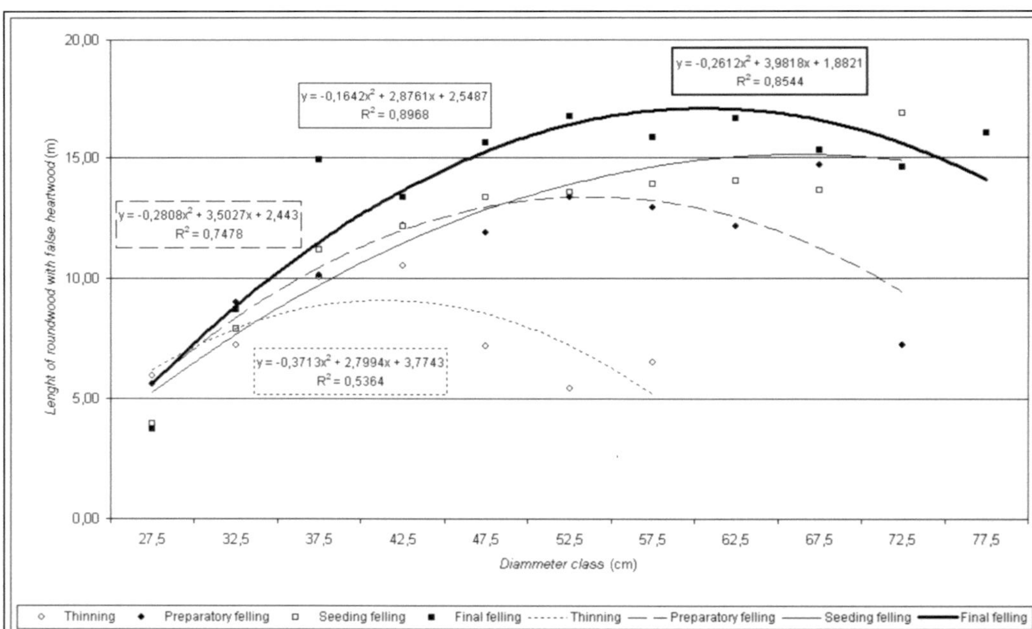

Fig. 6. Mean length of industrial roundwood with false heartwood

Mean values of the length of false heartwood and the trend line are presented in Figure 6 only for diameter classes containing three or more trees of diameter classes with false heartwood. The increasing trend of the length of industrial roundwood with false heartwood can clearly be seen from thinnings to final cut. The reasons of such trend may be explained by the above stated factors that cause the formation of false heartwood. This fact affects the absolute value of false heartwood volume in the volume of industrial roundwood of a certain type of cut. Lower values of the average lengths of false heartwood in the volume of industrial roundwood in larger diameter classes of thinnings (and even in preparatory cut) imply that the diameter at breast height is not a decisive factor in the formation of false heartwood.

The data presented in Table 7 and Figure 8 show the increase of the average absolute values of the volume of false heartwood in industrial roundwood by diameter classes and from thinnings to final cut. The lowest mean values were recorded with trees with false heartwood in thinnings and then trees in preparatory cut. With trees in preparatory cut the increase of heartwood volume of industrial roundwood of larger diameter classes is not as significant as with trees in seeding and final cut. In this respect, trees with false heartwood in seeding and final cut show very close values and an almost linear dependence.

Mean percentage values of the share of false heartwood volume of industrial roundwood, presented in Table 7 and Figure 7, show considerably different characteristics, almost contrary to absolute values. The more or less regular increase of mean percentage values of false heartwood volume in the volume of industrial roundwood from thinning to final cut is a common feature with absolute values of false heartwood volume in the volume of industrial roundwood.

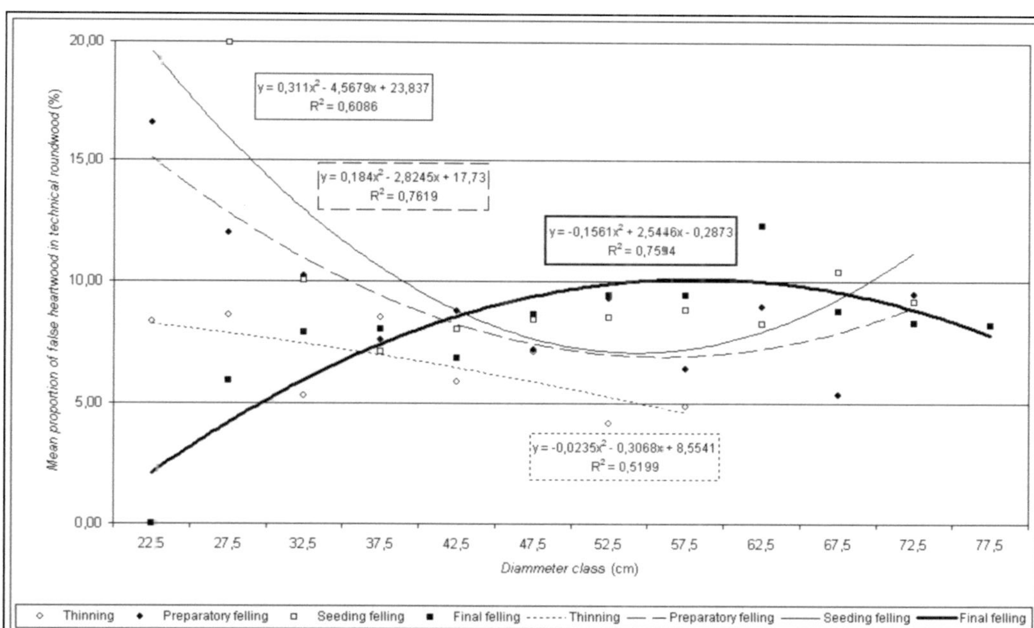

Fig. 7. Mean share of false heartwood in the volume of industrial roundwood of beech trees

A weaker correlation between the mean absolute values of false heartwood volume in the volume of industrial roundwood and the trend line is the effect of a considerably higher range of stand age of thinnings, where the measurements were carried out. This age ranges between 50 and 91 in thinnings, between 96 and 111 in preparatory cuts, between 101 and 112 in seeding cuts, while in final cuts it ranges between 98 and 112. As the above mentioned researches outlined the older age of beech trees (from 60 to 90 years) as the beginning of intensive formation of false heartwood, the sample of trees from thinning stands is the least homogenous in that respect (stand age). Mean values of absolute volume of false heartwood by diameter classes in seeding and final cuts show a great similarity as well as strong correlation, as well as trend lines of these types of cut.

Diameter class cm	Thinning		Preparatory felling		Seeding felling		Final felling	
	Proportion of heartwood							
	m^3	%	m^3	%	m^3	%	m^3	%
22,5	0,01	8,31	0,04	16,58	-	-	-	-
27,5	0,03	8,56	0,04	12,00	0,06	19,94	0,01	5,92
32,5	0,02	5,30	0,06	10,20	0,06	10,03	0,05	7,88
37,5	0,06	8,49	0,08	7,59	0,09	7,11	0,11	8,01
42,5	0,08	5,85	0,12	8,74	0,13	8,00	0,12	6,81
47,5	0,08	7,08	0,12	7,19	0,17	8,39	0,19	8,61
52,5	0,03	4,14	0,22	9,28	0,21	8,48	0,25	9,43
57,5	0,08	4,86	0,19	6,38	0,23	8,79	0,28	9,41
62,5	0,06	3,25	0,26	8,93	0,29	8,24	0,43	12,31
67,5	-	-	0,20	5,36	0,36	10,38	0,35	8,75
72,5	0,19	4,03	0,23	9,47	0,37	9,14	0,40	8,27
77,5	-	-	-	-	0,37	5,67	0,43	8,20
82,5	-	-	-	-	-	-	0,27	4,47

Table 7. Volume of false heartwood in industrial roundwood of beech trees

For all types of cut, except for final cut, the percentage shares of false heartwood volume in the volume of industrial roundwood show a decreasing trend by diameter classes. This trend may be explained by the fact that absolute volume of industrial roundwood increases with the increase of the diameter at breast height, and consequently the percentage share of industrial roundwood with false heartwood decreases. On the other hand, trees with smaller diameter at breast height with false heartwood show larger shares of false heartwood in the volume of industrial roundwood due to a lower share of volume of industrial roundwood. This all leads to a more or less decreasing trend of percentage shares of false heartwood in the volume of industrial roundwood. In this respect mean values of thinning stands show a linear correlation.

False heartwood affects most significantly the structure of beech assortments of the highest quality in terms of lowering their quality and market value. International (EU) standards allow up to 20 % of sound false heartwood in A quality class and up to 30 % in B quality class, while there are no limits for C and D classes. Star-like heartwood is not allowed in A class, in B class it may be present up to 10 %, and in C class up to 40 %, while in D class there are no limits. The influence of false heartwood on the quality and value of wood assortments can be primarily determined through the share of sub-classes in the classes of the highest quality (A and B class) of beech industrial roundwood. The Croatian standard

HRN EN 1316-1:1999 for beech provides the possibility of application of sub-class *A-red (A-s) and B-red (B-s)* depending on trade agreements. In these sub-classes, unlimited presence (up to 100 %) of homogenous and sound false heartwood (red heartwood) is allowed. In other words, industrial roundwood is classified into these sub-classes (A-s and B-s) if by its dimensions and other criteria it meets the requirements of A and B class, and however contains an excessive share of homogenous and sound false heartwood.

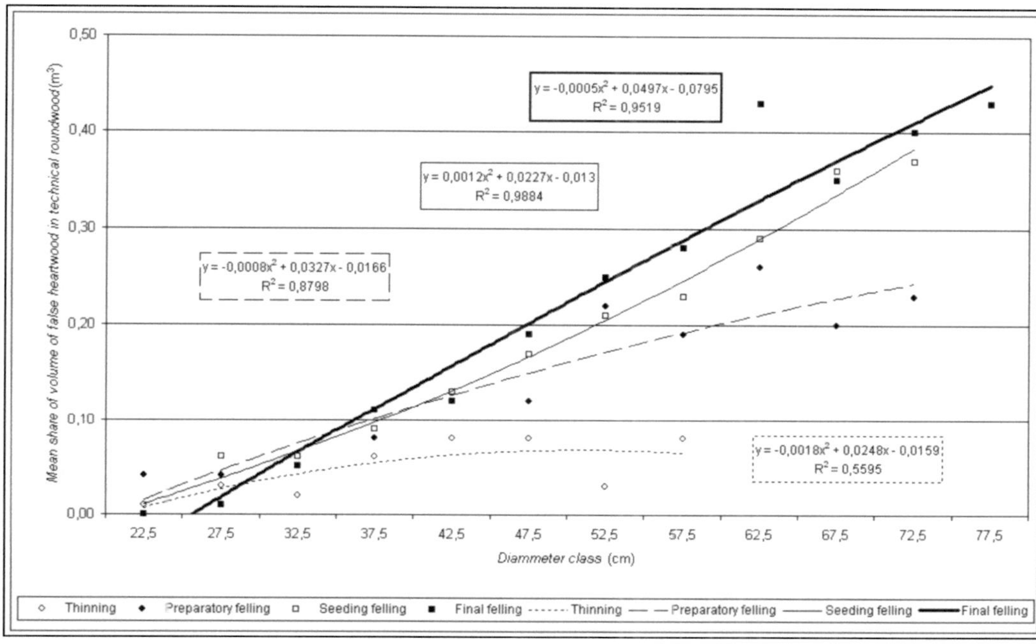

Fig. 8. Volume of false heartwood in industrial roundwood of beech trees

As the appearance, development and share of false heartwood in thinnings and preparatory cuts is not very significant, which was determined by previous researches (Prka, 2003, 2005, Krpan et al., 2006), these sub-classes of wood assortments of the highest quality have no significant effect on the assortment structure of thinnings and preparatory cuts. The presence of these sub-classes was investigated on 519 trees of seeding and final cut. Table 8 shows the percentage share of sub-class A-s and B-s according to number of pieces of industrial roundwood and share in the volume of industrial roundwood of quality class A and B. A-s logs account, on average, for 14,36 % in total volume of A class logs, and B-s logs account for 15,27 % in total volume of B class logs.

Type of felling	Forest block	A		A-s				B		B-s			
		Number of roundwood	Volume of roundwood m³	Number of roundwood	Volume of roundwood m³	Share in number %	Share in volume %	Number of roundwood	Volume of roundwood m³	Number of roundwood	Volume of roundwood m³	Share in number %	Share in volume %
Seeding	66a	47	32,54	4	2,99	8,51	9,20	122	79,79	8	6,14	6,56	7,70
Final	11a	25	21,61	6	5,31	24,00	24,59	76	62,56	7	7,48	9,21	11,06
Final	38a	69	79,45	10	12,70	14,49	15,98	108	103,4	18	19,80	16,67	19,14
Final	94b	82	104,8	14	19,47	17,07	18,58	140	141,6	24	33,78	17,14	23,86
Final	95b	47	65,78	3	3,22	6,38	4,90	70	67,33	2	2,25	2,86	3,34
Total		270	304,18	37	43,69	13,70	14,36	516	454,68	59	69,45	11,43	15,27

Table 8. Share of logs of A-s and B-s in A and B class according to number and volume

Obstacles related to processing, precision and practical application of assortment tables, and to the increase of reliability of business decision-making in planning the assortment structure of managed beech stands, usually arise out of the following facts:

- quality of trees and of the whole stand is the result of continuing impact of different abiotic and biotic factors,
- total volume of the stand cannot be used as the basis for planning felling, processing and extracting, and most of all not as the basis for calculating the financial income,
- usable volume of trees and stands varies in a wide range of values from approximately 30 % to 80 % (or more) compared to the total volume,
- distribution of wood assortments in individual trees is conditioned by diversity of their habitus and occurrence of faults on and in the tree,
- occurrence of faults, their size and number on and in the tree, is of random character and it cannot be correlated with measurable tree parameters,
- wood assortments of the same quality are not always produced from trees with the same dimensions and equal quality characteristics,
- there are differences between the classification of wood assortments in different countries, and the classification standards are subject to changes with time,
- in determining the quality of wood assortments, apart from measurable parameters, there is also a series of subjective assessments,
- analysis of the structure of wood assortments achieved in the process of wood production provides no possibility to make final conclusions primarily because there is no correlation between breast-height diameters of individual trees and the produced assortment structure, and due to the effects of the market and other relationships on the production process.
- assortment structure of managed stands is partly the result of man's impact, and such impacts have not been sufficiently researched nor recognized.

Due to the above reasons, the method for determining the assortment structure that would be relatively quick, simple and accurate is still to be found. In all methods used so far, model trees were used for determining the total volume and volume assortments by sectioning of standing and felled trees.

Tables showing shares of wood assortments determined in accordance with the *Croatian standards on forest harvesting products of 1995* and in accordance with the Croatian standard *HRN EN 1316-1:1999 Hardwood Round Timber – Qualitative classification, Part 1: Oak and beech* were developed separately for thinning and preparatory felling, and separately for seeding and final felling. This was done due to numerous reasons stated and explained above, and due to results of researches published before. The reasons for separating thinning and preparatory felling trees into special assortment tables are as follows:

- thinning sites and preparatory felling sites have an exceptionally high share of undamaged trees of abnormal growth and generally a higher percentage share of trees with negative impact on assortment structure of the felling site in the total number of marked trees compared to seeding and final felling (Prka, 2005, 2006a),
- marked trees of thinning and preparatory felling have on average a lower trunk height and consequently a lower share of technical roundwood is made of tree trunks compared to seeding and final felling (Prka, 2005, 2006b),

- validation analysis of trees by type of cut shows that thinning and preparatory felling trees have lower index values compared to trees from final and seeding felling (Prka, 2003a, 2005),
- total percentage share of technical roundwood in the net volume of trees is lower with thinning sites compared to other types of cut and it increases from thinning towards final felling (Prka, 2005),
- analysis of total deviations of percentage shares of wood assortments of the highest quality from the plan (analysis carried out in a three-year period in the research area) shows that tables of wood assortments, currently in use, overestimate the percentage share of veneer logs and peeling logs in thinning and preparatory felling sites (Prka, 2003a),
- in thinning and preparatory felling, the occurrence of trees with the highest quality assortments of technical roundwood is less probable (F - veneer and L – peeling logs – A and B quality class) and consequently the percentage share of wood assortments of the highest quality in the volume of large wood is also smaller compared to trees of seeding and final felling (Prka, 2005, 2008, Prka & Krpan, 2007),
- occurrence of false heartwood is not significant in felling sites up to the age of approximately 90 years due to the fact that in older thinnings around 15 % of trees with false heartwood may be expected. On the other hand in felling sites aged from 100 to 110 false heartwood is quite significant as it can be expected with more than 50 % of marked trees (Prka, 2003b, 2005, Prka et al., 2009, Krpan et al., 2006),
- number of trees with false heartwood increases from thinning towards final felling, as well as the length of technical roundwood with false heartwood and shares of technical roundwood affected by false heartwood (Prka, 2005, Prka et al., 2009, Krpan et al., 2006),
- seeding and final felling compared to thinning and preparatory felling show higher shares of the highest quality wood assortments depending on the diameter class by: approximately 8 to 14 % (veneer logs and peeling logs) with the application of the *Croatian Standards for Forest Harvesting Products (1995)*, and approximately 11 to 13 % (A and B – quality class) with the application of Croatian standard *HRN EN 1316-1:1999 Hardwood Round Timber – Qualitative classification, Part 1: Oak and beech* (Prka, 2005, 2008a, Prka & Krpan 2007),
- percentage shares of wood assortments by quality classes retain the same ratios (of course not the same percentage shares) regardless of the applied system of standards (Prka, 2005, 2008a, Prka & Poršinsky, 2009).

All the above reasons for dividing the sample of model trees and differences between these two groups of felling types are the result of our decisions. A common feature of the marked trees in thinning and preparatory felling is that they are chosen by selection criteria, which becomes irrelevant when the preparatory felling is completed because the key role in selecting trees for felling is then played by seeding, state of young growth and space distribution of the remaining trees. The share of wood assortments by individual types of felling is largely the result of our decisions in selecting trees for felling, based on which the objectives and guidelines of stand management are implemented.

In short, when determining the mathematical model for the development of volume of wood assortments, it should be taken into account that it depends on natural laws of development, which are not well known and which are affected during rotation by changing directly the stand structure (by silvicultural activities and natural regeneration through preparatory and finishing felling) and by applying the system of standards and trade conventions, both changeable with time.

In implementing sustainable management of beech forests, the factors that affect the assortment structure of even-aged beech felling sites, determined by this research, are as follows:

- selection criteria of trees for felling (marked) beech trees by which the prescribed type of cut is performed in managing natural beech stands,
- reached technological level of wood production which involves both technical equipment and development of forest infrastructure, as well as professional competence of all participants in wood production and the whole forest management,
- faults of beech wood formed as the consequence of natural development of beech stands and man's impact, among which false heartwood is the most conspicuous,
- procedures with beech technical roundwood during and after operations of wood production and the prescribed ways of measuring and calculating the volume of beech roundwood as well as their operational application,
- market relationships, besides demand and supply, also greatly conditioned by the development of capacities for processing beech into products such as wood products and other products (energy and similar),
- reached level of knowledge of the possibility to achieve the quantity and quality of beech from natural beech stands, and operational application of such knowledge.

4. Conclusion

False heartwood of beech affects considerably the quality of wood assortments in beech felling sites. The impact of individual types of cut on the assortment structure of beech felling sites will primarily depend on the number of trees affected by the process of development of false heartwood. This number ranges between 11,7 % of trees in thinnings, more than 54,7 % of trees in preparatory cut and 71,3 % in seeding cut and up to 84,6 % of trees with false heartwood in final cut. Hence, with the increase of the diameter at breast height of the tree, the number of tree with false heartwood increases, as well as the length of industrial roundwood with false heartwood and the volume of red heartwood in the volume of industrial roundwood of the tree. Contrary to that, the percentage share of false heartwood in the volume of industrial roundwood decreases with the increase of diameter at breast height of the tree, except in final cut.

The appearance of false heartwood has no special significance in planning assortment structures in thinnings of even-aged beech stands, considering the fact that about 10 % to 15 % of trees with false heartwood may be expected in older thinnings. On the other hand, in planning assortment structures of preparatory, seeding and final cuts, the appearance of false heartwood may be expected in approximately 55 % to 85 % of marked trees. In seeding and final cuts approximately 15 % of the volume of wood assortments of the highest quality

(A and B quality class) has an excessive share of sound false heartwood in A and B quality class. With respect to the volume of large wood, the share of A-s sub-class, depending on the diameter class, ranges between 0,3 and 2 %, and the share of B-s quality class between 1,1 and 3,2 % of the volume of large wood.

The distribution of trees in the sample with false heartwood, by type of cut and diameter class, fits fully into the latest researches of the cause of formation of false heartwood in beech trees (Torelli, 1984, 1994). Therefore, these results should be interpreted as the disruption of balance of water regime within the tree and dehydration of the central part of the tree trunk due to the disturbance of physiological balance in older trees. In investigating assortment structure of beech felling sites, the frequency of occurrence and volume of false heartwood in main fellings is the factor that affects considerably the quality of industrial roundwood.

Also, there are many other factors affecting the assortment structure of managed beech stands, and their impact is very complex. Some of these factors affect directly, and others indirectly, the possibility of achieving the quality of beech forest assortments. These factors are partly the result of natural laws of development of beech stands and trees, and partly the result of human and environmental impact. Some of these factors are objective and their impact cannot be avoided, while others are of a subjective nature and influence.

The effects on quantity and quality of beech forest assortments of even-aged management stands that can be achieved by performing a certain type of cut can be operationally related to:

- abiotic factors shown through climate, edaphic and orographic phenomena, and having a continuous and interdependent impact, whose adverse effect can usually be seen in extreme values,
- biotic factors mostly shown in human activities (beneficial and harmful) although there are also others (bacteria, fungi, insects, wild game, etc.),
- historic development of beech stand management, i.e. development of the organization of forestry science, operating practice and education of forestry staff, which finally resulted in the current state and potential of managed beech stands,
- comprehensive management of beech stands and organization of forestry economic activities, as well as position of these activities in a wider business (social) environment,
- achieved technological level of wood production that provides optimal implementation of operations in the production of beech forest assortments, including technical equipment and professional competence of all participants in the process of wood production,
- wood faults whose occurrence and number, although they are of random nature, can be partly affected (as well as total habitus of beech trees) by beech stand management,
- rotation of even-aged beech stands, i.e. determination of time of felling of individual trees, which is particularly emphasized with this species due to the occurrence and pattern of development of false heartwood,

- bucking of beech technical roundwood, precision of measurement and consistent use of applicable regulations (standards) for the classification of forest wood assortments into quality class,
- research of these issues and application of research results with the aim of developing more precise and operationally more applicable assortment tables, and more objective planning and control of production of beech assortments,
- development (changes) of standards for the classification of forest assortments into quality classes, technology of production and use of beech forest assortments, as well as development of market relationships.

Production possibilities of management stands are not without limits. From our point of view (the view of the currently active forestry staff), they are more or less permanent. We are not in a position to make, in a short term, a considerable increase (if any) of the production of natural stands. We can suggest where (on which trees) the increment of wood volume will be accumulated and when it should be sold. By our short-term operations (silvicutural operations, wood production) we can only make optimal use of and preserve the production possibilities of the stand, and leave to future generations stable stands of a potentially higher quality.

By managing beech stands, meeting scientific and professional requirements, the said impacts on the quantity and quality of forest wood assortments may be directed to a certain extent, through the life cycle, to the target production. With the application of tending procedures and regeneration of beech stands, with passing of time, the value of the stand and wood assortments produced by individual type of cut increases. In biological (physiological) and economic context, the development of false heartwood with beech trees has an opposite trend, and the positive selection of trees that carry the development of the stand and natural regeneration provide the transfer of the best properties to future generations. This is a compromise that must be accepted by proper management and determination of harvesting rotation of beech stands.

The economic development of each region in the Republic of Croatia should be based on its natural resources. Forests, forestry and wood production are significant natural (and economic) resources of the region of Bjelovar and Bjelovar and Bilogora County, where the share of beech and beech forest assortments are of the highest significance. Unfortunately, during the last turbulent 20 years, wood production in this area has been losing the significance expected based on traditional and natural resources. This has a negative effect both on forestry and the development of this region as a whole. The reasons are many, and one of the most important is the lack of concept of forestry and wood industry development based on a comprehensive understanding of the ways of using forest wood assortments applied before and on objective planning of the achievable assortment structure of the allowable cut in this area in future. Although such analyses are part of decision making at a strategic level, in the Republic of Croatia they have not been implemented satisfactorily and by using the relevant (true, accurate) data. In the past, the results of such unreliable and incomplete analyses were used by specific circles for the purpose of retaining favored market positions, rather than for the purpose of long-term development of forestry, wood industry and economy as a whole. At the end of 2008, negative global economic trends (recession) started in this segment of Croatian economy, too, and most past omissions have been disclosed and the position of economic entities still dealing with forestry and wood production has become even more difficult.

Regardless of the degree of (de)centralization of management of these resources, the profit of forestry activities, whose duration corresponds to the duration "of water droplets on a hot stove", should not be the basic interest of the state - the owner of most forests and enterprises that manage them. It is not hard to conclude that the permanent interest of the owner (society as a whole - state) should be to employ as many people as possible based on sustainable and economic principles of forestry, and use (process) forest products, especially in regions having such resources.

The use of beech-wood, especially processing of beech technical roundwood, is specific in the area of Bjelovar due to its high share in the allowable cut, as well as somewhat higher processing requirements and lower profits that can be made right after the primary processing of beech logs, and after a low degree of wood processing. Besides, compared to wood assortments of some other species, the market demand of beech assortments is less stable. Therefore, beech technical roundwood (e.g. compared to technical roundwood of pedunculate oak) is often treated as necessary evil in commercial and processing context. On the other hand, the quality of beech-wood and its share in the allowable cut of this area show that such attitude should be changed. This requires considerable changes in the approach to this problem of all involved in this segment of Croatian economy, and especially those connected in any way with this area. As it is highly probable that this region will become a "raw material basis" due to further decay of wood processing capacities, unfortunately this is not a matter of wishes or possibilities, but rather a necessity requiring urgent action. To put it mildly, it is highly unreasonable to expect solutions from those who have won (and are still fighting for) their position in the global market.

5. References

Anon (1995). Croatian Standards for Forest Harvesting Products, Zagreb.

Anon (1997). EN 1310, 1997: Round and sawn timber – Method of measurement of features. CEN, Brussels, 1–22.

Anin (1999). EN 1316-1, 1999: Hardwood round timber – Qualitative classification, Part 1: Oak and beech. CEN, Brussels, 1–7.

Anon (2006). EN 1309-2, 2006: Round and sawn timber – Method of measurement of dimensions, Part 2: Round timber – Requirements for measurement and volume calculation rules. CEN, Brussels, 1–15.

Baylot, J. & Vautherin, P. (1992). *Classement des bois ronds feuillus*. Departement BOIS et SCIAGES, CTBA, Paris, 1–76.

Benić, R. (1987). Assortment, *Encyclopedia of Forestry*, III, pp. 244-246, Zagreb.

Bosshard, H. H. (1965). Aspects of the aging process in cambium and xylem. *Holzforschung,* 19: 65–69.

Bosshard, H. H. (1966). Mosaikfarbkernholz in *Fagus silvatica* L. Schweiz. Zeitsch. F. Forstwesen, 116: 1–11.

Glavaš, M. (1999). *Fungal Diseases of Forest Trees*, pp. 45-57, ISBN 953-6307-39-1, Faculty of Forestry, Zagreb University, Zagreb.

Glavaš, M. (2003).Red heart and beech rot, *Common beech in Croatia*, pp. 561-573, ISBN 953-98571-1-2, Academy of Forestry sciences, Zagreb.

Govorčin, S.; Sinković T.; Trajković, J. & Despot, R. (2003). Beech, *Common beech in Croatia*, pp. 652-669, ISBN 953-98571-1-2, Academy of Forestry sciences, Zagreb.

Horvat, I. (1944). Beech with red heart, *Journal of forestry* 68(5–6), pp. 100–104, ISSN 0373-1332, Croatian Forestry Society, Zagreb.

Kelvin, L. (1953). Stabler, *Mathematical thought Cambridge*, 118.

Krpan, A. P. B. & Šušnjar, M. (1999). Standardization of forest timber products in Croatia. *Journal of forestry* 123(5–6), pp. 241–245, ISSN 0373-1332, Croatian Forestry Society, Zagreb.

Krpan, A. P. B. & Prka, M. (2001). Quality of beech trees from regeneration fellings of Bilogora region. *Wood industry,* 52(4), pp. 173–180, ISSN 0012-6772, Faculty of Forestry, Zagreb University, Zagreb.

Krpan A. P. B. (2003). Beech Forest Products and Timber Technologies Harvesting from Beech Stands, *Common beech in Croatia*, pp. 625-640, ISBN 953-98571-1-2, Academy of Forestry sciences, Zagreb.

Krpan, A.; Prka, M. & Zečić, Ž., (2006). Phenomenon and characteristic of false heartwood in the beech thinnings and regenerative fellings in mangement unit »Bjelovarska Bilogora«, *Annales experimentis silvarum culturae provehendis* 5 (2006), pp. 529-542, ISSN 0352-3861, Faculty of forestry, University of Zagreb, Zagreb.

Krpan, A.P.B. & Prka, M. (2008). Defining assortment structure of even-aged beech stands according to standard HRN EN 1316-1:1999, Formec 2008, KWF, Gro-Umstad: KWF, 2008., 236-237.

Nečesany, V., 1958: The change of parenchymatic cells vitality and the physiological base for the formation of beech heart. Drev. Vyskum, 3: 15–16.

Nečesany, V., (1966). Die Vitalitatsveranderung der Parenchymzellen als physiologischer Grundlage der Kernbildung. Holzforschung u. Holzverwetung, 18(4): 61–65.

Nečesany, V., (1969). Forstliche Aspekte bei der Entstechung des Falschkerns der Rotbuche. Holz - Zbl., 95(37): 563.

Matić, S & Skenderović J. (1992). Forest tending, *Forest of Croatia*, pp. 81-95, Faculty of Forestry, Zagreb University & "Croatian Forest" ltd., Zagreb.

Matić, S. (2009). The relationship between nature-based forest management and life stages in the development of a virgin forest, *Virgin forest ecosystems of Dinaric karst and nature-based forest management in Croatia,* Croatian Academy of Sciences and Arts, pp. 9-19, ISBN 978-953-154-856-4, Zagreb.

Matić, S. & Anić, I. (2009). *Virgin forest ecosystems of Dinaric karst and nature-based forest management in Croatia,* Foreword, Croatian Academy of Sciences and Arts, pp. 5-6, ISBN 978-953-154-856-4, Zagreb.

Pranjić, A. & Lukić, N. (1997). *Forest measurements*, pp. 65-66, ISBN 953-6307-26-X, Faculty of forestry, University of Zagreb, Zagreb.

Prka, M. (2003a). Valuably Characteristic of Common Beech Trees with Regard to the Type of Feeling in Cutting Areas of Bjelovar Bilogora, *Journal of forestry* 127(1–2), pp. 35–44, ISSN 0373-1332, Croatian Forestry Society, Zagreb.

Prka, M. (2003b). Occurrence of false heartwood in beech trees and technical beech roundwood coming from thinning and preparatory felling in the area of Bjelovar Bilogora, *Journal of forestry* 127(9–10), pp. 467–474, ISSN 0373-1332, Croatian Forestry Society, Zagreb.

Prka, M., (2005). Quality characteristics of beech's trees and assortment structure from thinnings and regeneratory fellings in area of Bjelovarska Bilogora, pp. 1-171, PhD Thesis, Forestry Faculty of Zagreb University, Zagreb.

Prka, M. (2006a): Features of Assigned Beech Trees According to the Type of Felling in the Felling Areas of Bjelovar Bilogora and their Influence on the Assortment Structure, *Journal of forestry* 130(7–8), pp. 319–329, ISSN 0373-1332, Croatian Forestry Society, Zagreb.

Prka, M. (2006b). Height and Purity of Beech Tree Trunks According to the Type of Felling and the Percentage of Technical Roundwood in the Trunks and Tree Tops in Relation to Applied Standard, *Journal of forestry* 123(11–12), pp. 511–522, ISSN 0373-1332, Croatian Forestry Society, Zagreb.

Prka, M. & Krpan, A. (2007). The problem of establishing the assortment structure of even-aged beech stands. *Journal of forestry* 131(5–6), pp. 219–236, ISSN 0373-1332, Croatian Forestry Society, Zagreb.

Prka, M. (2008a). Defining Assortment Structure of Even-Aged Beech Stands According to Standard HRN EN 1316-1:1999, *Journal of forestry* 132(5–6), pp. 223–238, ISSN 0373-1332, Croatian Forestry Society, Zagreb.

Prka, M. (2008b). Forestry in the Area of the Bjelovar-Bilogora Country from its Beginnings until Today, Croatian Academy of Sciences and Arts, pp. 143-167, ISSN 1846-9787, Zagreb - Bjelovar.

Prka, M. & Poršinsky, T. (2009). Structure Comparison of Technical Roundwood in Even-Aged Beech Cutblocks by Assortment Tables with Application of Standards HRN (1995) and HRN EN 1316-1:1999, *Journal of forestry* 133(1–2), pp. 15–25, ISSN 0373-1332, Croatian Forestry Society, Zagreb.

Prka, M.; Zečić, Ž.; Krpan, A.P.B. & Vusić, D. (2009). Characteristic and Share of European Beech False Heartwood in Felling Sites of Central Croatia, *Croatian Journal of Forest Engineering*, 30(2009)1, pp. 37-49, ISSN 1845-5719, Zagreb.

Prka, M. (2010). *Beech Forest and Beech Roundwood of Bjelovar Area*, Croatian Forestry Society, pp. 1-252, ISBN 978-953-56470-0-3, Bjelovar.

Tomaševski, S. (1958). Share and distribution of false heartwood in beech trees in Ravna Gora, *Journal of forestry* 82(11–12), pp. 407–410, ISSN 0373-1332, Croatian Forestry Society, Zagreb.

Torelli, N. (1984). The Ecology of discoloured Wood as illustrated by Beech (*Fagus sylvatica* L.). *IAWA Bulletin* n.s., Vol. 5 (2), 121–127.

Torelli, N. (1994). Relationship between Tree Growth Characteristics, Wood Structure and Utilization of Beech (*Fagus sylvatica* L.). *Holzforschung und Holzverwertung,* 45. Jahrgang, Heft 6, 112–116.

12

Use of the Pilodyn for Assessing Wood Properties in Standing Trees of *Eucalyptus* Clones[1]

Wu Shijun[1], Xu Jianmin[1*], Li Guangyou[1], Risto Vuokko[2], Lu Zhaohua[1], Li Baoqi[1] and Wang Wei[1]
[1]Research Institute of Tropical Forestry, Chinese Academy of Forestry, Guangdong Guangzhou
[2]Guangxi StoraEnso Forestry Corporation Ltd., Nanning, Guangxi
China

1. Introduction

As one of the major timer production species in the world, eucalyptus is characterized as fast-growing, high-yielding, and well adapted to different flat and mountainous environments with extreme low temperature of -5°C ((Qi 2007)). Most of eucalypts species are naturally distributed in the continental Australia of Oceania, and a few native to the Timor Island of Indonesia and Papua New Guinea (Qi 2007).Identification and selection of superior trees in forest management and breeding programs provide a means to improve the properties and value of future wood products (Knowles et al. 2004). In recent years, breeding objectives in tree improvement have moved from volume per hectare alone, to include also wood properties and their impact on industrial end products (Wei and Borralho 1997). Wood basic density is considered as one of the most important wood properties which has a major impact on the freight costs, chipping properties, pulp yield per unit mass of wood and paper quality (Pliura et al. 2007; Laurence et al 1999). Wood basic density generally shows a high heritability and responds well to genetic improvement. But the genetics of wood density has not been studied in great detail (Macdonald et al. 1997). Currently, published information on genetic variation in wood basic density in eucalyptus is limited with a few studies conducted in China (Kien et al. 2008; Lu 2000; Luo 2003).

Measurement of wood density is expensive and time consuming and also create varying degrees of damage to experimental materials, and that has restricted the number and accuracy of the studies published (Hansen 2000; Wei et al. 1997). However, pilodyn sampling is faster, cheaper, and not destructive, thus resulting in overall higher expected gains for selection of

---

[1] This study belongs to the project of National Eleventh Five-Year Science and Technology" Breeding of High yield and High Quality Fast-growing Wood Species of Eucalyptus" (2006BAD01A15-4)
Author: Wu Shijun (1984--) male, Shandong Weifang, Under Post-graduate Student, wushijun0128@163.com
*Corresponding Author: Xu Jianmin Professor

trees or culling of seedling seed orchards in comparison with the more destructive direct assessment of density (Greaves et al. 1996). Kube and Raymond (2002) reported that core sampling for basic density is assumed to cost $10.5 per tree, which includes field collection and laboratory processing, whereas the cost of pilodyn measurements is assumed to be $1.5 per tree.

The primary objective of this study is to test the effectiveness of pilodyn for evaluating wood basic density, modulus of elasticity (MoE) and other traits of eucalyptus clones in standing trees. This information will be used to develop appropriate selection strategies for eucalyptus breeding programs in southern China.

2. Materials and methods

2.1 Trial description

The trial was established at Shankou town in Guangxi (21°34' N, 112°42 E, 29m asl.), and is affected by the north tropical monsoon with annual mean temperature of 23°C and annual mean rainfall of 1589mm. The lateritic red earth was derived from sandstone and contains 0.15% of organic matter (0-20cm). Previous vegetation was a plantation of Eucalyptus. Indigenous vegetation was found on site. 22 eucalyptus clones (table 1) were planted in April 2004. Field design was randomized complete blocks with 7 replications and 5-tree plot in a spacing of 4m × 2m. Measurements and increment cores were collected in December 2008, at which time the trial was aged 56 months.

Clone number	Clone Identity.	Parental Combination	Style of Seedling
1	GRDH32-26	*E.urophylla* ×*E.grandis*	Cuttings
2	W5	ABL 12×Unknown	Tissue culture
3	GRDH32-29	*E.urophylla* ×*E.grandis*	Cuttings
4	M1	*E.urophylla* ×*E.grandis*	Tissue culture
5	GRDH32-28	*E.urophylla*×*E.grandis*	Cuttings
6	SH1	Leizhou NO.1×Unknown	Tissue culture
7	GRDH33-9	*E.urophylla* ×*E.grandis*	Cuttings
8	U6	*E.urophylla* x *E.tereticornis*	Tissue culture
9	GRDH32-25	*E.urophylla*×*E.grandis*	Cuttings
10	DH32-29	*E.urophylla*×*E.grandis*	Tissue culture
11	GRDH42-6	*E.grandis* ×*E.urophylla*	Cuttings
12	RGD3	*E.urophylla*×*E.camaldulensis*	Tissue culture
13	DH196	*E.urophylla*×*E.grandis*	Cuttings
14	DH32-28	*E.urophylla*×*E.grandis*	Tissue culture
15	GRDH30-10	*E.urophylla*×*E.grandis*	Cuttings
16	TH9224	*E.urophylla*×*E.camaldulensis*	Tissue culture
17	GRDH33-27	*E.urophylla*×*E.grandis*	Cuttings
18	LH1	*E.urophylla*×*E.tereticornis*	Tissue culture
19	TH9224	*E.grandis* x *E.camaldulensis*	Cuttings
20	DH32-22	*E.urophylla*×*E.grandis*	Tissue culture
21	DH32-13	*E.urophylla*×*E.grandis*	Tissue culture
22	DH32-25	*E.urophylla*×*E.grandis*	Tissue culture

Note: Male parents of U6, W5 and SH1 were not clear.

Table 1. Details of clones in the analysis

2.2 Assessments of wood properties

2.2.1 Pilodyn penetration

The pilodyn wood tester is an instrument originally developed in Switzerland for determining the degree of soft rot in wooded telephone poles (Raymond et al. 1998; Hansen 2000). Pilodyn penetration (PP), an indirect method for determining wood basic density, has been effective in assessing large number of trees in eucalyptus (Wei et al. 1997; Kien et al. 2008; Macdonald et al. 1997; Raymond et al. 1998) and other species (Ishiguri et al. 2008; Pliura et al. 2006). PP was measured using a 6-J Forest pilodyn with 2.5mm steel needle, by over the bark and removing a small section of bark (approximately 40mm × 20mm) at 1.3m respectively and taking two pilodyn shots on each of four aspects (north, south west and east) from an average tree per plot. The pilodyn is attractive in that it is rapid, does not require the use of an increment borer (destructive sampling), and is, in principle, free of operator bias (Cown 1978; Hansen 2000). To avoid introducing additional sources of error, all clones were sampled by the same team of people, minimizing the potential for operator error (Raymond et al. 1998).

2.2.2 Modulus of elasticity

FAKOPP microsecond timer is able to measure acoustic velocity in standing trees, by timing the acoustic wave as it travels along the stem between points a known distance apart (Knowles et al. 2004; Chauhan et al. 2006). The results signals were engendered by start and stop transducers and recorded on an oscilloscope. Stress wave velocity (SWV) was then calculated by dividing the test span by the measurement stress wave transmission time (Wang et. al. 2000).

$$SWV = L / t \quad (1)$$

Where L=1500 mm is the distance between two probes, t is the transmission time in microseconds (μs).

The SWV is combined with density measurements to give an estimated of dynamic MoE (Knowles et. al., 2004).

$$MoE = \rho\omega^2 \quad (2)$$

Where MoE is the dynamic modulus of elasticity, ρ is the average green density of the stem, ω is the SWV.

2.2.3 Wood basic density

Wood basic density was defined as oven-dry wood mass per unit volume of green wood, and was measured using the water displacement method (Kube and Raymond 2002; Tappi 1989). Five mm increment cores from pith to bark were extracted at a height of 1.3 m in the south-north orientation from an average tree per plot, immediately stored in plastic tubes with both ends sealed (Kien et al. 2008).Wood basic density was determined using the water displacement method, with two weights for every sample: weight of water displaced by immersion of wedge (w_1) and oven dry weight (w_2) (Kien et al. 2008). Basic DEN was then calculated as:

$$\text{Basic DEN } (g \bullet cm\text{-}3) = w_2 / w_1 \quad (3)$$

Wood basic density, outer wood basic density and heartwood basic density were tested respectively.

2.3 Statistic analysis

The SAS software package was used to analyze the variance of different Pilodyn penetration and the relationship between the Pilodyn penetration and wood density or MOE, respectively.

The mean by ramet at each clone of sampling was submitted to a variance and a covariance analysis according to the following linear model (Hansen et. al. 1997):

$$y_{ij} = \mu + \alpha_i + \beta_i + \varepsilon_{ij} \quad (4)$$

where y_{ij} is the performance of ramet of ith clone within j^{th} block, μ is the general mean, α_i is the random effect of the ith clone, β_j is the random effect of the jth block, ε_{ij} is the random error.

3. Results and discussion

3.1 Comparison between Pilodyn penetration and wood properties

The mean values of Pilodyn penetration and wood properties of 22 clones are listed in Table 2. The mean value ranged from 9.44 to 15.41 mm for Pilodyn penetration, 0.3514 to 0.4913 $g.cm^{-3}$ for wood basic density and 3.94 to 7.53 GPa for MoE, which were smaller than previous studies on the same species (Knowles et al. 2004; Kien et al. 2008; Wei et al. 1997) as well as other species (Jacques 2004; Zhu et al. 2008; Zhu et al. 2009). The most suitable range of basic density for pulpwood in eucalyptus is 0.48 to 0.57 $g \bullet cm^{-3}$ and pulp yield decrease sharply when basic density falls below 0.4 exceeds 0.60 (Dean 1995; Ikemori et. al. 1986). There were considerably lower density values than those found in this study. Consequently, wood basic density should be improved substantially to about 0.55 $g \bullet cm^{-3}$, and this would benefit pulp production in southern China (Kien et al. 2008). Clones of M1, RGD3 and TH9224 had higher basic density and MoE, meanwhile, clones of DH32-28, GRDH42-6 and DH196 had lower basic density and MoE.

The variation coefficient of Pilodyn penetration over the bark was ranged from 9.15% to 11.83%, whereas those measured by removing the bark was ranged from 13.40% to 14.45% (Table 3). One possible explanation could be that bark thickness and branch cluster frequency could affect this value. This agreed well with previously published results by Wei (1997) and Yin (2008).

The analysis of variance of pilodyn is presented in Table 4. There were significant (1% level) differences between pilodyn penetration of different treatment, different directions and different clones, indicating that selection of clones for pilodyn would be effective.

The regression equations and phenotypic correlations between pilodyn penetration and wood properties are given in Table 5 and Table 6, respectively. Generally strongly negative correlations were found between pilodyn and wood properties, ranging from -0.433 to -0.755, slightly lower than previously published study (Wei et al. 1997; Chapola 1994). The possible explanation could be at least in part to the relatively small age of materials or less pilody penetration and pith taken from clones. The results indicated that PP was generally reliable as an indirect measure of wood basic density. The correlations between pilodyn and MoE were significantly and negative. However, the relationship between pilodyn and MoE does not seem to be documented. And further research is needed to clarify in further. The correlations between pilodyn and heartwood density were slightly positive to strongly positively, lower than the correlations between pilodyn and other wood properties because of the short length of steel needle.

Clone number	Mean value of PP (mm)	wood basic density ($g.cm^{-3}$)	MoE (GPa)
16	9.44	0.4395	6.48
4	10.03	0.4913	7.42
12	10.28	0.4638	7.53
2	10.66	0.4145	4.93
15	10.81	0.4384	6.14
19	11.03	0.4302	6.31
8	11.15	0.362	3.94
20	11.41	0.4236	5.76
6	11.44	0.4295	5.25
18	11.47	0.4371	5.85
21	11.47	0.4262	5.84
10	11.63	0.4627	6.33
1	12	0.4614	5.88
3	12.22	0.4237	5.52
14	12.69	0.3938	5.48
9	12.97	0.4106	5.32
22	13.09	0.4172	5.65
17	13.5	0.4266	5.92
7	14.03	0.4164	5.47
11	14.94	0.3924	4.45
13	15.34	0.3899	4.35
5	15.41	0.3514	4.29

Variance analysis of pilodyn

Table 2. The mean value of Pilodyn penetration and wood properties

Treatment	PP over the bark					PP with bark removal				
Index	East	West	South	North	Mean	East	Wes	South	North	Mean
Mean PP (mm)	14.50	14.99	14.59	14.62	14.67	12.04	12.33	12.15	12.02	12.14
C V (%)	10.42	11.83	9.15	10.94	10.41	14.40	14.45	13.40	14.06	13.57

Table 3. The mean value and variation coefficient of pilodyn penetration on four directions

Source	DF	F Value	Pr≥F
Treatment	1	16.47	< 0.0001
Directions	6	21.13	< 0.0001
Clones	21	8.10	< 0.0001

The correlations between pilodyn and wood properties

Table 4. Variance analysis of pilodyn

Directions	Wood properties	Regression equation	R^2	R
East	MoE	$y = 0.0341x^2 - 1.3743x + 18.331$	0.365	-0.604**
	Basic density	$y = 0.0003x^2 - 0.0189x + 0.6422$	0.281	-0.530*
	Outer wood density	$y = 0.0007x^2 - 0.0356x + 0.8051$	0.417	-0.646**
	Heartwood density	$y = -0.0003x^2 + 0.0008x + 0.4794$	0.188	-0.433*
West	MoE	$y = -0.0037x^2 - 0.1913x + 9.3565$	0.374	-0.611**
	Basic density	$y = -2E\text{-}05x^2 - 0.0103x + 0.581$	0.363	-0.603**
	Outer wood density	$y = 0.0009x^2 - 0.0409x + 0.8485$	0.482	-0.695**
	Heartwood density	$y = -0.0005x^2 + 0.0076x + 0.4308$	0.274	-0.523*
South	MoE	$y = 0.05x^2 - 1.9367x + 23.169$	0.424	-0.651**
	Basic density	$y = 0.0019x^2 - 0.0735x + 1.0769$	0.395	-0.629**
	Outer wood density	$y = 0.0031x^2 - 0.1139x + 1.4198$	0.521	-0.722**
	Heartwood density	$y = 0.0011x^2 - 0.0459x + 0.8543$	0.289	-0.538**
North	MoE	$y = -0.0105x^2 - 0.0191x + 8.1845$	0.357	-0.597**
	Basic density	$y = -0.0003x^2 - 0.0042x + 0.5399$	0.357	-0.598**
	Outer wood density	$y = 0.0014x^2 - 0.0579x + 0.9693$	0.464	-0.681**
	Heartwood density	$y = -0.0012x^2 + 0.0255x + 0.3069$	0.284	-0.533*
Mean value	MoE	$y = 0.0187x^2 - 0.9296x + 15.216$	0.389	-0.624**
	Basic density	$y = 0.0006x^2 - 0.0313x + 0.7468$	0.359	-0.599**
	Outer wood density	$y = 0.0022x^2 - 0.0823x + 1.1665$	0.493	-0.702**
	Heartwood density	$y = -0.0003x^2 - 7E\text{-}05x + 0.4975$	0.262	-0.511*

Table 5. Regression analysis of wood properties (y) to pilodyn penetration (x) over the bark on four directions

Directions	Wood properties	regression equation	R^2	R
East	MoE	$y = 0.0429x^2 - 1.3872x + 15.992$	0.431	-0.656**
	Basic density	$y = 9E\text{-}05x^2 - 0.0133x + 0.5688$	0.357	-0.598**
	Outer wood density	$y = 0.0006x^2 - 0.0304x + 0.7061$	0.529	-0.727**
	Heartwood density	$y = -0.0005x^2 + 0.0031x + 0.4548$	0.235	-0.484*
West	MoE	$y = 0.0082x^2 - 0.5428x + 11.063$	0.431	-0.656**
	Basic density	$y = 1E\text{-}05x^2 - 0.0117x + 0.5661$	0.414	-0.644**
	Outer wood density	$y = 0.0008x^2 - 0.0367x + 0.7538$	0.530	-0.728**
	Heartwood density	$y = -0.0005x^2 + 0.0033x + 0.4582$	0.313	-0.560**
South	MoE	$y = -0.0063x^2 - 0.171x + 8.6605$	0.365	-0.604**
	Basic density	$y = -0.0016x^2 + 0.0291x + 0.3081$	0.356	-0.596**
	Outer wood density	$y = 0.0002x^2 - 0.0204x + 0.6545$	0.528	-0.727**
	Heartwood density	$y = -0.0027x^2 + 0.0589x + 0.1055$	0.263	-0.513*
North	MoE	$y = 0.0054x^2 - 0.4438x + 10.18$	0.344	-0.587**
	Basic density	$y = -0.0007x^2 + 0.0061x + 0.453$	0.375	-0.613**
	Outer wood density	$y = 0.0012x^2 - 0.045x + 0.7988$	0.541	-0.736**
	Heartwood density	$y = -0.0018x^2 + 0.0353x + 0.2603$	0.282	-0.531*
Mean value	MoE	$y = 0.0079x^2 - 0.5492x + 11.117$	0.418	-0.646**
	Basic density	$y = -0.0008x^2 + 0.0084x + 0.4432$	0.404	-0.635**
	Outer wood density	$y = 0.0006x^2 - 0.0326x + 0.7318$	0.569	-0.755**
	Heartwood density	$y = -0.0018x^2 + 0.0341x + 0.2696$	0.295	-0.543**

Table 6. Regression analysis of wood properties (y) to pilodyn penetration(x) with bark removal on four directions

4. Conclusion

In the present study, the effectiveness of pilodyn for assessing wood properties of eucalyptus clones in standing trees was discussed. The results obtained are as follows:

1. The mean value of Pilodyn penetration, wood basic density and MoE ranged from 9.44 to 15.41 mm, 0.3514 to 0.4913 g.cm^{-3} and 3.94 to 7.53 GPa, respectively.
2. There were significant differences between pilodyn penetration of different treatment, different directions and different clones. The coefficient of variation ranged from 9.15% to 11.83% for Pilodyn penetration over the bark and ranged from 13.40% to 14.45% for Pilodyn penetration with bark removal.
3. The correlations between pilodyn and wood properties were generally strongly negative, and the coefficients ranged from -0.433 to -0.755. The results indicated that wood basic density and MoE can be predicted by using pilodyn. Results from this study also tend to confirm those of Cown (1981) who concluded that Pilodyn is not an accurate equipment for measurement, but it does provide an effective and efficient means of estimating wood properties.

5. Acknowledgements

We thank Huang Hongjian, Tan Peitao and Hu Yang from Xinhui Forest Bureau for their assistance. Comments from K. Harding, D. Pegg, Dr. Zeng Jie and an anonymous reviewer are appreciated.

6. References

Chapola Gbj. 1994. Assessment of some wood properties of eucalyptus species grown in Malawi using pilodyn method. Discovery and Innovation. 6(1):98-109

Chauhan S.S., Walker J.C.F. 2006. Variation in acoustic and density with age, and their interrelationships in radiation pine. Forest Ecology and Management. 229:388-394

Cown, D.J.. 1981. Use of the pilodyn wood tester for estimating wood density in standing trees – in fluce of site and tree age. World Forestry Conference.

Cown D J . 1978. Comparison of the Pilodyn and torsiometer methods for the rapid assessment of wood density in living trees. N ZJ For Sci ,:384 – 3911

Dean, G.H. 1995. Objectives for wood fibre quality and uniformity. In: pott, B.M., Borralho, N.M.G., Reid, J. B., Cromer, R. N., Tibbits, W.N. and Raymond, C.A. (eds). Eucalypts plantations: improving fibre yield and quality. CR-IUFRO Conf., Hobart, 19-24 Feb. 483 pp.

Ikemori Y. K., Martins F. C. G. and Zobel, B. J.. 1986 The impact of accelerated breeding on 31 wood Properties. In proceedings of the 18th IUFRO World 32 Conference Division 5: Forest products. Ljubljana, Yugoslavia. p.358-368

Ishiguri, F., Matsui, R., Lizuka, K., Yokota, S. and Yoshizawa, N. 2008. Prediction of the mechanical properties of lumber by stress-wave velocity and Pilodyn penetration of 36-year-old Japanese larch trees. Oiginal Arbelten · Originals. 66: 275–280

Greaves, B.L., Borralho, N.M.G, Raymond, C.A. and Farrington, A. 1996. Use of a pilodyn for the indirect selection of basic density in *Eucalyptus nitens*.Canadian Journal of Forest Research. 26(9):1643-1650

Hansen C. P. 2000. Application of the Pilodyn in forest tree improvement. DFSC Series of Technical Notes. TN55. Danida Forest Seed Centre. Humlebaek, Denmark.

Hansen J.K., Roulund H. 1996. Genetic parameters for spiral grain, stem form, pilodyn and growth in 13 years old clones of *Sitka Spruce* (Picea sitchensis (Bong.) Carr.). Silva Genetica. 46 (2-3):107-113

Jacques D., Marchal M. and Curnel y. 2004. Relative efficiency of alternative methods to evaluate wood stiffness in the frame of *hybrid larch* (Larix × *eurolepis* Henry) clonal selection. Ann. For. Sci. 61. pp:35-43

Knowles R. Leith, Hansen Lars W., Wedding Adele, Downes Geoffrey. 2004. Evaluation of non-destructive methods for assessing stiffness of Douglas fir trees. New Zealand Journal of Forestry Science. 34(1):87-101

Kube P., Raymond C. 2002. Selection strategies for genetic improvement of basic density in *Eucalyptus nitens*. Tcchnical repoet 92.

Knowles, L. R., Hansen, L.W., Wendding, A. and Downes, G.. 2004. Evaluation of non-destructive methods for assessing stiffness of douglas fir trees. New Zealand Journal of Forestry Science. 34(1)::87-101

Laurence, R Schimleck, Anthony J Michell, Carolyn A Raymond, Allie Muneri. 1999. Estimation of basic density of *Eucalyotus globulus* using near-infrared spectroscopy. Canadian Journal of Forest Research. Feb. 29:194-202

Lu, Z., Xu, J., Bai J. and Zhou,W.. 2000. A study on wood property variation between Eucalyptus *tereticornis* and Eucalyptus *camalduensis*. Forest Research. 13(4):370-376

Luo, J. 2003 Variation in growth and wood density of *Eucalyptus urophylla*. In turnbull, J. W.(Ed.) "*Eucalypts in Asia" ACIAR proceedingsNo. 111*. Zhanjiang, Guangdong, China.

Macdonald A.C., Borralho N.M.G.and Potts B.M. 1997. Genetic variation for growth and wood density in *eucalyptus globulus ssp.globulus* in Tasmania (Australia). Silva Genetica. 46 (4):236-241

Kien N. D., Jansson G., Harwood C., Almqvist C., Thinh H. H. 2008. Genetic variation in wood basic density and pilodyn penetration and their relationships with growth, stem straightness and branch size for *eucalyptus urophylla* S.T.Blake in Northern Vietnam. New Zealand Journal of Forestry Science. 38 (1):160-175

Pliura A., Zhang S. Y., Mackay J. and Bousquet. 2007. Genotipic variation in wood density and growth traits of poplar hybrids at four clonal trials. Forest Ecology and Management. 2007 238: 92-106

Qi S. 2007. Applied Eucalypt cultivation in China. Beijing: China Forestry Publishing House

Raymond, C.A. and MacDonald, A.C. 1998. Where to shoot your pilodyn: within tree variation in basic density in plantation *eucalyptus globulus* and *E. nitens* in Tasmania. New Forests. 15: 205–221.

Tappi.1989. Basic density and moisture content of pulpwood. TAPPI no. T258 om-98

Wang X., Ross R. J., Mcclellan M., Barbour R. J., Erickson J. R., Forsman J.W. and Mcgin G. 36 D. 2000. Strength and stiffness assessment of standing trees using a 37 nondestructive stress wave technique. Res. Pap. FPL-RP-585. U.S. Department of 38 Agriculture, Forest Service, Forest Products Laboratory, Madison, WI.

Wei X., N.M.G.Borralho. 1997. Genetic control of basic density and bark thickness and their relationships with growth traits of *Eucalyptus urophylla* in south east China. Silvae Genetica. 46(4):245-250

Yin Y., Wang L., Jiang X.. 2008. Use of Pilodyn tester for estimating basic density in standing trees of hardwood plantation. Journal of Beijing Forestry University 30 (4): 7-11

Zhu J., Wang J., Zhang S., Zhang J., Sun X., Liang B.. 2008. Wood property estimation and selection of *populus tomentosa*. Scientia Silvae Science. 44(7): 23-28

Zhu J., Wang J., Zhang S., Zhang J., Sun X., Liang B., Zhao K. 2009. Using the pilodyn to assess wood traits of standing trees *Laix kaempfri*. Forest Research. 22(1): 75-79

13

An Overview on Spruce Forests in China

Zou Chunjing[1], Xu Wenduo[2], Hideyuki Shimizu[3] and Wang Kaiyun[1]
[1]Shanghai Institute of Urban Ecology and Sustainability
School of Life Science, East China Normal University, Shanghai
[2]Institute of Applied Ecology, Chinese Academy of Sciences, Shenyang
[3]National Institute for Environmental Studies, 16-2 Onogawa, Tsukuba, Ibaraki
[1,2]P. R. China
[3]Japan

1. Introduction

The genus *Picea* A. Dietrich (spruce), which is a relative isolated group under evolution, belongs to Pinaceae family (Ran et al., 2006; Bobrow, 1970; Buchholz, 1929, 1931; Alvin, 1980; Mikkola, 1969). It includes 28–56 species depending on different systems of classification (Farjón, 1990; Ledig et al., 2004), most of which are in Eastern Asia, while many researchers thought that there were about 40 species in *Picea* genus and were only found in the north hemisphere (Budantsey, 1992, 1994; Wolfe, 1975, 1978; Tiffney and Manchester, 2001). The distribution range is from 21°N (Huanglian Mountains of Vietnam) to 70°N (Far Eastern area of Russia) (Fig. 1). Spruce forests are the main dominant vegetation in alpine coniferous forest in subtropical zone and temperate zone, and they are only found in alpine area, subalpine area and plateau from 21°N to 46°N (Li, 1995). In cold temperate zone and its adjacent regions (47°N to 57°N), spruce forests are the zonal vegetation types in boreal coniferous forest. From 57°N to 70°N, spruce forests transform from horizontal (latitudinal) zonal distribution to vertical (altitudinal) zonal distribution and from continuous distribution to discontinuous distribution.

In the north of Euro-Asia continent, the main spruce species are *Picea abies* (L.) H. Karst. and *P. obovata* Ledeb., which form the continuous boreal coniferous forest (Ferguson, 1967; Florin, 1954, 1963; Guerli et al., 2001). *P. abies* is found in Alps of France, the Balkan Peninsula or the Balkan Mountains in the west, Germany and Scandinavian Peninsula in the north, Poland and the north and middle region of Russia in the east. In Siberian area of Russia, *P. obovata* takes the place of *P. abies*, it is found until to Lena River Valley and Okhotsk. But in the east Siberian area, *P. obovata* retreats from the dominant position, and is taken the place by *Larix sibirica* Ledeb. due to the rigorous continental climate (Colleau, 1968; Corrigan et al., 1978; Harris, 1979; Hart, 1987).

In North America, spruce species are abundant, including *P. glauca* (Moench) Voss, *P. mariana* (Mill.) Britton and al., *P. engelmannnii* Parry ex Engelm. and *P. sitchensis* (Bong.) Carriére (Barbour and Bilings, 1988; Klaus, 1987). *P. glauca* is distributed extensively in Canada and North USA, from Labradorian Peninsula and Alaska to Montana, North Dakota, Minnesota, Wisconsin, Michigan, to Massachusetts near Atlantic coast. *P. mariana* is distributed almost in the whole Canada, and extensively in the eastern provinces and Newfoundland, to Alaska across Rocky Mountains in the west, and to Pennsylvania, north Virginia, Wisconsin, and Michigan in the south. Britain Colombia Province of Canada is the west border of *P. mariana*. *P. engelmannnii* is found in the west of North America, from Alberta Province and Britain Colombia Province of Canada to Arizona and New Mexico of USA along Rocky Mountains, it is also distributed in Cascade Range in Washington and Oregon. *P. sitchensis* distributes in the northwest of North America, and can be found from Aleutian Islands to Pacific coast of the northwest of California too (Delevoryas and Hope, 1973; Hsu, 1983; Weng and Jackson, 2000).

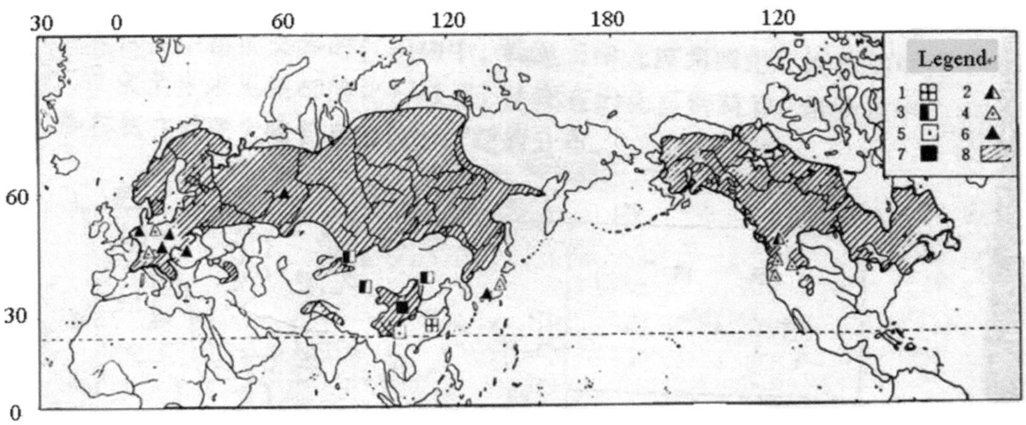

Fig. 1. The modern distribution range and fossil localities of *Picea* spp. in the world (based on Li, 1995, Lű et al., 2004, and McKenna, 1975) (1-7. Fossil localities: 1. Eocene; 2, 3. Oligocene; 4, 5. Miocene; 6, 7. Pliocene; 8. Modern distribution)

In China, the distribution range of spruce forests is very large, from Daxinganling Mountains (north) to Gaoligong Mountains (south), and Tianshan Mountains (west) to Central Mountains of Taiwan Province (east) (Fang, 1995, 1996; Fang and Liu, 1998). The spruce forests are found as long as there are site conditions of cold-temperate moisture types. In China the spruce forests belong to vertical zonal distribution with 17 species and 8 variations of *Picea* genus and take more than 40% of the species in the world. Furthermore, the almost all of the species are endemic in China, except for those in Daxinganling Mountains which belongs to East Siberian area and Arertai Mountains (belonging to West Siberian area). In China, spruce forests are distributed in Northeast, North, Northwest and Southwest.

In the mountains of Daxinganling, Xiaoxinganling and Changbai of Northeast China, *P. koraiensis* Nakai, *P. jezoensis* var. *microsperma* (Lindl. Cheng et L. K. Fu) and *P. jezoensis* var. *komarovii* (V. Vassil.) Cheng et L. K. Fu are the edificators of upland dark coniferous forests, which are extended partition of dark coniferous forests of Far East Area of Russia (Editorial Committee of Forest of China, 1997; Li, 1980; Li and Zhou, 1979).

The distribution range of spruce forest is restricted for drought in North and Northwest China. *P. meyeri* Rehder, E. H. Wilson and *P. wilsonii* Mast. are found in Jibei Mountains, Xiaowutai Mountains of Hebei Province, Guanqin Mountains, Wutai Mountains, Guandi

Mountains of Shanxi Province in North China. In Northwest China, Arertai Mountains are the south border of *P. obovata*. *P. schrenkiana* var. *tianshanica* (Rupr.) Cheng et S. H. Fu is found in Tianshan Mountains, the west border of spruce forests in China. *P. crassifolia* Kom. Is distributed extensively in Qilian Mountains, Helan Mountains and Yinshan Mountains of Qinghai Province, Gansu Province and Ningxia Hui Nation Autonomous Region (Editorial Committee of Forest of China, 1997).

In Southwest China, including West Sichuan Province, north Yunnan Province and south Tibet Autonomous Region, there are 17 spruce species, which takes 43.3% of spruce species in the world. The important species are *P. likiangensis* (Franch.) E. Pritz., *P. likiangensis* var. *linzhiiensis* Cheng et L. K. Fu, *P. likiangensis* var. *balfouriana* (Rehd. et Wils.) Hillier ex Slavin, *P. purpurea* Mast., *P. brachytyla* (Franch.) E. Pritz., and so on, they form subalpine dark coniferous forests in Southwest China (Kuan, 1981; Sun, 2002; Wu et al., 1995).

P. spinulosa (Griff.) A. Henry and *P. smithiana* (Wall.) Boiss. are found in moist area in Himalayas in south Tibet Autonomous Region, and they always form small pure forest or mixed forest (Kuan, 1981).

P. morrisonicola Hayata forms the dominant pure coniferous forest in Central Mountains in Taiwan, which is the only subalpine coniferous forest of the east China in subtropical zone (Liu, 1971).

Monophyly of *Picea* has never been debated (Wright, 1955; Prager et al., 1976; Frankis, 1988; Price, 1989; Sigurgeirsson and Szmidt, 1993), but infrageneric classification of the genus remains quite controversial (Liu, 1982; Schmidt, 1989; Farjón, 1990, 2001; Fu et al., 1999), owing to morphological convergence and parallelism (Wright, 1955), and high interspecific crossability (Ogilvie and von RudloV, 1968; Manley, 1972; Gorden, 1976; Fowler, 1983, 1987; Perron et al., 2000). In addition, little is known about phylogenetic relationships of most species, especially the geographically restricted species growing in the montane regions of southwest China (LePage, 2001). Moreover, the origin and biogeography of *Picea* have drawn great interest from both geologists and biologists (Wright, 1955; Aldén, 1987; Page and Hollands, 1987; LePage, 2001, 2003), but they are still far from being resolved.

Spruce species are fine trees for lumbering, so researches on spruce were conducted very early in China (Editorial Committee of Vegetation of China, 1980). However, basic characteristics, flora, distribution types, and evolution relationship of the spruce species in China, and the relationship among spuce in China and abroad need more concern. There are many data about the topics above, but they are always scattered.

The aim of this study was (1) to summarize systematically the researches on spruce in China, and (2) to try to clarify the relationship among Chinese spruces, and among spruce in China and abroad.

2. Characteristics of species composition in spruce forests in China

2.1 The edificators in spruce forests in China

Picea spp. is distributed extensively in China (Editorial Committee of Forest of China, 1997). It is difficult to expatiate on the characteristics of edificators in spruce forests, so we divided China into five parts according to their districts, including Northeast China, Northwest

District	Species	F	D	RD	P	RP
Northeast China	*P. koraiensis*	96.00	11.52	47.50	3.482	48.60
	P. jezoensis var. *microsperma*	82.50	7.65	31.80	2.239	28.50
	Pinus korainensis Sieb.	47.50	1.56	4.29	0.626	5.21
	Populus davidiana Dode.	28.00	2.62	3.24	0.248	2.78
	Quercus mongolica Fischer ex Ledebour	16.50	2.13	1.87	0.104	1.67
	Betula platyphylla Suk.	32.50	1.19	2.61	0.182	2.36
	Betula ermanii Cham.	10.50	0.87	0.99	0.167	2.13
	Betula costata Trautv.	11.90	1.23	1.32	0.098	1.98
	Abies nephrolepis (Trautv.) Maxim.	31.75	2.95	2.45	0.204	3.11
	Larix gmelini (Rupr.) Rupr.	15.38	1.85	1.98	0.128	2.52
	Pinus sylvestriformis Taken.	9.82	0.62	0.23	0.075	0.85
Northwest China	*P. schrenkiana* Fisch. et Mey.	88.50	11.00	35.65	4.457	42.80
	P. schrenkiana var. *tianshanica*	75.00	9.88	29.54	4.023	32.60
	P. obovata	62.50	5.00	22.26	3.285	15.93
	Larix sibirica	28.00	1.61	3.28	0.241	2.75
	Betula pendula Roth.	15.50	0.95	1.06	0.113	0.82
	Sorbus tianschanica Mast.	9.20	0.43	0.95	0.108	0.36
	Populus talassica	11.42	0.87	0.62	0.168	1.55
	Betula tianschanica Cheng et S. H. Fu	8.75	0.65	0.53	0.201	1.10
North China	*P. meyeri*	84.00	12.25	24.21	4.114	29.61
	P. wilsonii	81.30	11.64	22.80	4.109	27.52
	P. asperata Mast.	67.50	9.58	15.21	2.628	14.20
	P. crassifolia	48.00	6.62	13.45	2.124	12.74
	Abies ernestii Rehd.	26.70	3.15	5.82	1.100	3.69
	Larix principis-rupprechtii Maryr.	22.58	2.69	4.63	0.482	2.37
	Pinus tabulaeformis Carr.	30.50	2.89	4.97	0.368	2.15
	Acer davidii Franch.	11.20	1.25	1.35	0.191	1.26
	Populus ningshanica L.	14.70	1.96	1.45	0.207	1.18
	Betula platyphylla	11.36	1.82	0.95	0.158	1.51
	Betula albo-sinensis Burk.	6.88	1.12	0.73	0.079	0.08
	Quercus liaotungensis Koidz	2.35	0.53	0.82	0.045	0.09
Southwest China	*P. likiangensis*	65.40	12.27	17.23	3.885	17.62
	P. likiangensis var. *balfouriana*	60.37	12.62	15.81	4.001	14.56
	P. purpurea	61.50	11.57	16.20	3.629	14.28
	P. asperata	45.22	5.68	9.45	2.156	9.76
	P. brachytyla	66.75	10.13	15.87	4.107	18.19
	P. brachytyla var. *complanata* Mast.	62.53	9.69	14.64	3.982	15.33
	Abies faxoniana Rehd. et Wils.	25.50	2.85	2.96	1.003	1.17
	Abies spectabilis (D. Don) Mirb.	21.23	4.23	2.35	0.691	2.21
	Pinus griffithii McClelland	17.79	3.91	1.46	0.268	1.28
	Pinus tabulaeformis	11.33	2.86	1.09	0.551	1.54
	Pinus armandii Franch.	7.83	2.15	0.78	0.079	0.08
	Populus davidiana	9.33	3.56	0.89	0.104	0.11
	Acer flabellatum Rehd.	6.74	3.54	0.34	0.073	0.08
	Quercus semicarpifolia Smith	8.12	2.31	0.28	0.097	0.08
	Betula platyphylla	9.92	1.75	0.34	0.162	0.09
	Betula utilis var. *prattii* D. Don	2.47	1.10	0.11	0.052	0.06
	Juglans cathayensis Dode	2.24	0.59	0.05	0.038	0.04

Taiwan	*P. morrisonicola*	100.00	14.87	77.77	5.451	84.72
	Tsuga chinensis (Franch.) Pritz.	62.50	1.87	9.80	0.233	3.62
	Pinus armandii var. *mastersiana* (Hay.) Hay.	37.50	1.00	5.22	0.256	3.97
	Chamaecyparis obtuse var. *formosana* Matsum.	25.00	0.62	3.26	0.241	3.74
	Chamaecyparis formosensis Matsum.	12.50	0.12	0.65	0.002	0.02
	Cunninghamia konishii Hayata	12.50	0.12	0.65	0.085	1.31
	Trochodendron aralioides Sieb. et Zucc.	12.50	0.50	0.61	0.166	2.57

Table 1. Species characters in spruce forest in different districts in China (F-Frequency (%), D-Density (/100m^2), RD-Relative density (%), P-Predominance, RP-Relative predominance (%))

China, North China, Southwest China and Taiwan. The characteristics of spruce forest in different districts in China are as shown in Table 1. In Northeast China, Northwest China and Taiwan, there are few edificators in spruce forests, while in North China and Southwest China, many species of spruce forests are found (Editorial Committee of Forest of China, 1997; Editorial Committee of Vegetation of China, 1980; Zhou, 1988; Chou, 1986, 1991).

2.2 Flora characters of spruce forests in China

In different districts in China, the floristic and geographical elements of spruce forests are complex (Table 2) (Wu, 1991; Wang, 1992, 2000). Generally speaking, there are more species belong to temperate zone element (Northeast China (83.39%), Northwest China (81.25%), North China (77.43%), Southwest China (72.50%), and Taiwan (70.66%)). In tropical China, spruce forest takes the following proportions: in North China (16.38%), Southwest China (22.92%), and Taiwan (24.50%). However, in temperate zone, spruce distribution in the three districts are as follows: North China (3.45%), Southwest China (2.08%), and Taiwan (2.28%)) are relatively less than the other two districts (Northeast China (8.15%) and Northwest China (10.41%)). China endemic elements in the three districts (North China (23.49%), Southwest China (37.92%), and Taiwan (32.19%)) are distinctly more than the other two districts (Northeast China (10.66%) and Northwest China (7.99%)), due to these two districts are connected with other districts, such as northeastern Asia, Siberian, and Far East of Russia (Editorial Committee of Forest of China, 1997).

3. Section grouping based on cytogenetical studies

3.1 Karyotype of 17 *Picea* species in China

Karyotype is the basis of cladistics. We collected all the pictures on chromosome of different *Picea* species (Sudo, 1968; Hizume, 1988; Taylor and Patterson, 1980; von RudloV, 1967; Wang et al., 2000; Wu, 1985, 1987; Xu et al., 1994, 1998; Mehra, 1968). The pictures were treated by using the software Motic Images Advanced 3.0 to get the length of arms of chromosome. Researchers have found karyotype characters of 17 *Picea* species in China up to now (Table 3). Karyotype equations of these species include four types: 2n=24m, 2n=22m+2sm, 2n=20m+4sm, and 2n=16m+8sm. B chromosome is found only in *P. meyeri, P. wilsonii, P. jezoensis* var. *microsperma,* and *P. obovata.* There is no variation of chromosome number.

Floristic types	Geographical elements	Northeast China	Northwest China	North China	Southwest China	Taiwan
I	(1)	15 (4.70)	15 (5.21)	22 (4.74)	18 (2.50)	9 (2.56)
II	(2)	21 (6.58)	13 (4.51)	12 (2.59)	14 (1.94)	8 (2.28)
	(3)	5 (1.57)	17 (5.90)	4 (0.86)	1 (0.14)	0 (0.00)
III	(4)	45 (14.11)	28 (9.72)	19 (4.09)	21 (2.92)	3 (0.85)
	(5)	46 (14.42)	36 (12.50)	54 (11.64)	43 (5.97)	32 (9.12)
	(6)	27 (8.46)	17 (5.90)	53 (11.42)	58 (8.06)	36 (10.26)
	(7)	42 (13.17)	12 (4.17)	32 (6.90)	49 (6.81)	18 (5.13)
	(8)	18 (5.64)	12 (4.17)	27 (5.82)	28 (3.89)	25 (7.12)
	(9)	34 (10.66)	23 (7.99)	109 (23.49)	273 (37.92)	113 (32.19)
	(10)	13 (4.08)	25 (8.68)	18 (3.88)	29 (4.03)	18 (5.13)
	(11)	9 (2.82)	25 (8.68)	11 (2.37)	8 (1.11)	0 (0.00)
	(12)	17 (5.33)	22 (7.64)	14 (3.02)	5 (0.69)	0 (0.00)
	(13)	15 (4.70)	34 (11.81)	13 (2.80)	8 (1.11)	3 (0.85)
IV	(14)	12 (3.76)	9 (3.13)	76 (16.38)	165 (22.92)	86 (24.50)
Total 5	14	319 (100.00)	288 (100.00)	464 (100.00)	720 (100.00)	351 (100.00)

Table 2. The floristic geographical elements of spruce forests in different districts in China (I World element ((1) World element), II Cold zone element ((2) North temperate zone-arctic element, (3) Siberia element), III Temperate zone element ((4) North temperate zone element, (5) Ancient world temperate zone element, (6) Temperate zone-Asia element, (7) East Asia element, (8) Sino-Japan element, (9) China endemic element, (10) Middle Asia element, (11) Aertai-Mongolia-Dahuri element, (12) Dahuri-Mongolia element, (13) Mongolia steppe element), IV Tropical zone element ((14) North temperate-tropical zone element))

No.	Species	Karyotype equation	Arm ratio	Chromosome length ratio	Karyotype type
1	*P. asperata*	20m+4sm	1.31±0.04	1.71±0.32	1A
2	*P. retroflexa* Mast.	22m+2sm	1.24±0.14	1.89±0.56	2A
3	*P. koraiensis*	20m+4sm	1.42±0.23	1.72±0.12	2A
4	*P. meyeri*	22m+2sm+2B	1.36±0.15	1.77±0.29	2A
5	*P. wilsonii*	20m+4sm+1B	1.27±0.28	1.89±0.31	2A
6	*P. schrenkiana*	20m+4sm	1.38±0.29	1.83±0.42	2A
7	*P. schrenkiana* var. *tianshanica*	16m+8sm	1.42±0.18	2.12±0.18	2B
8	*P. smithiana*	20m+4sm	1.31±0.47	1.85±0.21	1A
9	*P. morrisonicola*	16m+8sm	1.50±0.33	1.87±0.34	2A
10	*P. likiangensis*	20m+4sm	1.27±0.18	1.60±0.19	2A
11	*P. likiangensis* var. *balfouriana*	20m+4sm	1.34±0.26	1.77±0.45	2A
12	*P. purpurea*	20m+4sm	1.28±0.16	1.86±0.16	2A
13	*P. jezoensis* var. *microsperma*	22m+2sm+1B	1.35±0.28	1.82±0.53	2A
14	*P. brachytyla*	20m+4sm	1.34±0.32	1.84±0.37	2A
15	*P. brachytyla* var. *complanata*	20m+4sm	1.36±0.17	1.99±0.42	2A
16	*P. mongolica*	20m+4sm	1.33±0.18	1.83±0.15	1A
17	*P. obovata*	24m+3B	1.35±0.27	1.87±0.28	2A

Table 3. Karyotype characters of 17 spruce species in China

For karyotype type, there are 3 1A types (including *P. asperata, P. smithiana* and *P. mongolica* W. D. Xu) and 1 2B type (*P. schrenkiana* var. *tianshanica*). The others are 2A types.

3.2 Structure variation of chromosomes and evolution hierarchy

We took arm ratio as x-coordinate, and chromosome length ratio as y-coordinate. All *Picea* spp. were drawn as shown in (Fig. 2a, b). The change range of arm ratio is from 1.23 to 1.50, and most of species (22) are from 1.25 to 1.35. The change range of chromosome length ratio is from 1.60 to 2.12, and only 14 species are from 1.75 to 1.85 (Wang et al., 1990).

Structure variation of chromosomes of Chinese *Picea* spp. (Fig. 2a) is more obvious than *Picea* spp. found in other parts of the world (Fig. 2b).

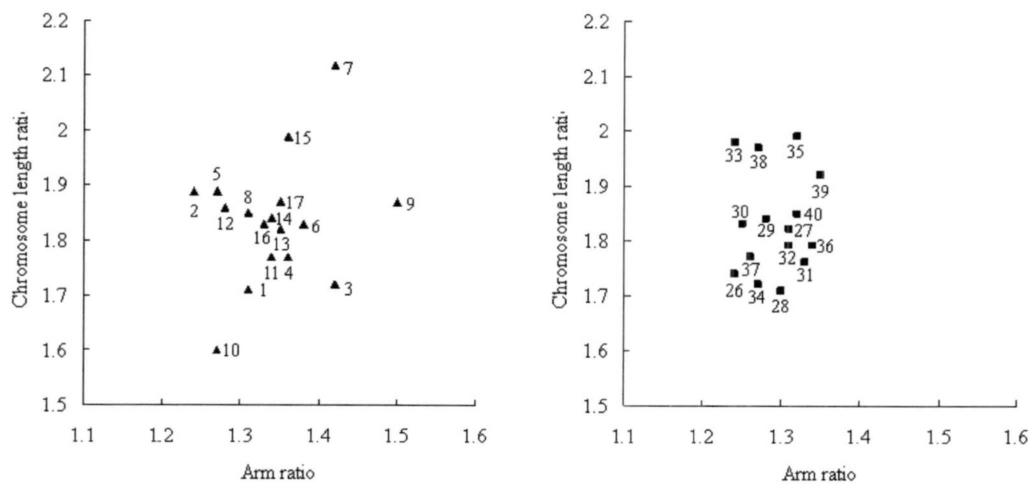

a. Chinese *Picea* spp. b. world *Picea* spp.

Fig. 2. Chromosomes structure of Chinese *Picea* spp. and other world *Picea* spp. (data based on Table 5)

Some researchers (Wang et al., 1990) thought a coefficient *k* (Karyotypic asymmetry in both average arm ratio and ratio of longest / shortest of chromosomes) was a good index for expressing the evolution hierarchy of certain species and genus.

$$k = \frac{A_i + L_i}{A_{\max} + L_{\max}} \times 100\% \quad (1)$$

Where A_i – average arm ratio of species (or genus), L_i – chromosome length ratio of species (or genus), A_{max} – maximum arm ratio in genus (or family), L_{max} – maximum chromosome length ratio in genus (or family).

According to value of k of *Picea* spp., the evolution hierarchy of 17 *Picea* spp. in China (Fig. 3) and 15 *Picea* spp. abroad (Fig. 4) were determined.

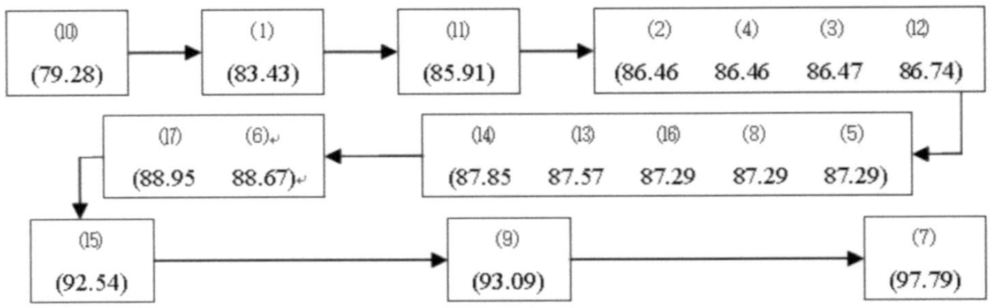

Fig. 3. Evolution hierarchy of 17 *Picea* spp. in China

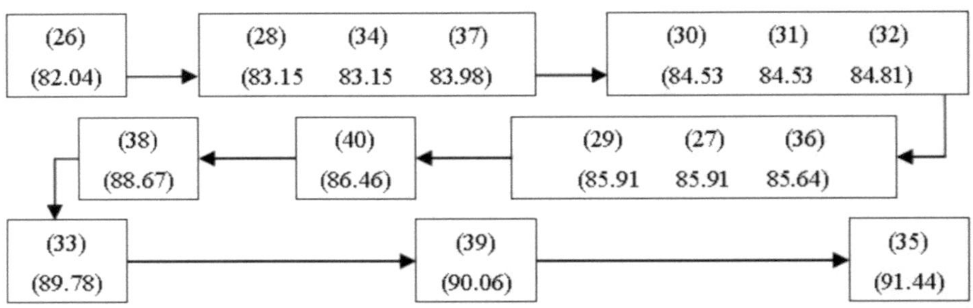

Fig. 4. Evolution hierarchy of 15 *Picea* spp. abroad

3.3 Section grouping

In taxonomy, *Picea* genus in China can be divided into three sections according to their karyotypes and the coefficient *k*. These sections are Sect. *Casicta*, Sect. *Omirica*, and Sect. *Picea*. Furthermore, we can determine their evolution hierarchy as in (Fig. 5) (Ran et al., 2006; Wu, 1991).

4. Distribution types of *Picea* spp. in China

4.1 Distribution range and niches of *Picea* spp. in China

The data of some *Picea* species (including *P. koraiensis, P. jezoensis* var. *microsperma, P. jezoensis* var. *komarovii,* and *P. mongolica*) are based on our previous field investigation. And we conducted the interpretation of TM image of some pivotal regions (including Tianshan Mountains in Xinjiang Weiwuer Autonomous Region, Hengduan Mountians in Sichuan Province and Tibet Autonomous Region, Qilian Mountains in Shaanxi Province and Gansu Province) (Liu et al., 2002; Yang et al., 1994; Editorial Committee of Forest of China, 1997; Cen, 1996).

4.2 Grouping of distribution types

Principal Components Analysis (PCA) was performed to compress the autocorrelated metric environmental variables by creating a reduced number of compounds (principal components) that explain the observed variation of distribution type (Jolliffe, 2002; Norusšis,

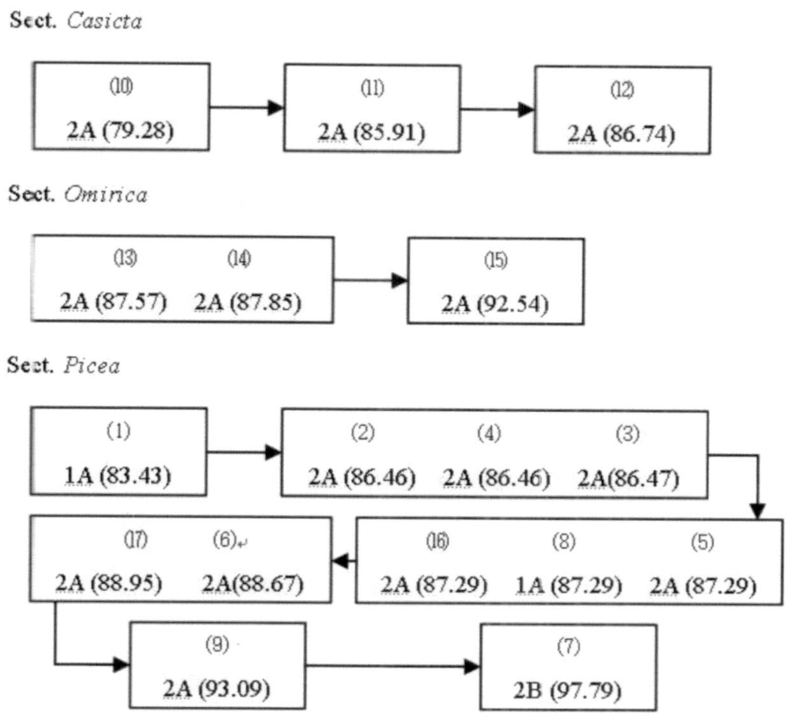

Fig. 5. Section grouping and evolution hierarchy of 17 *Picea* spp. in China

1990). Only compounds that accounted for more variation than any individual variable (eigenvalue > 1) were used in the final model. A 'varimax' rotation was applied to the reserved components to redistribute the variance among factors to obtain factor scores. Fuzzy clustering was then applied to the sample scores from the PCA ordination to identify the main distribution types. The fuzzy clustering specification used 3–6 clusters, a fixed fuzziness criterion of 2 and a convergence coefficient of 0.001. Then we obtained three categories of distribution types.

The first category is based on species adaptability to climate (mainly temperature, precipitation, and moisture). There are three types, including cold-moist type (10, 11, 12, 14, 15, 20, 22, 23, 25), cold-drought type (3, 6, 7, 13, 16, 17, 18, 19, 21, 24) and warm-moist type (1, 2, 4, 5, 8, 9).

The second category is based on environmental factors (particularly altitude). There are four types, including upland type (1, 2, 4, 5, 6, 7, 8, 9, 10, 11, 12, 14, 15, 18, 19, 21, 22, 23, 25), valley type (3, 13, 20), plain type (17, 24), and sandy land type (16).

The third category is based on distribution range (longitude and latitude) of species. There are three types, including narrow-distribution type (8, 9, 16, 17, 18, 22, 23, 25), medium-distribution type (1, 2, 4, 10, 11, 12, 14, 15, 21, 24) and broad-distribution type (3, 5, 6, 7, 13, 19, 20).

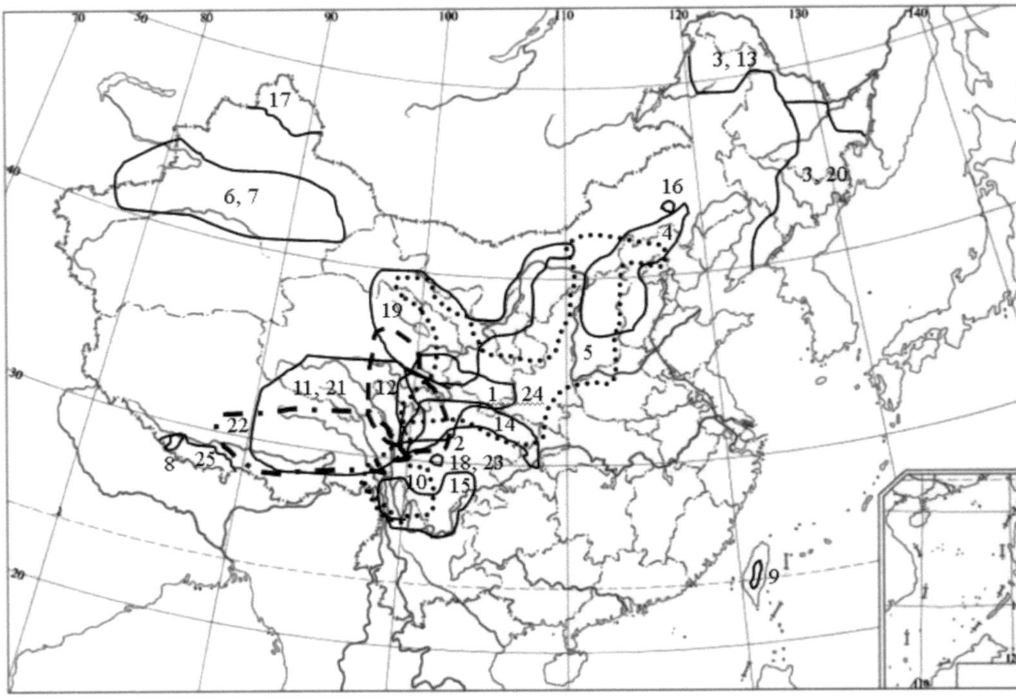

Fig. 6. The main distribution range of 25 *Picea* species in China

5. Discussion

5.1 The origin of *Picea* genus

Severe climatic oscillations associated with glacial cycles in the arctic during the late Tertiary and throughout the Quaternary era resulted in great changes in species distribution and population structure (Böhle et al., 1996; Qian and Ricklefs, 2000; Liu et al., 2002; Petit et al., 2003; Hewitt, 2004; Thomas, 1965). Meanwhile, descendent sea level created land connections for intercontinental exchanges of flora and fauna, especially boreal species (Tiffney, 1985a, b; Wen, 1999; Xiang et al., 2005). Spruce, as a kind of gymnosperm, is an archaic group under evolution, although pioneer reliable fossils of *Picea* genus are not available so early in Oligocene (Miller, 1975, 1977). Later in Oligocene and Miocene, fossils of *Picea* genus appear widely in Europe, North America and Japan (Page, 1988; Axelord, 1986, 1976; Ferguson, 1967). According to the fossils and modern distribution range, it can be concluded that ancestor of *Picea* genus might be a branch differentiate from Pinaceae during evolvement metaphase. But until Tertiary, ancient *Picea* spp. became the same as modern *Picea* spp.

Where does *Picea* genus originate from? There are many hypotheses in botanical science. Wright (1955) thought *Picea* genus might originate from northeastern Asia, and moved to North Arctic or diffused towards south along mountains. It seems logical because there are many *Picea* spp. there, including *P. jezoensis* var. *hondoensis* Mayr., *P. polita* Sieb. et Zucc., *P. jezoensis* Carr., *P. jezoensis* var. *microsperma*, *P. jezoensis* var. *komarovii*, *P. korainensis*. However, the hypothesis can not give a reasonable explanation to the phenomenon that there are many *Picea* spp. in lower latitudinal regions in eastern Asia (Miller, 1988; Hopkins et al., 1994).

Wu (1991) thought the distribution center of *Picea* spp. was in East Asia, particularly in Hengduan Mountains according to his research findings. Li (1995) reported that in Hengduan Mountains, *Picea* spp. belonged to almost all of the subgroups except for an obvious evolutional subgroup – *P. pungens* subgroup originated from Rocky Range. In Hengduan Mountains in western Sichuan (Sun, 2002), northern Yunnan and eastern Tibet, there are more interspecific differentiations in *Picea* genus. Some species from Sect. *Picea* (*P. asperata, P. jezoensis*, and so on), Sect. *Casicta* (*P. likiangensis, P. likiangensis* var. *balfouriana*, and so on) and Sect. *Omirica* (*P. brachytyla, P. brachytyla* var. *complanata*, and so on) are found there. For example, in relatively ancient subgroup – Sect. *Picea*, there are 30 species and 2 variations in Sect. *Picea*, and 13 species and 1 variation are found here, which take 43.3% of the total species of Sect. *Picea*. So many researchers thought Hengduan Mountain was the original center of *Picea* genus, at least it was one of the most important differentiated centers.

It proved has been proven according to analysis on fossils and pollens of *Picea* genus (Jain, 1976; Miller, 1972, 1974, 1985; Schall and Pianka, 1978; Shi et al., 1998) that during the ice age in Quaternary, the forest composed of *Picea* spp. and *Abies* spp. and the two species were distributed widely in the mountains and plains of Southwest China, Northwest China, North China, East China, and Taiwan. During that time, cold-temperate coniferous forests have wider distribution range than present. It's well known that glacier activity was active in Quaternary, and vegetation zone moved in both horizontal and vertical directions. With the advance and retreat of ice sheets, species went extinct over large parts of their range, and some populations were dispersed to new locations or survived in refugia and then expanded again (Hewitt, 2000; Stewart and Lister, 2001; Abbott et al., 2000). This repeated process would on the one hand stimulate adaptation and allopatric speciation (Hewitt, 2004), whereas, on the other, provide the opportunities for hybridization between recolonized populations, even reproductively unisolated species (Abbott and Brochmann, 2003). During interglacial time, because of climate warming, some cold-temperate coniferous forests retreated to north, and others moved towards the mountains when the glacier melted, which formed the modern distribution range shrinking again and again. In Hengduan Mountains, there are more spaces and diverse habitats for cold-temperate coniferous trees moving upwards to the high environments (Sun, 2002). However, it's latitudinal is lower, so the cold-temperate coniferous species such as *Picea* spp. are distributed in the medium and top parts of mountains, which detached the distribution area into many parts, and some *Picea* spp. differentiate into many subspecies. The place became the center of geographical distribution and differentiation of *Picea* genus. The reticulate evolution and biological radiation resulted from climatic, ecological and geological changes broght many difficulties to the evolutionary and biogeographical studies of some taxa with long generation times, widespread distributions and low morphological divergence.

5.2 The relationship among Chinese *Picea* spp. and other world *Picea* spp.

Karyotype equations of 17 species of spruce in China include four types (2n=24m, 2n=22m+2sm, 2n=20m+4sm, and 2n=16m+8sm) (Table 3). Karyotype data of 15 species spruce abroad (Table 5) are shown (Hizume, 1988; Kinlaw and Neale, 1997; Niemann, 1979; Rushforth, 1987; Hilis and Ogllvie, 1970; Doyle, 1963), three of which are included in

Karyotype equations (2n=24m (*P. sitchensis*), 2n=22m+2sm (12 species), 2n=20m+4sm (2 species)).

No.	Species	Longitude (°)	Latitude (°)	Altitude (m)	Precipitation (mm)	Temperature (°C)	Moisture (%)
1	*P. asperata*	100.1-106.8	30.2-34.6	2400-3600	550-850	2-12	60-80
2	*P. retroflexa*	100.1-103.7	30-33.1	2100-4100	550-800	2-10	55-70
3	*P. koraiensis*	116.4-129	40.7-52.5	300-1800	600-900	2-4	60-80
4	*P. meyeri*	111.4-117.5	37.5-40.6	1400-2700	500-900	2-10	60-70
5	*P. wilsonii*	101.7-117.5	30-42.2	1400-2800	500-900	5-11	50-70
6	*P. schrenkiana*	75.2-95	37.7-45.6	1200-3000	500-600	-3-6	50-65
7	*P. schrenkiana* var. *tianshanica*	77-94.5	37-46	1250-3000	500-600	-3-5	50-70
8	*P. smithiana*	85.3	29	2300-3200	700-1000	6-13	50-70
9	*P. morrisonicola*	120.8-121.5	23.2-24.5	2500-3000	1000-1400	10-20	70-85
10	*P. likiangensis*	98.9-102.1	26.5-30.2	2500-3800	500-1100	0-9	60-80
11	*P. likiangensis* var. *balfouriana*	93.7-102.5	29.5-33.8	3000-4100	700-1100	2-8	70-80
12	*P. purpurea*	100.4-105.2	30.6-36.3	2600-3800	450-1100	0-6	60-80
13	*P. jezoensis* var. *microsperma*	124-134	41-52.5	300-1800	700-900	0-6	60-80
14	*P. brachytyla*	100.4-112	29.2-35.2	1500-3300	700-1100	2-9	60-80
15	*P. brachytyla* var. *complanata*	92-103.7	24.5-31.9	2000-3800	600-1100	0-9	60-80
16	*P. mongolica*	117.5	44.6	1100-1300	200-400	-2-2	30-60
17	*P. obovata*	86.5-90.5	46.7-48.6	1300-1800	400-600	-2-6	40-70
18	*P. aurantiaca* Mast.	102.1	30.2	2600-3600	600-700	-3-5	55-70
19	*P. crassifolia*	98.4-111.2	32.5-41	1600-3800	400-600	0-5	60-75
20	*P. jezoensis* var. *komarovii*	124-134	41-52.5	600-1800	700-900	0-6	60-80
21	*P. 1ikiangensis* var. *hirtella* Cheng et L. K. Fu	96.4-107.2	28.8-31.5	3000-4100	600-900	2-7	50-65
22	*P. likiangensis* var. *linzhiensis*	90.8-100.2	27.1-30.2	2900-3700	600-1000	4-9	55-70
23	*P. likiangensis* var. *montigena* Cheng ex Chen	102.1	30.2	3300	600-700	-3-5	55-70
24	*P. neoveitchii* Mast.	102.5-110.8	31-34.6	1300-2000	400-600	3-8	50-70
25	*P. spinulosa*	85.2-89.1	27.8-29	2900-3600	450-900	0-8	60-75

Table 4. The distribution range and environmental factors of 25 spruce species in China

In Chinese *Picea* spp., B chromosome is found only in *P. meyeri, P. wilsonii, P. jezoensis* var. *microsperma,* and *P. obovata*. In **abroad** *Picea* spp., B chromosome is found only in *P. sitchensis*. There is no variation of chromosome number.

According to karyotypic asymmetry in both average arm ratio and length ratio of chromosomes, Chinese *Picea* spp. are more than that of abroad *Picea* spp. (Fig. 2, 10, 3, 9, 7). 2B karyotype is a relative evolutional type, and this type is only found in Chinese *Picea* spp..

Karyotype equation (16m+8sm) is a relative evolutional type, this type is not found in abroad *Picea* spp.. On the contrary, karyotype equation (24m), which is a relatively primordial chromosome, is found in them (*P. sitchensis*). We can conclude from karyotype structure that Chinese *Picea* spp. are relatively evolutional than abroad *Picea* spp.

No.	Species	Karyotype equation	Arm ratio	Chromosome length ratio	Karyotype type
26	*P. abies*	22m+2sm	1.24±0.33	1.74±0.34	2A
27	*P. orientalis* (L.) Link.	22m+2sm	1.31±0.42	1.82±0.24	2A
28	*P. glauca*	22m+2sm	1.30±0.21	1.71±0.19	2A
29	*P. mariana*	22m+2sm	1.28±0.09	1.84±0.53	1A
30	*P. rubens* Sarg.	22m+2sm	1.25±0.23	1.83±0.18	2A
31	*P. engelmannii*	20m+4sm	1.33±0.50	1.76±0.43	2A
32	*P. pungens* Engelm.	22m+2sm	1.31±0.11	1.79±0.12	2A
33	*P. bicolor* (Maxim.) Mayr.	22m+2sm	1.24±0.12	1.98±0.24	2A
34	*P. glehnii* (F. Schmidt) Mast.	22m+2sm	1.27±0.24	1.72±0.18	2A
35	*P. koyamae* Shirasawa	20m+4sm	1.32±0.08	1.99±0.45	2A
36	*P. polita*	22m+2sm	1.34±0.48	1.79±0.29	2A
37	*P. sitchensis*	24m+2B	1.26±0.23	1.77±0.18	1A
38	*P. omorika* (Pančić) Purk	22m+2sm	1.27±0.12	1.97±0.57	2A
39	*P. jezoensis*	22m+2sm	1.35±0.45	1.92±0.32	2A
40	*P. jezoensis* var. *hondoensis*	22m+2sm	1.32±0.09	1.85±0.18	2A

Table 5. Karyotype characters of 15 spruce species abroad

Hu et al. (1983) reported the differences of interspecific zymogram distances of genus *Picea* (Table 6). Firstly, concerning abroad *Picea*, *P. abies* is similar to the Chinese *Picea*, but it has long zymogram distance with *P. polita*. The zymogram distance between *P. polita* and other *Picea* except for *P. wilsonii* is long. The zymogram distance between *P. pungens* and other *Picea* except for *P. schrenkiana* and *P. wilsonii* is long. About the relationship between the Chinese *Picea*, the zymogram distances are short except for the following three pairs, those are *P. koraiensis* and *P. meyeri*, *P. meyerii* and *P. crassifolia*, *P. wilsonii* and *P. likiangensis*.

Species	*P. abies*	*P. koraiensis*	*P. meyeri*	*P. crassifolia*	*P. schrenkiana*	*P. wilsonii*	*P. polita*	*P. likiangensis*	*P. pungens*
P. abies	0	0.13	0.22	0.13	0.13	0.22	0.50	0.33	0.30
P. koraiensis		0	0.40	0.13	0.13	0.30	0.30	0.38	0.44
P. meyeri			0	0.43	0.14	0.22	0.50	0.29	0.50
P. crassifolia				0	0.25	0.30	0.45	0.25	0.44
P. schrenkiana					0	0.13	0.40	0.14	0.22
P. wilsonii						0	0.20	0.43	0.20
P. polita							0	0.55	0.45
P. likiangensis								0	0.44
P. pungens									0

Table 6. Interspecific zymogram distances of *Picea* genus (Hu et al., 1983)

In zonal distribution, there are close contact among abroad *Picea* spp. and Chinese *Picea* spp., particularly, in Northwest China and Northeast China. In Northwest China, *P. schrenkiana* and *P. schrenkiana* var. *tianshanica* are distributed widely in Tianshan

Mountains. They diffuse towards west along Tianshan Mountains into mountains of Pakistan and Afghanistan. *P. obovata* distributes in Aertai Mountains in northern Xinjiang, and it is connected with Siberian region of Russia. In Northeast China, *P. koraiensis* is found in Da Xinganling Mountains, Xiao Xinganling Mountains, Wanda Mountains, Zhangguangcailing Mountains, and Changbai Mountains. It is also found in Korean Peninsula, and Far East of Russia (Zheng and Fu, 1978). *P. jezoensis* distributes widely in Northeastern Asia, including Far East of Russia, Korean Peninsula, and North Japan (Ying, 1989). When it extends into Northeast China, it differentiates into some variations, such as *P. jezoensis* var. *microsperma* (in Da Xinganling Mountains, Xiao Xinganling Mountains, Zhangguangcailing Mountains) and *P. jezoensis* var. *komarovii* (in Changbai Mountains). In North China, Southwest China, and Taiwan, the *Picea* spp. have few connection with abroad *Picea* spp., so there are many China endemic species in spruce forests of these regions.

6. Acknowledgements

We gratefully acknowledge the financial support from the National Nature Science Foundation of China under the grants Nos. 39900019, 30070129, and 31170388, a grant from Shanghai Institute of Urban Ecology and Sustainability (SHUES2011A03), and Global Environmental Research Fund by Ministry of the Environment of Japan. We would like to thank Dr. Zhenzhu Xu, Dr. Yasumi Yagasaki and Dr. Shoko Ito for many valuable comments on the earlier versions of the manuscript and English corrections.

7. References

Abbott R.J., Brochmann C. 2003. History and evolution of the arctic flora: in the footsteps of Eric Hultén. Mol Ecol 12, 299-313.

Abbott R.J., Smith L.C., Milne R.I., Crawford R.M.M., Wolff K., Balfour J. 2000. Molecular analysis of plant migration and refugia in the Arctic. Science 289, 1343-1346.

Aldén B. 1987. Taxonomy and geography of the genus. Int. Dendr. Soc. Yearb. 1986, 85-96.

Alvin K.L. 1960. Further conifers of Pinaceae from the wealden formation of Belgium. Mem Inst Roy Sci Nat Belgium, 146, 1-39.

Axelord D.I. 1976. History of the coniferous forests, California and Nevada. Univ Calif Publ Bot 70.

Axelord D.I. 1986. Cenozoic history of some western American pines. Ann Missouri Bot Gard 73, 565-641.

Barbour M.G., Bilings W.D. 1988. North American Terrestrial Vegetation. Cambridge University Press.

Bobrow E.G. 1970. Generis *Picea* Historia et Systematica. Nov Syst Pl Vasc 7, 7-39.

Böhle U.R., Hilger H.H., Martin W.F. 1996. Island colonization and evolution of the insular woody habit in *Echium* L. (Boraginaceae). Proc Natl Acad Sci USA 93, 11740-11745.

Boqrzan Z., Papes D. 1978. Karyotype analysis in *Pinus*: A contribution to the satandardization of the karyotype analysis and review of some applied techniques. Silvae Genetica 27, 144–149.

Buchholz J.T 1929. The embryogeny of the conifers. Pro Intern Congr Plant Sci 1, 359–292.

Buchholz J.T. 1931. The pine embryo and embryos of related genera. Trans Ill Acag Sci 23, 117–125.

Budantsey L.Y. 1992. Early stages of formation and dispersal of the temperate flora in the boreal region. Bot Rev 58 (1), 1–48.

Budantsey L.Y. 1994. The fossil flora of the Paleogene climatic optimum in northeastern Asia. In: Boulter M.C., Fisher H.C. (Eds.), Cenozoic Plants and Climates of the Arctic. Springer, Berlin, pp. 297–313.

Cen Q.Y. 1996. The study of subfamily abietoidese (Pinaceae) flora from China. Supplement to the Journal of Sun Yatsen University 2, 87–92.

Chou Y.L. 1986. Ligneous Flora of Heilongjiang. Heilongiiang Science Press, Harbin.

Chou Y.L. 1991. Vegetation of the Greater Xingan Mts., Science Press, Beijing.

Colleau C. 1968. Anatomie comparée des feuilles de *Picea*. Cellule 67, 185–253.

Corrigan D., Timoney R.F., Donnelly D.M.X. 1978. N-Alkanes and ω-hydroxyalkanoic acids from the needles of twenty-eight *Picea* species. Phytochemistry 17, 907–910.

Delevoryas T., Hope R.C. 1973. Fertile coniferophyte remains from the late Triassic Deep River Vasin, Morth Carolina. Amer J Bot 60, 810–818.

Doyle J. 1963. Proembryogeny in Pinus in relations to that in other conifers - A survey. Proc Roy Ir Acad 62, B. 13, 181–216.

Editorial Committee of Forest of China. 1997. Forest in China (Vol 1). Beijing: China Forestry Press. 513–574.

Editorial Committee of Vegetation of China. 1980. Vegetation of China. Beijing: Science Press.

Fang J.Y. 1995. Three-dimension distribution of forest zones in east Asia. Acta Geographica Sinica 50 (2), 160–167.

Fang J.Y. 1996. The distribution pattern of Chinese natural vegetation and its climatologic and topographic interpretations. In: Researches on Hotspots of Modern Ecology. Beijing: China Science and Technology Press 369–380.

Fang J.Y., Liu G.H. 1998. Ecology of plant distribution: historical review and modern advance. Scientific Foundation in China 12, (special issue), 48–53.

Farjón A. 1990. Pinaceae: Drawings and Descriptions of the Genera *Abies, Cedrus, Pseudolarix, Keteleeria, Nothotsuga, Tsuga, Cathaya, Pseudotsuga, Larix* and *Picea*. Koeltz ScientiWc Books, Königstein, Germany.

Farjón A. 2001. World Checklist and Bibliography of Conifers, Second edn. Royal Bot. Gard., Kew, England.

Ferguson S.K. 1967. On the phytogeography of coniferales in the European Cenozoic. Palaeogeogr Palaeoclim Palaeoecol 3 (1), 73–100.

Florin R. 1954. The female reproductive organs of conifers and taxads. Biol Rev Cambridge Phil Soc 29, 367–389.

Florin R. 1963. The distribution of conifer and taxad genera in time and space. Acta Hort Verg 20, 194–256.

Fowler D.P. 1983. The hybrid black × sitka spruce, implications to phylogeny of the genus *Picea*. Can J For Res 13, 108–115.

Fowler D.P. 1987. The hybrid white × sitka spruce: species crossability. Can J For Res 17, 413–417.

Frankis M.P. 1988. Generic inter-relationships in Pinaceae. Notes Roy Bot Gard Edinb 45, 527–548.

Fu L., Li N., Mill R.R. 1999. *Picea*. In: Wu, Z.Y., Raven, P.H. (Eds.), Flora of China (4). Science Press, Beijing and Missouri Botanical Garden Press, St. Louis, pp. 25–32.

Gorden A.G. 1976. The taxonomy and genetics of *Picea rubens* and its relationship to *Picea mariana*. Can J Bot 54, 781–813.

Gugerli F., Sperisen C., Büchler C., Magni F., Geburek T., Jeandroz S., Senn J. 2001. Haplotype variation in a mitochondrial tandem repeat of Norway spruce (*Picea abies*) populations suggests a serious founder effect during postglacial re-colonization of the western Alps. Mol Ecol 10, 1255–1263.

Harris T.M. 1979. The Yorkshire Jurassic flora 5, Coniferales. London: British Museum (Natural Hostory).

Hart J.A. 1987. A cladistic analysis of conifers: preliminary results. J Arm Arb 68, 269–306.

Hewitt G.M. 2000. The genetic legacy of the Quaternary ice ages. Nature 405, 907–913.

Hewitt G.M. 2004. Genetic consequences of climatic oscillations in the Quaternay. Philos Trans R Soc Lond B 359, 183–195.

Hilis L.V., Ogllvie R.T. 1970. Picea banks in Beaufort formation (Tertiary), Northwestern Banks Island, Arctic Canada. Can J Bot 48, 457–463.

Hipkins V.D., Krutovskii K.V., Strauss S.H. 1994. Organelle genome in conifers: structure, evolution, and diversity. For Genet 1, 179–189.

Hizume M. 1988. Karyomorphological studies in the family Pinaceae. Natural Science 8, 1–108.

Hsu J. 1983. Late cretaceous and cenozic vegetation in China emphasizing their connections with North America. Ann Missouri Bot Gard 70, 490–508.

Hu Z.A., Wang H.X., Yan L.F. 1983. Biochemical systematics of gymnosperm: POD of Pinaceae. Acta Phytotaxonomica Sinica 4, 423–434.

Jain K.K. 1976. Note evolution of wood structure in Pinaceae. Israel Bot 25, 28–33.

Kinlaw C.S., Neale D.B. 1997. Complex gene families in pine genomes. Trends Plant Sci 2, 356–359.

Klaus W. 1987. Mediterranean pines and their history. Pl Syst Evol 162, 133–162.

Kuan C.T. 1981. Fundamental features of the distribution of coniferae in Sichuan. Acta Phytotaxonomica Sinica 19 (4), 393–407.

Ledig F.T., Hodgskiss P.D., Krutovskii K.V., Neale D.B., Eguiluz-Piedra T. 2004. Relationships among the spruces (*Picea*, Pinaceae) of southwestern North America. Syst Bot 29, 275–292.

LePage B.A. 2001. New species of *Picea* A. Dietrich (Pinaceae) from the middle Eocene of Axel Heiberg Island, Arctic Canada. Biol J Linn Soc 135, 137–167.

LePage B.A. 2003. The evolution, biogeography and palaeoecology of the Pinaceae based on fossil and extant representatives. Acta Hort 615, 29–52.

Li N. 1995. Studies on the geographic distribution, origin and dispersal of the family Pinacease Lindl. Acta Phytotaxonomica Sinica 33 (2), 105–130.

Li W.H. 1980. Community Structure and Succession of Valley Spruce-Fir Forests of the Less Xingan Mts. J Natural Resources 4, 17–29.

Li W.H., Zhou P.C. 1979. Study on distributional general rules and mathematical models of dark coniferous forests in Asian-European Continent. J Natural Resources (1), 21–34.

Liu J.Q., Gao T.G., Chen Z.D., Lu A.M. 2002. Molecular phylogeny and biogeography of the Qinghai-Tibet Plateau endemic *Nannoglottis* (Asteraceae). Mol Phylogenet Evol 23, 307–325.

Liu T.S. 1971. A monography of the genus *Abies*. Taipei, The Department of Forestry College of Agriculture National Taiwan Univeristy.

Liu T.S. 1982. A new proposal for the classification of the genus *Picea*. Acta Phyt Geobot 33, 227–244.

Liu Z.L., Fang J.Y., Piao S.L. 2002. Geographical distribution of species in genera *Abies, Picea* and *Larix* in China. Acta Geogrophica Sinica 57 (5), 577–586.

Lü H., Wang S., Shen C., Yang X., Tong G., Liu K. 2004. Spatial pattern of modern *Abies* and *Picea* pollen in the Qinghai-Xizang Plateau. Quat Sci 24, 39–49.

Manley S.A.M. 1972. The occurrence of hybrid swarms of red and black spruce in central New Brunswick. Can J For Res 2, 381–391.

McKenna M.C. 1975. Fossil mammals and early Eocene North Atlantic land continuity. Ann Mo Bot Gard 62, 335–353.

Mehra P.N. 1968. Cytogenetical evolution of conifers. Indian Journal Genetics Plant Breeding 28, 97–111.

Mikkola L. 1969. Observations on interspecific sterility in *Picea*. Ann Bot Fenn 6, 285–339.

Miller C.N. 1972. Pityostrobus palmeri, A new species of petrified conifer cones from the Late Cretaceous of New Jersey. Amer J Bot 59 (4), 325–358.

Miller C.N. 1974. Pityostrobus hallii, A new species of structurally preserved conifer cones from the Late Cretaceous of Maryland. Amer J Bot 61, 798–804.

Miller C.N. 1975. Early evolution in the Pinaceae. Rev Palaeobot Palyno 21, 101–117.

Miller C.N. 1977. Mesozoic conifers. Bot Rev 43, 217–218.

Miller C.N. 1985. Pityostrobus pubescens, A new species of petrified conifer cones from the Late Cretaceous of New Jersey. Amer J Bot 92 (4), 520–529.

Miller C.N. 1988. The origin of modern conifer families. In: Beck CB ed. Origin and Evolution of Gymnosperms. New York: Colombia Univ Press, 448–487.

Niemann G.J. 1979. Some aspects of the chemistry of Pinaceae needles. Acta Bot Neerl 28 (1), 73–88.

Ogilvie R.T., von RudloV E. 1968. Chemosystematic studies in the genus *Picea* (Pinaceae). IV. The introgression of white and Engelmann spruce as found along the Bow River. Can J Bot 46, 901–908.

Page C.N. 1988. New and maintained genera in the conifer families Podocarpaceae and Pinaceae. Notes RBG Edinb 45 (2), 377–395.

Page C.N., Hollands R.C. 1987. The taxonomic and biogeographic position of Sitka spruce. Proc R Soc Edinb 93b, 13–24.

Perron M., Perry D.J., Andalo C., Bousquet J. 2000. Evidence from sequence-tagged-site markers of a recent progenitor-derivative species pair in conifers. Proc Natl Acad Sci USA 97, 11331–11336.

Petit R.J., Aguinagalde I., de Beaulieu J.L., Bittkau C., Brewer S., Cheddadi R., Ennos R., Fineschi S., Grivet D., Lascoux M., Mohanty A., Muller-Starck G., Demesure-Musch B., Palme A., Martin J.P., Rendell S., Vendramin G.G. 2003. Glacial refugia: hotspots but not melting pots of genetic diversity. Science 300, 1563–1565.

Prager E.M., Fowler D.P., Wilson A.C. 1976. Rates of evolution in conifers (Pinaceae). Evolution 30, 637–649.

Price R.A. 1989. The genera of Pinaceae in the Southeastern United States. J Arnold Arbor 70, 247–305.

Qian H., Ricklefs R.E. 2000. Large-scale processes and the Asian bias in species diversity of temperate plants. Nature 407, 180–182.

Ran J.H., Wei X.X., Wang X.Q. 2006. Molecular phylogeny and biogeography of *Picea* (Pinaceae): Implications for phylogeographical studies using cytoplasmic haplotypes. Molecular Phylogenetics and Evolution 41, 405–419.

Rushforth K. 1987. Conifers. Christopher Helm, London.

Schall J.J., Pianka E.R. 1978. Geographical trends in numbers of species. Science 201, 679–686.

Schmidt P.A. 1989. Beitrag zur Systematik und Evolution der Gattung *Picea* A. Dietr. Flora 182, 435–461.

Shi Y.F., Li J.J., Li B.Y. 1998. Uplift and Environmental Changes of Qinghai-Tibetan Plateau in the Late Cenozoic. Guangdong Science and Technology Press, Guangzhou.

Sigurgeirsson A., Szmidt A.E. 1993. Phylogenetic and biogeographic implications of chloroplast DNA variation in *Picea*. Nordic J Bot 13, 233–246.

Stewart J.R., Lister A.M. 2001. Cryptic northern refugia and the origins of the modern biota. Trends Ecol Evol 16, 608–612.

Sudo S. 1968. Anatomical studies on the wood of species of *Picea*, with some considerations on their geographical distribution and taxonomy. Bull Gov For Exp Stat 215, 39–130.

Sun H. 2002. Evolution of Arctic-Tertiary flora in Himalayan–Hengduan Mountains. Acta Bot Yunnan 24, 671–688.

Taylor R.J., Patterson T.F. 1980. Biosystematics of Mexican spruce species and populations. Taxon 29, 421–469.

Thomas G. 1965. The saccate pollen grains of Pinaceae mainly of California. Grana Palynologica 6, 270–289.

Tiffney B.H. 1985a. Perspectives on the origin of the floristic similarity between Asia and eastern North America. J Arnold Arbor 66, 73–94.

Tiffney B.H. 1985b. The Eocene North Atlantic land bridge: its importance in the Tertiary and modern phytogeography of the Northern Hemisphere. J Arnold Arbor 66, 243–273.

Tiffney B.H., Manchester S.R. 2001. The use of geological and paleontologyical evidence in evaluating plant phylogeographyic hypotheses in the Northern Hemisphere Tertiary. Int J Plant Sci 162, s3–s17.

von RudloV E. 1967. Chemosystematic studies in the genus *Picea* (Pinaceae). Can J Bot 45, 891–901.

Wang H.S. 1992. Floristic Geography. Beijing: Science Press.

Wang H.S. 2000. The distribution pattern and floristic analysis of family Pinaceae of China. Bull Bot Res 20 (1), 12–19.

Wang M.L., Shi D.X., Zeng P.A. 1990. The situation of karyomorphological analysis and biological signification of *Picea* plants in China. J Sichuan Agric Univ 15 (1), 74–81.

Wang X.Q., Tank D.C., Sang T. 2000. Phylogeny and divergence times in Pinaceae: evidence from three genomes. Mol Biol Evol 17, 773–781.

Wen J. 1999. Evolution of eastern Asian and eastern North American disjunct distributions in flowering plants. Ann Rev Ecol Syst 30, 421–455.

Weng C., Jackson S.T. 2000. Species differentiation of North American spruce (*Picea*) based on morphological and anatomical characteristics of needles. Can J Bot 78, 1367–1383.

Wolfe I.A. 1978. A paleobotanical inter pretation of Tertiary climates in the Northern Hemisphere. Amer Sci 66, 694–703.

Wolfe J.A. 1975. Some aspects of plant geography of the Northern Hemisphere during the Late Cretaceous and Tertiary. Ann Mo Bot Gard 62, 264–279.

Wright J.W. 1955. Species crossability in spruce in relation to distribution and taxonomy. For Sci 1, 319–340.

Wu D.X. 1985. Artificial Cultivatied *Picea koraiensis* Forests. Newsletter of Forestry Science and Techniques 1: 16–18.

Wu H.Q. 1987. Spruce Forests in the Northeast China. Dynamics of Spruce Forests. Doctoral Thesis, Northeast Forestry University, Harbin.

Wu S.G., Yang Y.P., Fei Y. 1995. On the flora of the alpine region in the Qinghai-Xizang (Tibet) Plateau. Acta Bot Yunnannica 17, 233–250.

Wu Z.Y. 1991. The areal – types of Chinese genera of seed plants. Acta Bot Yunnannica (S), 1–139.

Xiang Q.Y., Manchester S.R., Thomas D.T., Zhang W., Fan C. 2005. Phylogeny, biogeography, and molecular dating of cornelian cherries (*Cornus*, Cornaceae): tracking Tertiary plant migration. Evolution 59, 1685–1700.

Yang G.T., Erich H., Sun B., Zhang J. 1994. *Picea-Abies* forest of the Northeast China. Bull Bot Res 14 (3), 313–328.

Ying T.S. 1989. Areography of the gynmosperms of China (1). Acta Phytotaxonomica Sinica 27 (1), 27–38.

Xu W.D., Li W.D., Zheng Y. 1994. The taxonomy of *Picea mongolica* in Inner Mongolia. Bull Bot Res 14 (1), 59–68.

Xu W.D., Liu G.T., Duan P.S., Zou C.J. 1998. Study on *Picea mongolica* forest ecosystem in Baiyinaobao Natural Reserve, Inner Mongolia. China Forestry Publishing House, Beijing..

Zheng W.J., Fu L.G. 1978. Flora Reipublicae Popularis Sinicae. (Tomus 7). Beijing: Science Press.

Zhou Y.L. 1988. Some vegetation types of China (VI): deciduous coniferous forests. Biological Bulletin 51, 6–10.

14

Chemical Defenses in Eucalyptus Species: A Sustainable Strategy Based on Antique Knowledge to Diminish Agrochemical Dependency

S. R. Leicach[1*], M. A. Yaber Grass[1], H. D. Chludil[1], A. M. Garau[2], A. B. Guarnaschelli[2] and P. C. Fernandez[1,3]

[1]*Química de Biomoléculas*

[2]*Dasonomía, Facultad de Agronomía (FA), Universidad de Buenos Aires (UBA) Ciudad Autónoma de Buenos Aires C1417DSE*

[3]*National Institute of Agricultural Technology INTA, EEA Delta del Paraná Paraná de las Palmas y Canal Comas s/n Campana*

Argentina

1. Introduction

A large number of tree species from different genera have being used world over for their timber resources. Most of them produce roundwood for sawmill and commercial valuable derivatives such as those related to pulp and paper, hardboard and particleboard industries (FAO 2011b). Species within Fabaceae, Pinaceae, Myrtaceae, Cupressaceae, Araucariaceae, Meliaceae, Fagaceae, and Proteaceae families are exploited by those industries.

Wood is characterized by a quite heterogeneous structure based on cell walls mainly composed by cellulose (41-43%), hemicellulose (20-30%), and lignin (27%). Phenylpropanoid derivatives are also contained within lignocellulosic wood structure (Baucher et al 2003, Boerjan et al 2003).

Besides timber uses, some wood particular components are considered adequate resources for other kind of industries. Cellulose derivatives are currently used as a source for natural adhesives. New hydrocolloids are being obtained from cellulose derivatives; some of them have been applied to improve cohesion of wound bandages. Lignins have also industrial applications in fiber-board and paste applications in plywood (Otten et al 2007).

Different wood properties are considered as key characteristics depending on the industrial utilization of forest species. Wood density is the most useful parameter when measuring wood quality. For solid wood purposes wood density is positively correlated to mechanical strength and shrinkage; other properties related to solid wood uses or structural applications are modulus of rupture, modulus of elasticity, percentage of tension, dimensional stability, grain and texture (Hoadley 2000, Wiedenhoeft 2010).

* Corresponding Author

Current concerns about environmental stability and conservation of natural resources have increased most countries interest in integrated exploitation of forestry species encouraging the use of non-timber products; they have proved to be the source of a wide range of bioactive chemicals. Most of them are secondary metabolites that can be used for nutraceutical, pharmaceutical, and medicinal purposes, also exhibiting potential in integrated pest management to diminish agrochemicals use or even as agrochemicals substitutes (Willfor et al 2004, Krasutsky 2006, Fernándes and Cabral 2007, Domingues et al 2011).

Timber plantations can produce besides bioactive chemicals, many highly valuable non-timber products, including an array of goods like seeds, nuts, oils and fragrances. Some of them also produce other industrial materials such as latexes, tannins, gum exudates, dyes and resins (Wong et al 2001, Jones and Lynch 2007, FAO 2011a).Timber trees producing economically valuable non-timber products have been named 'timber plus trees' (Mull 1993).

Acacia, Pinus and *Eucalyptus* genera include some of the most commonly cultivated species in the tropics and South America. Non-timber tropical forest products have been grouped into four categories: (i) fruits and seeds, with plant parts harvested mainly for fleshy fruit bodies, nuts and seed oil; (ii) plant exudates such as latex, resin and floral nectar; (iii) vegetative structures such as apical buds, bulbs, leaves, stems, barks and roots, and (iv) small stems, poles and sticks harvested for housing, fencing, fuel wood, and craft and furniture materials such as carvings and stools (Cunningham 1996, Dovie 2003).

Gums, resins, and latexes have been the most widely used categories of non-wood forest products (Copper et al. 1995; Willfor et al., 2004; Krasutsky, 2006; Fernandes and Cabral, 2007). Species within *Acacia, Sterculia, Ceratonia, Prosopis, Agathis,* and *Shorea* genera are among the most important supply sources for such industries. Gums and resins are employed for a wide range of food and pharmaceutical purposes. Pharmaceutical industry uses gums as binding agents for tablets production, and as emulsifying agents in creams and lotions. Food and printing industries have also taken advantage of their thickening, emulsifying, and stabilizing characteristics (Mbuna and Mhinzi 2003, Gentry et al 1992, Anderson 1993). Rosin derivatives from pine trees are being used alone or in combination with acrylic resin solutions and emulsions to produce better quality water-based flexographic inks and varnishes (Vernardakis 2009). Resins are used for paper, wood paints production, and also to prepare varnishes and lacquers, whereas soft resins and balsams are used to produce fragrances (Mbuna and Mhinzi 2003, Messer 1990). Latexes have been early used in the chewing gum industry (Williams 1963).

From an ecological standpoint, non timber products are considered more valuable than timber ones as they can be harvested through the years without cutting down the trees, with almost no perturbation on ecosystems. Moreover, plantations may provide environmental benefits across diverse areas including the atmosphere (replacing ozone-depleting substances), agriculture (land rehabilitation, phytoremediation), and carbon sequestration (Barton 1999). From a social point of view integrated exploitation of forest resources has been considered within the group of multiple-use strategies, which increases the range of income generating alternative options for forest-dependent communities, while avoiding some of the ecological costs of timber cutting (Olajide et al 2008).

2. Eucalypts

Eucalyptus species, belonging to Myrtaceae family, are native to Australia although there are only a few native species from Papua New Guinea, Indonesia and the Philippines. Eucalypts represent a very important group of tree species in Australia, playing a dominant role in continental vegetation and being extremely important for biodiversity conservation. There are more than 800 species, divided in 13 subgenera, and hybrids (Boland et al 2006, Brooker et al 2006). Classification of the genus has been made based on a wide range of attributes. Among genetic, anatomical and chemical features, leaf surface characteristics and the presence of specific secondary metabolites, particularly terpenoids and phenolic derivatives (i.e., renantherin is present in most *Monocalyptus* section), have been recognized as fundamental in eucalypts interaction with specific groups of insects and mammals.

Eucalyptus species grow in natural forests and have been successfully introduced worldwide, being cultivated as one of the main biomass sources under different environmental conditions, depending on the species. Most of them are commercially used to obtain timber and fiber. Eucalypts can play useful environmental roles in natural forests, plantations being also useful for ornamentation and shade purposes in cities green spots, being also useful in lowering water tables. They represent one the most important fiber sources for pulp and paper production in European, African, and Asian countries, and in South America, particularly in Argentina, Brazil, and Chile. Even when the genus is rich in secondary metabolites only a few species are used for essential oil production.

Eucalyptus species grow faster than most other commercially exploited tree species, usually producing high yields (Binkley and Stape 2004). Their abilities to grow rapidly when environmental conditions are suitable and to survive and recover easily from fire and other damaging factors are related to the bud system characterizing their trees. They do not need resting period during winter, growing whenever water is available and temperatures are mild, regardless the time of the year. If they do stop growing (due to a summer drought or a winter cold period) they are able to resume growth whenever stressful factors disappear (Williams and Woinarski 1997, Wei and Xu 2003, Florence 2004).

Eucalypts show better growth performance outside their native environment, it has been suggested that at their natural habitats tree regeneration is affected by insects and diseases commonly absent in other environments. Trees seem to have less access to light, water and nutrients due to neighboring competition in their native land, where soil conditions may have became less adequate and chances of fire higher.

Outside native habitat volume yields can be quite different depending on each particular species genetic potential, resources availability and plantation strategy, i.e. mean annual increments varying from 10 to 90 m^3/ha/year have been obtained (depending on the use of wood, and whether clear-cut or thinning is used) in sample plots as a result of rotation periods from 5 to 25 years (Eldridge et al 1994, Wei and Xu 2003). One of the most important advantages of the genus is its capacity to coppice which allows a new production of wood from the same plantation. *E. grandis*, *E. camaldulensis*, *E. globulus*, *E. tereticornis*, *E. urophylla*, *E. dunni*, and *E. nitens* are among the most successful species. Iberian countries use around 1.29 million ha for timber plantations, predominantly *E. globulus*, whereas in South America *E. grandis*, the hybrid *E. urograndis* and *E. globulus* are among the preferred species, being cultivated in approximately 6.64 million ha (Iglesias-Trabado at al 2009). It has been estimated that pulp production by both these regions may currently represent more than 80% *Eucalyptus* spp. pulp produced worldwide (BRACELPA 2009).

Tolerances to abiotic stresses such as drought, cold, and salinity have been reported for different species. *E. camaldulensis* seems to be a successful one in arid and semi-arid regions, also exhibiting a better performance under moderate to severe soil salinity. *E. nitens* is a cold adaptable species while *E. globulus* grows better in soils characterized by slight to moderate salinity, mainly under temperate climates, and *E. grandis* is mainly cultivated in subtropical and regions (Florence 1996, Teulieres et al 2007).

Several research and breeding programs have been successful improving timber yields, obtaining significant changes in fast growing eucalypt plantations that incremented production volume in more than 40 m^3/ha/year (Binkley and Stape 2004, Shani et al 2003). They have reported the development of 25 transgenic lines in *E. camaldulensis*, *E. grandis* and its hybrids with the ability to produce considerably higher biomass amounts.

Wood quality traits, particularly those needed to produce solid wood products have been investigated by different research groups. Pilodyn penetration is an indirect method for determining wood basic density, Muneri and Raymond (2000) have analyzed genetic parameters and genotype-by-environment interactions for basic density, pilodyn penetration, and stem diameter in *E. globulus*. Genomics and markers linked to wood properties were also studied by other authors, Barros et al (2002), Moran et al (2002), and Thamarus et al (2004).

Even when commercial exploitation of transgenic eucalypts has not been developed in a significant degree, research work has been performed in order to obtain genetically improved clones producing higher yields of high quality woods, also exhibiting higher levels of resistance to environmental stress (Girijashankar 2011). Eucalypt fibers are particularly appreciated for manufacturing high-quality grades of tissue paper, writing and printing papers (Foelkel 2009). Advances have been made related to microfibrillar orientation of secondary cell wall in *E. grandis* (Thumma et al 2005, Spokevicius et al 2007) and *E. nitens* (MacMillan et al 2010), and a reduction of lignin content has been achieved in *E. grandis* x *E. urophylla* (Tournier et al 2003) and in *E. camaldulensis* modified clones (Kawaoka et al 2006, Chen et al 2001). Improved plant raw materials for pulp and paper purposes have been obtained by modifying lignin biosynthetic pathway; these changes can affect lignin content, composition, or both (Baucher et al 2003).

Water deficit, a common constraint in forestry, is the main cause of plant stress during eucalypts establishment, when seedling growth can be also compromised by herbivory and disease. Genetic changes to improve plant survival chances enhancing resistance levels towards environmental adversities would contribute to obtain higher yields in forestry plantations. The ability to successfully grow under different kinds of abiotic stress represents an important trait in eucalypts plantations. Transgenic lines in several eucalypt species with tolerance to salinity, low temperatures and drought were obtained by Yamada-Watanabe et al (2003) and Navarro et al (2011).

Higher resistance levels to plagues and herbicides have been obtained in transgenic *E. camaldulensis* (Harcourt et al 2000) and *E. urophylla* (Shao et al 2002), commercial potential of this feature being an important management option during establishment period.

Although successful development of most species depends on their ability to compete for resources, it has been also related to their production of bioactive defensive chemicals. Research on genetic control of secondary metabolites production by forest trees has been increased in recent years. Several studies performed on *E. globulus* (Freeman et al 2008, O´Reilly-Wapstra et al 2011, Külheim et al 2011) and *E. nitens* (Henery et al 2007) allowed the identification of the quantitative trait loci for terpenes and formylated phloroglucinol derivatives, compounds considered to be the most significant defensive treats against herbivory. Essential oils have been accepted as the main defensive trait in eucalypts. Research work on genetic modulation of essential oil production has been performed by Ohara et al (2010). These authors developed a transgenic *E. camaldulensis* clone introducing genes associated to limonene synthesis; they demonstrated that plastidic and cytosolic expression of PFLS (Perilla frutescens limonene synthase) resulted in 2.6 to 4.5-fold increments in limonene concentrations, also emphasizing a synergistic increase in 1,8-cineole and α-pinene biosynthesis.

High hemicellulose content represents a desirable pulp characteristic for paper production, since hemicellulose levels are positively correlated to pulp resistance; surprisingly it seems to be also related to essential oil production. Correlations between leaf volatile organic compounds and pulp properties may be useful in genetic improvement programs. Positive correlations between leaf volatiles and hemicellulose content in pulp have been previously reported by Zini et al (2003) for 14 eucalypt clones.

3. Plants natural defenses

While primary metabolism has been related to biomass development and reproduction, exhibiting very slight differences among living organisms; secondary metabolism includes a wide range of chemical families, making it possible to classify plants in different species through chemical taxonomy, according to each particular profile.

During evolution, terrestrial plants have developed new biosynthetic pathways to produce flavonoids, terpenoids, alkaloids, cyanogenic glycosides, glucosinolates, and numerous phenolic compounds (including polymers such as lignins and tannins), secondary metabolites that have provided advantages to the producing species allowing their successful survival despite invasive organisms and other environmental stresses.

Benzoic and cinnamic acids and their phenolic derivatives, flavonoids, and long chain hydrocarbon compounds including derivated alcohols, carbonylic compounds and fatty acids, are among the most common defensive chemicals in plant kingdom. Alkaloids, terpenoids, glucosinolates, hydroxamic acids, tiophenes, cyanogenic glycosides, disulfures and sulfoxides are less distributed and restricted to particular genera, (Leicach et al, 2009a, Leicach et al 2010, Yaber Grass et al 2011).

Most of them are natural products also known as **allelochemicals**, which were originally defined as secondary metabolites involved in plant-plant and plant-microorganism interactions, until International Allelopathy Society (IAS) extended in 1998 its definition to every natural product playing a role in plant-environment interactions (Leicach et al 2009b). They have proved to provide protection towards competition by weeds and other plants and to avoid detrimental action of herbivores, fungi, bacteria, and viruses. Insects and mammals feeding on leaves have co-evolved in such a way that some secondary metabolites also play important roles in host selection by herbivores (Matsuki et al 2011). Secondary metabolites seem to represent the chemical language in plant-environment interactions, continuously growing in number and changing as co-evolution takes place.

Terpenoids are physiologically, ecologically, and commercially important, protecting plants against herbivores and pathogens and having important roles in allelopathic interactions, nutrient cycling and attraction of pollinators. Different woody species produce volatile mixtures of terpenoids known as essential oils that are responsible for their particular odours and fragrances. They are produced by species within different families such as Asteraceae (*Matricaria*), Labiatae (*Mentha*), Myrtaceae (*Eucalyptus*), Pinaceae (*Pinus*) and Rutaceae (*Citrus*) and contain structurally related terpenes as main components (Harborne 1998, Leicach et al 2009a). Essential oils are usually obtained from non-woody parts of the plant, particularly foliage, through steam distillation also known as hydrodistillation. They are complex mixtures of monoterpenes (C_{10}) and sesquiterpenes (C_{15}), with minor abundances of phenylpropanoids, and acyclic hydrocarbon derivatives such as oxides, ethers, alcohols, esters, aldehydes, and ketones.

Figure 1 shows the chemical structure of most distributed essential oil mono and sequiterpenic compounds.

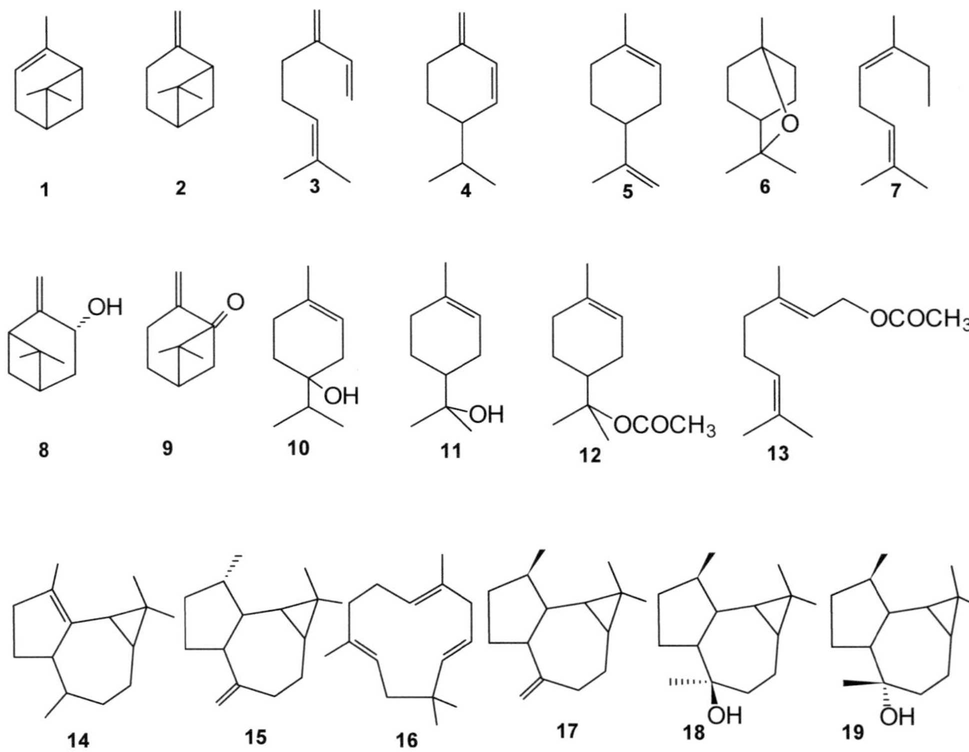

1-α–pinene, **2**- β–pinene, **3**- myrcene, **4**- α–phellandrene, **5**- limonene, **6**- 1,8-cineole, **7**- *cis*-ocymene, **8**- *trans*-pinocarveol, **9**- pinocarvone, **10**- 1-terpinen-4-ol, **11**- α–terpineol **12**- α-terpinyl acetate, **13**- geranyl acetate. Sesquiterpenes: **14**- α-gurjenene, **15**- aromadendrene, **16**- β–humulene, **17**- allo-aromadendrene, **18**- globulol, **19**- epiglobulol.

Fig. 1. Essential oils most common terpenic derivatives

Essential oils are commonly contained in glandular trichomes, a paradigmatic example of joint physicochemical mechanism that has demonstrated to be one of the most effective defences against noxious organisms (Hammerschmidt and Schultz 1996, Bowers et al 2000). Enzymes involved in the different metabolic steps associated to these terpenes biosynthesis are located within these structures (Gershenzon and Croteau 1990, Gershenzon and Dudareva 2007).

Essential oil blend depends on the relative amounts of their components, a feature that characterizes each species. Table 1 shows essential oil most important components for some *Eucalyptus* species (Juan et al 2011, Gilles et al 2010).

Oil constituents	1	2	3	4	5	6	7	8	9	10	11	12	13
α-Thujene		0.6		0.3	0.2					3.1		0.5	
α-Pinene	4.3	5.4	2.3	1.2	10.1	9.3	5.6	8.3		0.4	5.5	5.6	52.7
Camphene		1.6	0.3		0.3	23.1	0.3						
β-pinene	25.3	0.1	1.7	0.7	2.1	2.7		2.5	6.2				
Myrcene		0.2	0.6		0.4		1.7	1.3		1.1		0.5	
α-Phellandrene				7.2	1.2		2.3			17.4	1.4	7.6	
limonene	4.6	5.4		2.6	6.4	5.1	10.1		3.5				
β-Phellandrene										2.8			
1,8-Cineole	5.2	58.9	1.2	35.7	57.7	44.3	61.3			0.7	48.5	26.7	18.7
β-ocimene				0.1	4.4					0.3			
γ-Terpinene	1.2	2.8	0.3	2.8	0.2			0.8		0.8	13.0	3.0	5.0
para-cymene	7.4	2.1				1.6	7.2	28.6	27.3	8.5	4.4	13.6	9.7
Terpinolene										2.4			
citronellal			72.7										
Linalool	0.4		0.1		2.5	0.3				0.9			
β-caryophyllene	4.3		2.6	0.2		2.2							
Terpin-4-ol	1.7			1.2		0.2		1.7		4.7	3.6	1.8	1.1
α-Terpineol	6.2	2.7	0.7	1.4	1.3	0.3	3.1	5.6	6.3	1.0	3.6	1.5	5.7
Piperitone										40.5			
cryptone		1.1		25.4	0.4	1.3	3.7	17.8					
α–Terpinyl acetate		2.1	1.5			1.2		0.2		0.3	5.2		
citonellol	2.3		6.3			0.1							
Arommadendrene	1.7	2.1		1.3		1.3			0.3		3.4	2.3	
alloaromadendrene													
epiglobulol												7.5	
Spathulenol	4.1			0.2				1.8	1.1				
globulol	2.4	1.6		3.1	4.4	7.3	0.3	0.5			4.2	0.1	

***1**-E. alba, **2**- E. camaldulensis, **3**- E. citriodora, **4**- E. deglupta, **5**- E. urophylla, **6**- E. globulus, **7**- E. saligna, **8**- E. tereticornis, **9**- E. robusta, **10**- E. dives, **11**- E. dunnii, **12**- E. gunnii, **13**- E. grandis.*

Table 1. Essential oil components for some *Eucalyptus* species

1,8-cineole, also known as eucalyptol, is one of the most characteristic chemical features in *Eucalyptus* species, being the main component in many of their essential oils, with relative abundances varying from 20 to 70%, depending on the species. Some of them (*E. alba, E. camaldulensis, E. urophylla E. globulus* and *E. saligna*) produce high amounts of 1,8-cineole, the latter being the one with the higher relative abundance in Table 1. *E. sideroxylon* represents an extreme example as its essential oil can contain up to 90% 1,8-cineole (Alzogaray et al 2011).

Other species do not include 1,8-cineole in their essential oils (*E. tereticornis* and *E. robusta*) or produce it in much lower amounts. *E. grandis* essential oil contains α-pinene as the main component (52.7%) being 1,8-cineole the second one in relative abundance (18.7%), *E. alba* produces β–pinene (25.3%) as the main component and only 5.2% 1,8-cineole and *E. citriodora* characterized by a high amount of citronellal in its essential oil (72.7%), contains only 1.2% 1,8-cineole (Cimanga et al 2002).

Some *Eucalyptus* species have been studied by our research group. Essential oils obtained from fresh leaves of four species grown in Argentina within Agronomy School experimental field (34° 37′ S, 58° 20′ W, Buenos Aires University) were obtained by hydro-distillation using a Clevenger device to be further analyzed by GC and GC-MS. *E. globulus* showed the highest (1.5% fw) essential oil yield, followed *E. sideroxylon* (1.1% fw), *E. tereticornis* (0.9% fw), and *E. camaldulensis* (0.8%), adult leaves from the latter containing the highest 1,8-cineole relative abundance (60%), almost two-fold the value found in the other three species (unpublished results).

Eucalyptus species produce, besides essential oils, secondary metabolites that are ubiquitously distributed in plant kingdom such as hydrolyzable and condensed tanins, flavonoids, and others less distributed such as phloroglucinol derived compounds and cyanogenic glycosides (Moore et al 2004).

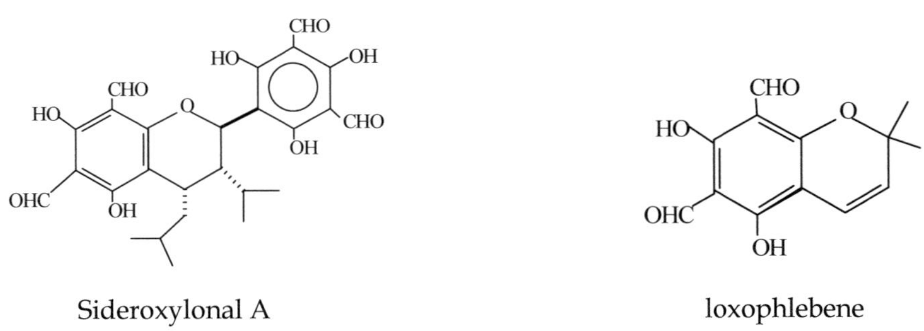

Fig. 2. Eucalypts formylated phloroglucinol compounds

Foley and coworkers have reported that there are species within *Eucalyptus* subgenera (e.g. *Monocalyptus*) lacking formylated phloroglucinols while others (e.g. *Symphyomyrtus*) contain a wide variety of them. They have also reported that there is a positive correlation between formylated phloroglucinols concentrations and those corresponding to terpenes in trees foliage, allowing koalas and other marsupial species to make food choices based on the way leaves smell (Foley et al 2009). Formylated phloroglucinol compounds are characterized by at least one fully substituted phenolic ring with one or two aldehyde groups. Phloroglucinol sesquiterpene derivatives have been identified in *E. globulus* leaves (Osawa et al 1996).

Dimeric phloroglucinol compounds such as sideroxylonals, have also been found in leaves and flower buds of several *Eucalyptus* species (Eschler and Foley 1999).

Recently Sidana et al. (2011) have reported the isolation of two new formylated phloroglucinols, loxophlebal B and loxophlebene in *E. loxophleba* ssp. *Lissophloia* leaves.

Cyanogenic glycosides, on contrary, are not common chemical features in eucalypts, however there are few species containing them. *E. cladocalyx* allocates up to 20% of leaf nitrogen to cyanogenic glycosides production, being prunasin the most representative compound (Gleadow et al 1998, Gleadow and Woodrow 2000). Gleadow et al. (2008) have described 18 cyanogenic species and one subspecies (e. g. *E. acaciiformis* Deane & Maiden, *E. leptophleba* F. Muell., *E. nobilis* Johnson & Hilld).

Most cyanogenic species produce mainly (*R*)-prunasin, although its epimer, sambunigrin, has been also detected. The diglycoside amygdalin has been reported to be produced by *E. camphora* (Neilson et al 2006).

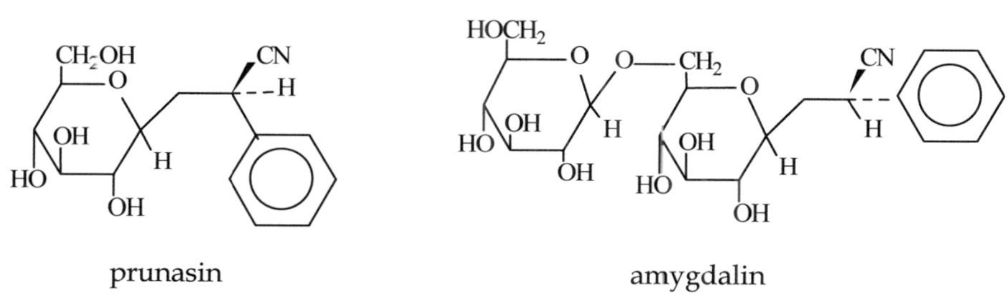

Fig. 3. Cyanogenic glycosides in some eucalyptus

4. Allelochemicals production

The amount and structural characteristics of biosynthesized defensive chemicals vary with the species and depend on each individual physiological and ontogenic state, also being strongly modulated by environmental conditions (Einhellig 1989, Einhellig 1995, Leicach 2009c). Essential oil yield and composition have demonstrated to change for a particular species depending on individual features as much as on geographical location, climate conditions and particular characteristics of soil. They can differ between individuals from the same species, even within a small population.

4.1 Individual features

The amount and structural characteristics of biosynthesized defensive chemicals vary with age in each species, and may dramatically change with changes in each individual physiological condition (Leicach 2009d). Even when it has been speculated that immature plant tissues lack of defensive chemicals since their production would be constrained by the lack of the corresponding enzymatic machinery, it has been demonstrated that concentration of some secondary metabolites (particularly low molecular weight phenolics, cyanogenic glycosides, terpenes, and alkaloids) reaches the highest value during early stages of seedling growth and leaf expansion, being only synthesized in young tissues (Herms and Mattson 1992). Most of them, including essential oil components, are powerful biocides that play a protective role against noxious organisms at this particularly sensitive stage of plant development. $^{14}CO_2$ incorporation rate values indicated maximum levels of camphor biosynthesis during *Salvia officinalis* leaf expansion (3-4 weeks), which declined to almost undetectable ones after 6 weeks (Genshenzon and Croteau 1990).

Essential oil yield and chemical composition have shown to be related to leaf age in *Eucalyptus* species (Silvestre et al 1997, Wildy et al 2000). Higher essential oil yields have been obtained from *E. camaldulensis* fully expanded but not fully lignified leaves (Doran and Bell 1994). Significant differences in essential oil yield and composition were also detected between *E. globulus* young and mature leaves (Chennoufi et al 1980, Silvestre et al 1997), and also among juvenile, mature, and senescent leaves from *E. citriodora* (Batish et al 2006).

We have studied differences in essential oil yield and composition between adult leaves and the new ones harvested three months after submitting *E. camaludulensis* trees to mechanical damage. We have found that young leaves produced higher yields, 66.3% more essential oil than adult ones. Non oxygenated terpenes were particularly abundant in young leaves, whereas higher relative abundances in many oxygenated terpenes, including 1,8-cineole, were found in adult leaves. Table 2 shows relative abundances of *E. camaludulensis* essential oil components exhibiting significant differences between adult and young leaves (unpublished results).

Terpene	**Adult leaves**	**Young leaves**
α-Pinene	4.4 ± 0.5	7.3 ± 0.6
Camphene	0.3 ± 0.09	0.1 ± 0.05
β-Pinene	3.1 ± 0.4	2.3 ± 0.2
Myrcene	0.7 ± 0.22	2.7 ± 0.5
α-Phellandrene	0.1 ± 0.04	0.5 ± 0.1
α-Terpinene	0.09 ± 0.02	0.4 ± 0.17
1,8 Cineole	59.6 ± 3.7	48.6 ± 3.3
β-Ocymene	0.5± 0.1	0.3 ± 0.1
γ-Terpinene	3.9 ± 0.3	5.7 ± 0.8
α–Terpinolene	traces	1.0 ± 0.2
Linalool	0.23 ± 0.08	0.55 ± 0.1
α–Fenchol	0.55 ± 0.2	0.31 ± 0.1
Trans-Pinocarveol	2.12 ± 0.70	0.52 ± 0.15
Pinocarvone	0.86 ± 0.14	0.55 ± 0.16
α-terpineol	8.35 ± 0.69	5.80 ± 0.61
geranil acetate	traces	0.43 ± 0.15
Isolongifolene	traces	0.33 ± 0.05
β–Humulene	0.67 ± 0.3	0.99 ± 0.24
Globulol	2.47 ± 0.63	4.25 ± 0.60
Epiglobulol	1.24 ± 0.47	3.65 ± 0.35
Oxigenated terpenes	80.27 ± 0.35	69.47 ± 0.46
Non-oxigenated terpenes	16.71 ± 0.17	24.44 ± 1.82

Table 2. Mean relative abundances of *E. camaldulensis* essential oil components in adults and young leaves. Data given as mean ± standard deviation.

Marzoug et al (2011) have also reported higher amounts of oxygenated derivatives in *E. oleosa* adult leaves (43.2%) compared to young ones.

Secondary metabolites can be found in almost all plant organs, their concentrations and relative abundances being characteristic of each kind. Particular chemical families are in general distributed in different proportions depending on the plant organ and tissue. Marzoug et al 2011 analyzed essential oil yield and composition in different *E. oleosa* organs, their results for most abundant components shown in table 3.

Oil component (%)	Stems	Adult leaves	Fruits	Flowers
α-Pinene	5.2	1.7	2.6	2.2
p-Cymene	6.8	10.6	9.0	9.2
Limonene	4.2	1.5	0.7	1.6
1,8-Cineole	31.5	8.7	29.1	47.0
cis-Sabinol	3.1	4.2	2.5	1.0
trans-Pinocarveol	9.9	-	0.1	0.1
Pinocarvone	3.5	1.8	1.0	0.3
Verbenone	2.1	3.7	1.4	0.8
α-Selinene	-	0.5	10.0	2.1
Spathulenol	3.5	16.1	3.4	-
γ-Eudesmol	5.6	15.0	16.4	12.5
Essential oil yields (%)	0.52	0.45	1.12	0.53

Table 3. Essential oil yield and composition in different *E. oleosa* organs.

Essential oils bioactivity depends on their particular composition, which changes from organ to organ; different biological activities have been reported for essential oils obtained from eucalypts different organs. Essential oils from *E. globulus* Labill leaves have been widely used as antiseptic and for the relief of cough symptoms, colds, sore throat and other infections (Kumar et al 2007). It has been early reported that the main compound of its fruit oil was 1,8-cineole (Basias and Saxena 1984), however aromadendrene was reported almost two decades later by other authors to be the main constituent for the same species (Pereira et al 2005); differences in environmental conditions might be responsible for the lack of agreement in the results. Mulyaningsih and coworkers (2010) have suggested that aromadendrene may significantly contribute to antimicrobial activity of its fruit oil. Combinations of aromadendrene and 1,8-cineole showed additive effects in most cases, but also synergistic behavior. *E. globulus* fruit essential oil has been proved to display a pronounced antimicrobial effect towards multidrug-resistant bacteria.

4.2 Environmental conditions

Biotic stresses (disease, herbivory and/or the presence of competitors) as much as abiotic stresses such as nutrient deficiency and drought can affect chemical defence's production, in particular essential oil yield and composition. Potential of nursery preconditioning to enhance survival chances of future trees by reducing their palatability or by attracting beneficial insects as a result of changes in chemical defenses may be an answer to overcome environmental adversities (Leicach et al 2010). It can affect leaves quality by modulating their chemical composition, a determinant feature, besides tissues hardness, towards herbivory. It has been demonstrated that a higher 1,8-cineol relative abundance in essential oil can lower defoliation level due to *Anoplognathus* beetles (Edwards et al 1993). In order to develop a more sustainable commercial forestry minimizing chemical input, studies have been performed by our research group to obtain more resistant seedlings to begin with. Controlled drought conditions at particular nursery stages have proved to be one of the possible ways to achieve it, particularly when they enhance seedlings chemical defensive potential (Leicach el al 2010).

4.2.1 Abiotic factors

Studies performed in the last fifty years confirmed that most important abiotic factors affecting allelochemicals biosynthesis are water, soil quality, light, and temperature, several pathways within secondary metabolism being often increased when plants are exposed to growth condition different from their optimal requirements. Spatial and temporal variation in resources availability may influence the relative magnitude of defence benefits in plants, each particular class of compounds within them being able to respond in different ways to changes in environmental conditions.

Soil degradation has proved to increase allelochemicals production in different species. We have previously reported data related to the modulating effect of soil deterioration on secondary metabolites production by two widespread weed species, *Chenopodium album* and *Senecio grisebachii*. *C. album* samples grown in continuously cultivated (deteriorated) Argentinean Rolling Pampa soils have demonstrated to produce higher levels of long chain hydrocarbon derivatives (known to inhibit plant germination) than those grown in pristine soil (Leicach et al 2003). We have also demonstrated that the invasive weed *S. grisebachii* produced higher levels of toxic pyrrolizidine alkaloids in soils with higher deterioration degree. Significant differences were found when reproductive organs were studied, inflorescences from samples grown in less deteriorated soil showing lower relative abundances of toxic alkaloids (Yaber Grass et al 2011).

Water deficiency has proved to be a usual stress affecting agriculture and forestry, drought conditions being able to change allelochemicals abundances, including essential oil yield and composition. As a result of previous studies we have reported qualitative changes in essential oils obtained from young *E. camaldulensis* leaves after submitting seedlings to drought during nursery period. We did not find significant changes in essential oil yield, but we found significant changes in essential oil relative composition. Leaves from seedlings submitted to drought developed an essential oil composition that has been previously reported to be characteristic of mature leaves (Silvestre et al 1997) and related to their higher resistance to herbivory. Total amount of oxygenated terpenes were significantly increased as a result of water deficiency; globulol, epiglobulol, and ledol abundances were doubled, and 1,8-cineole content was enhanced by 28.3%, whereas total amount of non-oxygenated terpenes was significantly decreased (44%) (Leicach et al 2010).

Light quality and nutrient availability are also determining factors. Gleadow and Woodrow (2002) have reported that *E. cladocalyx* cyanogenic glycoside concentration was increased in near 70% in fully expanded leaves in response to moderate water stress; another research group has demonstrated that light deficiency caused the opposite effect on this species, enhancing cyanogenic glycoside production (Burns et al 2002). In both reports total phenolics and condensed tannins remained unaffected. Studies related to fertilization effects showed that *E. globulus* seedlings reduced their condensed tannins content while significantly enhancing essential oil production when nutrients were added (O'Reilly-Wapstra et al 2005).

4.2.2 Biotic factors

Damage caused by biotic factors such as herbivory have often proved to enhance secondary metabolism, however response depends on the particular species. We have previously reported changes in quinolizidine alkaloids abundances in two *Lupinus* species (*L. albus* and *L. angustifolius*) that responded in different ways to mechanical damage and herbivory (Vilariño et al 2005, Chludil et al 2009).

Many plants species respond to wounding, herbivory or pathogens attack by increasing endogenous synthesis and releasing jasmonic acid and methyl jasmonate, starting in damaged organs. These volatile signals have proved to activate defence responses in intact neighbours from the same species. Even when *Eucalyptus* species produce a complex array of constitutive chemical defenses, no significant changes have been found in essential oils, polyphenolics and foliar wax related compounds from trees following foliar-chewing insect damage (Rapley et al 2007). These authors reported that only foliar tannins seemed to be affected, increasing their concentration three months after larval feeding. However they did not find such differences in wounded trees five months later, suggesting that increase of tannins production was most likely a rapid response, which proved to diminish further larval survival and branch defoliation. They have suggested that foliar tannins may operate as toxins and/or anti-feedants to *Mnesampela privata* larval feeding.

5. Ecological significance of natural defenses of plants

Both volatile and non-volatile secondary metabolites from eucalypts have been associated to a wide spectrum of roles associated to defensive strategies.

Host location by insects has proved to be related to leaves chemical composition, secondary metabolites involved in those interactions are known as **semiochemicals**. Research on semiochemicals has grown since the 1950s, parallel to that related to allelochemicals, in terms of isolation and identification of responsible secondary metabolites; the final goal in both areas being development of solutions for agriculture and forestry problems through applied research. Sex pheromones and kairomones represent the most important groups of chemical signals in intra and interspecies communication, having proved to affect host choice by a particular insect species, i.e. mountain pine beetles feed on pines avoiding alder trees as a result of their ability to detect kairomones of appropriate hosts and non-host species. Volatile organic compounds playing the role of kairomones, are detected by insects through olfactory receptors located usually in antennae hairs.

Similarities in secondary metabolites between *Eucalyptus* species and those belonging to other genera have proved to trigger utilization of novel hosts by insects feeding on leaves. Electroantennogram studies (EAG) have demonstrated that *M. privata* female responds to ubiquitous eucalypt monoterpenes using them as host location and assessment cues (Steinbauer et al 2004). These authors also suggested that epicuticular waxes can be used as a leaf age indicator. *M. private,* also known as autumn gum moth, seems to take oviposition decisions based on both nonstructural epicuticular wax and foliar monoterpene cues. They may lay their eggs on novel hosts if foliar chemistry resembles that of the primary host (Ostrand et al 2008), new expansion hosts sharing many terpenes with natural ones (Paine et al 2011). Behavioral assays showed that natural and novel hosts with high amounts of α-pinene received fewer eggs than those with lower amounts; the opposite occurred with α-terpineol that have shown to enhance *M. private* oviposition in eucalypts containing it higher concentrations (Ostrand et 2008). De Little (1989) has also suggested that similarities in foliar terpenes may partly explain host expansion of leaf beetles *Chrysophtharta bimaculata* onto *E. nitens* in Tasmania. Steinbauer and Wanjura (2002) have discussed the preferential selection by Christmas beetles (*Anoplognathus montanus* and *A. pallidicollis*) of exotic peppercorn trees (*Schinus molle*) over neighbouring previous eucalypt hosts, suggesting that it could be related to the lack of 1,8-cineole in *S. molle* combined with the presence of other attractant monoterpenes, since 1,8-cineole is a main constituent in eucalypt species resistant to Christmas beetles.

Comparative GC-EAG studies have shown that several volatile compounds, including 3-hydroxy-2-butanone, 3-methyl-1-butanol, ethyl 3-methylbutanoate, (*Z*)-3-hexen-1-ol, α-pinene, β-pinene, *p*-cymene, 1,8-cineole, limonene, guaiene, α-terpinene, and linalool are be detected as hosts signals by the woodborer *Phoracantha semipunctata* (Barata et al 2002). These authors suggested that α-cubebene and (*E*)-4,8-dimethylnone-1,3,7-triene may act as cues for avoidance of unsuitable hosts.

Eucalypts extracts contain an array of defensive allelochemicals exhibiting various biological effects, antibacterial, antioxidant, and antihyperglycemic, among them (Takahashi et al 2004), with essential oils playing a central role in many of these biological functions. Most essential oils, including those obtained from eucalypts, have shown to display some degree of antimicrobial activity that is in general related to the presence of terpenoid and phenylpropanoid compounds that have proved to exhibit individual antimicrobial effects. Biological activity of essential oils starts most likely, affecting cell membrane structure and functions. Because of their lipophilicity, many of their components can accumulate in membranes being able to disrupt its structure and to affect transport processes and other membrane-associated events such as signal transmission, and ATP synthesis particularly in microorganisms (Einhellig 1995).

Essential oils bioactivity has been often related to the presence of 1,8-cineole in eucalypts, however it has been suggested that several other components may eventually act additively or synergistically. Citronellal, citronellol, citronellyl acetate, *p*-cymene, eucamalol, limonene, linalool, α-pinene, γ-terpinene, α–terpineol, alloocimene, and aromadendrene are among them (Duke 2004, Batish et al 2006; Su et al 2006, Liu et al 2008). *E. camaldulensis* and *E. urophylla* antibacterial properties have been described by Cimanga et al (2002), their antifungal activities have been confirmed by Su (2006). Strong inhibiting effects on *S. aureus* and *E. coli* were described by Bachir Raho and Benali (2008) for *E. globulus* and *E. camaldulensis* essential oils. Another *Eucalyptus* species, *E. oleosa*, produces essential oils with proven antibacterial activity against Gram-positive bacteria, and inhibitory effect on yeast-like fungi. Both activities have been related to the presence of oxygenated monoterpenes, 1,8-cineole in particular (Marzoug et al 2011).

Different assays were also performed by our research group to determine *Eucalyptus* essential oil biological activities. Antifungal activity of four *Eucalyptus* species (*E. camaldulensis, E. globulus, E. sideroxylon, E. tereticornis*) was tested against common pathogens affecting crops production (*Aspergillus flavus, Aspergillus niger, Cladosporium cucumerinum*). Antifungal activity (8 μl/dot) was tested by means of bioautographic assay (Homans and Fuchs 1970), *E. camaldulensis* being more active than *E. globulus, E. sideroxylon* and *E. tereticornis*.

A. flavus and *C. cucumerinum* susceptibilities to essential oil decreased in the following order: *E. camaldulensis* > *E. globulus* > *E. sideroxylon* > *E tereticornis*, being *C. cucumerinum* the most affected species. At the same dose *A. niger* also showed differential susceptibility between four species essential oil. *E. camaldulensis* essential oil exhibited almost two-fold *E. globulus* activity and more than two-fold *E. tereticornis* activity, but it was only 25% more effective than *E. sideroxylon* essential oil. The fact that *E. camaldulensis* essential oil was the one containing higher amounts of oxygenated terpenoids may suggest a possible correlation to its higher activity (Marzoug et al 2011).

Eucalyptus essential oil components have also been associated to allelopathic activities, most of them inhibitory; few species have been able to develop successfully near these trees in natural habitats (Liu et al 2008).

Other *Eucalyptus* defensive chemicals such as phenolic derivatives have also proved to sensitize phospholipids bi-layer, increasing cell membrane permeability and leakage of vital intracellular constituents and also impairing microbial enzymes activity (Moreira et al 2005).

A different class of phenolic derivatives, formylated phloroglucinols, only produced by some *Eucalyptus* species have shown to display a wide range of ecologically significant biological activities, particularly as anti-feedant (Lawler et al 1998, Sidana et al 2010).

Ecological significance of formylated phloroglucinol derivatives in *Eucalyptus* species has been associated to marsupial folivores feeding behaviour (Eschler et al 2000, Marsh et al 2003, Sidana et al 2010). Interactions between eucalypts and mammalian herbivores that feed on them have also been investigated by other authors (Lawler et al 2000, Close et al 2003). O'Reilly-Wapstra et al (2002) analyzed the role of plant genotype in *E. globulus* resistance to browsing by a generalist marsupial folivore, the common brushtail possum. Formylated phloroglucinol compounds (Eschler et al 2000) seem to play a significant defensive leaf trait conferring resistance in *E. globulus* juvenile coppice foliage (O'Reilly-Wapstra et al 2004).

6. Potential uses

Secondary metabolites produced by forestry species might be used in Integrated Pest Management either as repellents or attractants. *Eucalytpus* species have shown not only to reduce atmospheric carbon dioxide levels and provide timber biomass, but also to perform a variety of indirect services through their chemical defenses, essential oils being their most characteristic defensive treat with a wide spectrum of biological effects.

Plant extracts and isolated natural compounds represent a wide range of possibilities to replace or at least diminish the use of synthetic products to control pests and diseases affecting plants, animals and/or human beings. Bioactive natural products should also be seriously considered as they have proved to be more specific in most of their biological activities.

Forestry-derived industries have been focused during last decades on development of breeding techniques to produce trees with higher timber yields and enhanced wood quality. An undesired feature observed in most improved cultivars was an enhancement of susceptibility to diseases and predators, suggesting that they were investing less metabolic energy in chemical defence (Wallis et al 2010). This fact and the development of pesticide resistance in previously controlled herbivores may explain the increment in pesticide amounts further used in forestry plantations. Continuous agrochemicals overuse seriously affected environmental sustainability also triggering pest resurgence and development of resistance/cross-resistance (Brunherotto and Vendramim 2001). Similar practices in agriculture associated to food demands of a permanently growing population, strongly contributed to enhance environmental degradation; soil and water pollution, and losses in biodiversity being the main deterioration symptoms faced by most countries in the world.

The fact that synthetic pesticides have proved to represent a significant hazard to mammals and mankind directly consuming plant material or derived foodstuff has been thoroughly confirmed. Current concerns about their presence contaminating grains, fruits, and vegetables, has limited the number of permitted synthetic fumigants, encouraging the search for friendlier pest control alternatives. Natural products represent one of the most important alternatives, to control pests and diseases that affect plants and animals without deleteriously effecting environmental safety (Isman 1997, Men and Hall 1999, Tripathi et al 2008). Plants natural ability to produce allelochemicals is currently revised to enhance the number of cultivated species, varieties, or clones with enriched chemical defences in order to diminish agrochemicals dependency. Selection of populations or individuals with particular morphological and/or chemical characteristics making them less susceptible to biotic or abiotic stress is one of the possible ways to achieve it (Gershenzon and Dudareva 2007, Yaber Grass et al 2011).

Chinese folk medicine has used *Eucalyptus* species for centuries, hot water extracts of dried leaves from *E. citriodora* have been, and are still used to prepare anti-inflammatory, analgesic and antipyretic formulas for respiratory infections, such as sinus congestion and flu. Essential oils are easily biodegraded and have proved to exhibit low toxicity against vertebrates also playing an important role as bioherbicide for weed management (Barton 1999, Batish et al 2007, Batish et al 2008).

E. camandulensis essential oil has also proved to be useful for pharmaceutical purposes. It has been used to treat lung diseases and cough in medicines like expectorants, also taking advantage of its antituberculosis, antibacterial, and antifungal properties.

The significant negative effects of *E. camaldulensis* and *E. urophylla* essential oils on *S. aureus* and *E. coli* development contribute to point out the potential of both *Eucalyptus* species for antiseptic, microbiostatic, or as disinfectant activities (Bachir Raho and Benali 2008).

Eucalyptus essential oils have been preferred over those obtained from other forestry exploited species because they have proved to be useful in perfumery, pharmaceutical and other industries playing multipurpose roles (FAO 1995). They have also proved to negatively affect virus development, Schnitzler et al (2001) reported *in vitro* activity against antiherpes virus. *E. globulus* essential oil components, alone or in combination with other antibacterial agents, may provide a promising new scheme in phytotherapy.

Besides essential oil components, other eucalypts defensive treats may also be used taking advantage of their biological effects. Formylated phloroglucinols have also proved to be therapeutically and/or pharmacologically significant (EBV inhibitory (Takasaki et al 1990), anti-bacterial (Satoh et al 1992), HIV-RT inhibitory (Nishizawa et al 1992), aldose reductase inhibitory (Satoh et al 1992), anti-protozoal (Bharate et al 2006). The dimeric derivative, sideroxylonal A, has proved to act as a potent marine anti-fouling agent comparable to the most active compound 2,5,6-tribromo-1-methyl-gramine (Singh et al 1996).

Chemical preservatives used during last decades to avoid losses in crops production and spoilage of packed, canned and bottled foodstuff, have proved to be responsible for several kinds of residual toxicity, carcinogenic and teratogenic effects, among them. *Eucalyptus* essential oil may represent a possible alternative to replace, at least partially, some agrochemicals currently used to control crops diseases and plagues, and as chemical preservatives in foodstuff. They have proved to be efficient as fumigants and contact insecticides in the control of stored-product insects (Batish et al 2008).

Eucalyptus essential oils have been early included in the list of Generally Regarded as Safe category by Food and Drug Authority of USA and classified as non-toxic (USEPA, 1993), and European countries have also accepted them as flavoring agents in foodstuff.

Eucalyptus oil and 1,8-cineole oral and acute LD_{50} have been reported to be 4440 mg/kg bodyweight (BW) and 2480 mg/kg BW (Regnault-Roger 1997), respectively, to rats, values that demonstrated they could be less toxic than pyrethrins (LD_{50}: 350–500 mg/kg BW; USEPA 1993) and even technical grade pyrethrum (LD_{50} values 1500 mg/kg BW) (Casida and Quistad, 1995). *Eucalypts* leaf extracts have also been approved as natural food additives because of their antioxidant properties and included in the List of Existing Food Additives in Japan (Amakura et al 2002, Tyagi and Malik 2011); some of them are used in cosmetic formulations (Takahashi et al 2004, Gilles et al 2010).

Fresh and dried *E. globulus* leaves are commonly used in Africa to control insects feeding on crops. *Eucalyptus* leaves have been also used in Brazilian grain stores to deter *Sithophilus zeamais* and *Rhysopertha dominica*. It has been demonstrated that secondary metabolites produced by *Eucalyptus* and other closely related species displayed high levels of repellency towards a variety of invertebrates (Thacker and Train 2010).

Some *Eucalyptus* essentials oils containing high 1,8-cineole amounts have also proved to be effective to control mites. They could be used as a natural acaricides, as they have shown to be effective against varroa mite, *Varroa jacobsoni*, an important parasite of honeybee, *Tetranychus urticae* and *Phytoseiulus persimilis* (Choi et al 2004) and *Dermatophagoides pteronyssinus* (Saad et al 2006).

Essential oils and their major constituents have shown toxicity against a wide range of microbes including bacteria and fungi, both soil-borne and post-harvest pathogens. Su et al. (2006) demonstrated the antifungal activity of essential oils from *E. grandis, E. camaldulensis,* and *E. citriodora* against the mildew and wood rot fungi viz, *Aspergillus clavatus, A. niger, Chaetomium globosum, Cladosporium cladosporioides, Myrothecium verrucaria, Penicillium citrinum, Trichoderma viride, Trametes versicolor, Phanerochaete chrysosporium, Phaeolus schweinitzii,* and *Lenzites sulphureus*.

Several authors have described *Eucalyptus* essential oils bioactivity against pathogenic and food spoilage bacteria and yeast (Papachristos and Stanopoulos 2002, Sartorelli et al 2007). It has been demonstrated that *E. globulus* essential oils can display significant bactericidal and bacteriostatic effects on *E. coli* (Moreira et al 2005). Citronellal, major constituent of *E. citriodora,* has been successfully used to control development of two fungal pathogens affecting rice crops, *Rhizoctonia solani* and *Helminthosporium oryzae*. Ramezani et al (2002) have reported the complete inhibition of both funguses by *E. citriodora* essential oil, emphasizing that citronellal was even more effective.

Tzortzakis (2007) has reported an alternative use of *E. globulus* essential oil, suggesting that its vapors represent a good choice when trying to maintain strawberry and tomato postharvest freshness and firmness during storage and transit. It was demonstrated that no changes occurred in their sweetness, organic acid and total phenolic content, after exposing fresh strawberries and tomatoes to oil vapors.

S. oryzae is one of the most distributed species deleteriously affecting stored grains. We have performed bioassays to analyze fagorrepellency, obtaining data on differential capacity of *E. camaldulensis* essential oils depending on leaves age. *E. camaldulensis* essential oils obtained by hydro-distillation from adult and young fresh leaves were tested at two doses (10 and 20 μl/dot) on *S. oryzae* as target insect. Essential oil higher repellency effects were observed for both kinds of leaves in the first 20 min; being always more active those obtained from adult leaves (50 y 60 %) compared to young ones (30-35%). After that period, essential oils from both, adult and young leaves, showed lower repellency values, with no significant differences between doses by the time bioassay was ending (unpublished results). The higher repellency levels of adult leaves essential oil might be related to its higher proportion of oxygenated terpenoids, as it has been previously suggested for other of its biological activities (Marzoug et al 2011). The following figure represents repellency percentages of *E. camaldulensis* adult and young leaves essential oils towards *S. oryzae*.

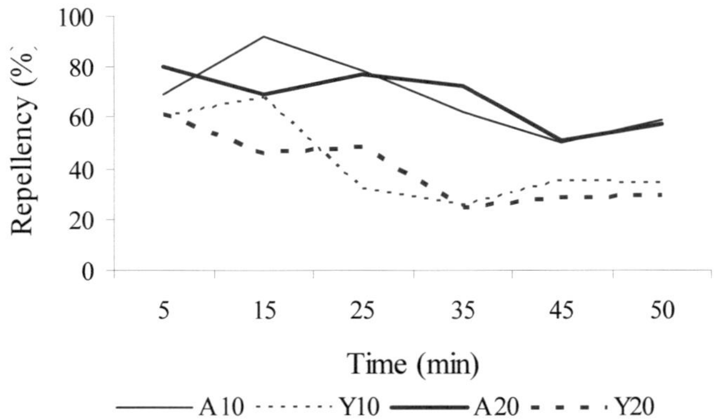

Fig. 4. Repellency of *E. camaldulensis* leaves essential oils towards *S. oryzae*. Adult leaves (A) 10 and 20 μl/dot, Young leaves (Y) 10 and 20 μl/dot.

Eucalyptus essential oil has been also used as antifeedant, particularly in formulas against biting insects (Chou et al 1997). Su et al (2006) have described insecticidal properties including larvicidal and mosquito repellent of different members of botanical Myrtaceae family. Trigg (1996) has earlier reported that products based on eucalypts essential oils used as insect repellent, can protect humans from biting insects up to 8 h depending on the concentration of the essential oil. Lucia et al (2007) have also demonstrated that *E. globulus* essential oil displays toxic effects on *Aedes aegypti* larvae, these authors determined its LC_{50}=32.4 ppm. In relation to this activity, Seyoum et al (2003) have reported that burning of *E. citriodora* leaves represents a cheap and effective method of household protection against mosquitoes in Africa.

Plant-parasitic nematodes represent another major plant plague that infesting different food crops such as vegetables and fruit plantations. They cause considerable economic losses related to reduced yields and unmarketable production features. *Eucalyptus* essential oils have also been shown to possess nematicidal activity. Pandey et al (2000) demonstrated that essential oils (at 250 ppm) obtained from *E. citriodora* and *E. hybrida* resulted highly toxic to *Meloidogyne incognita*, inhibiting growth of root-knot nematode at 250 ppm. More recently, Ibrahim et al (2006) confirmed those results as they reported that eucalypts essential oil demonstrated to be toxic to second-stage juveniles (J2s) of root-knot nematode *M. incognita*.

Native insects that became pests on *Eucalyptus* species in plantations outside Australia, are either highly polyphagous or have native trees belonging to Myrtaceae family as natural hosts. Insects in the latter group may be pre-adapted to shift hosts depending on host chemical composition (Kliejunas et al 2001, Wingfield et al 2008). Chemical identification of secondary metabolites used by insects as signals to select host trees can be useful in association to genetic programs to obtain clones lacking these substances, in order to turn them less attractive to them (Hall and Menn 1999). Potential of several essential oil components to diminish impact of most deleterious plagues should be considered in such programs.

Eucalyptus species produce, as other tree species, kairomones that have proved to modulate behavior of insects feeding on leaves. Semiochemicals, particularly host kairomones, can be useful tools to disrupt the location of food crops by pests or to design baited traps for monitoring programs (Thacker and Train 2010).

Traditional methods to detect pests in forest plantations were usually based on visual signs of damage, when plantation production was almost lost and there was not enough time to apply control methods to avoid it. Static traps have been developed in the last decades in order to prevent such losses in plantations yields. Baited traps, incorporating kairomone lures have proved to attract a range of insect families being able to detect the presence of target species even when populations are small, as in developing outbreaks or new incursions (Miller 2006). They can be applied to monitor periodical outbreak of established pests, their spread into new areas, and the emergence of new folivores.

As it was mentioned before, drought represents a common stress during eucalypts establishment and water stress has proved to diminish levels of resistance to insect attack in host trees. There are many reports on tree mortality caused by stem borers in mid-rotation *Eucalyptus* plantations that have been correlated with drought events. Damage by stem-boring insects has proved to kill or degrade pruned trees so severely that they cannot be used at all or can only be suitable for pulpwood (Bashford 2008).

Static traps containing *Eucalyptus* essential oil components have been developed for early detection of stem-boring insects in *Eucalyptus* plantations, they have been used to detect the presence of low (pre-outbreak) populations of stem-borers in *Eucalyptus* plantations where a range of species emerge at different times during summer months. Several host tree volatiles known to attract stem-boring beetle species have been studied by Bashford 2008. Eucalypt volatiles such as 1,8-cineole, phellandrene, α-pinene y α/β pinene have been successfully used to monitor almost all present stem borers in intercept panel traps. They were used to survey ambrosia beetles and other Scotylidae presence in *Eucalyptus* plantations in Brazil (Flechtmann et al 2000).

Phytotoxicity has been mentioned above as a characteristic biological activity of some eucalypts chemical components. *E. citriodora* essential oil has proved to be more effective towards broad-leaved (dicot) weeds than to grassy (monocot) ones (Singh et al 2006) most likely because citronellal, the main component of its essential oil, has proved to be more toxic towards to broad-leaved weeds. Though several authors have evaluated phytotoxic effects of eucalypts oils taking advantage of their herbicidal potential against weeds, there are several constraints mostly related to oil yield variability with season and climate, among many other environmental factors. Volatility and lipophilicity of oil components, difficulties in plant uptake, affectivity under field conditions and toxicity towards non-target plants are some other issues to be clarified before their use as commercial herbicides is accepted (Batish et al 2006).

The same constrains effecting their application as herbicide can be extended to the wide spectrum of other biological activities. The low amount of commercialized products based on eucalypts essential oils in spite of the huge scope and market for natural pesticides, is basically caused by the strict market regulations including toxicological evaluation against non-target organisms, need of product standardizations, demands related to the lack of reproducibility in plant material quality and regulatory approvals limiting their use.

Even when there is still much work to do to overcome those constrains, *Eucalyptus* essentials oils, which has proved to exhibit an environment friendly nature, can be considered a potential sustainable alternative for pest management in urban areas, homes and other sensitive areas such as schools, restaurants and hospitals. Moreover, farmers involved in organic crops and greenhouse production systems and those from developing countries who cannot afford costly synthetic pesticides, could also take advantage of their bioactivity using them as natural pesticides (Isman 2006).

7. References

Alzogaray, R. A.; Lucia, A., Zerba, E. N. & Masuh, H. M. (2011). Insecticidal activity of essential oils from eleven *Eucalyptus* spp. and two hybrids: Lethal and sublethal effects of their major components on *Blattella germanica*. *Journal of Economic Entomology*, Vol.104, No.2, (April 2011), pp. 595-600, ISSN 00220493

Amakura, Y.; Uminoa, Y.; Tsujia, S.; Itob, H.; Hatanob, T. & Yoshidab, T. (2002). Constituents and their antioxidative effects in eucalyptus leaf extract used as a natural food additive. *Food Chemistry*, Vol.77, No.1, (May 2002), pp. 47–56, ISSN 03088146

Anderson, D. (1993). Some factors influencing the demand for gum arabic (*Acacia senegal* (L.)) and other water-soluble tree exudates. *Forest Ecology and Management*, Vol.58, No.1-2, (April 1993), pp.1-18, ISSN 03781127

Bachir Raho, G. & Benali, M. (2008). Antibacterial activity of leaf essential oils of *Eucalyptus globulus* and *Eucalyptus camaldulensis*. *African Journal of Pharmacy and Pharmacology*, Vol.2, No.10, (December 2008), pp. 211-215, ISSN 19960816

Barata, E. N.; Mustaparta, H.; Pickett, J. A.; Wadhams, L. J. & Araujo, J. (2002). Encoding of host and non-host plant odors by receptor neurons in the *Eucalyptus* woodborer, *Phoracantha semipunctata* (Coleoptera: Cerambycidae). *Journal of Comparative Physiology A: Neuroethology, Sensory, Neural and Behavional Physiology*, Vol.188, No.2, (March 2002), pp.121–125, ISSN 03407594

Barros, E.; Verryn, S. & Hettasch, M. (2002). Identification of PCR-based markers linked to wood splitting in *Eucalyptus grandis*. *Annals of Forest Science*, Vol.59, No.5-6, (July-October 2002), pp. 675-678, ISSN 12864560

Barton, A. F. M. (1999). The Oil Malle Project. A multifaceted industrial ecology case study. *Journal of Industrial Ecology*, Vol.3, No.2-3, (April 1999), pp. 161-176, ISSN 10881980

Bashford, R. (2008). The development of static trapping systems to monitor for wood-boring insects in Tasmanian plantations. *Australian Forestry*, Vol.71, No.3, (June 2008), pp. 236-241. ISSN 00049158

Basias, R. & Saxena, S. (1984). Chemical examination of essential oil from the fruits of *Eucalyptus globulus* Labill. *Herba Hungarica*, Vol.23, pp. 21–23, ISSN 0018-0580

Batish D. R.; Singh H. P.; Setia N.; Kaur S.; Kohli R. K. (2006). Chemical composition and phytotoxicity of volatile essential oils from intact and fallen leaves of *Eucalyptus citriodora*. *Zeitschrift fur Naturforschung. Section C: Journal of Biosciences* Vol.61, No.7-8, (July-August, 2006), pp. 465–471, ISSN 09395075

Batish, D. R.; Singh, H. P.; Setia, N.; Kohli, R. K.; Kaur, S. & Yadav, S. S. (2007). Alternative control of littleseed canary grass using eucalypt oil. *Agronomy for Sustainable Development*, Vol.27, No.3, (September 2007), pp. 171–177, ISSN 17740746

Batish, D. R.; Singh, H. P.; Kohli, R. K. & Kaur, S. (2008). *Eucalyptus* essential oil as a natural pesticide. *Forest Ecology and Management*, Vol.256, No.12, (December 2008), pp. 2166-2174, ISSN 03781127

Baucher, M.; Halpin, C.; Petit-Conil, M.; Boerjan W. (2003). Lignin: Genetic engineering and impact on pulping. *Critical Reviews in Biochemistry and Molecular Biology*, Vol.38, No.4, (July-August 2003), pp. 305-350, ISSN 10409238

Binkley, D. & Stape, J. (2004). Sustainable management of *Eucalyptus* plantations in a changing world. Proceedings of an IUFRO Conference "*Eucalyptus* in a changing world", N. Borralho et al. (Ed.), Aveiro. Portugal, (October 11-15, 2004), pp. 11-17

Boerjan, W.; Ralph, J. ; Baucher, M. (2003). Lignin biosynthesis. *Annual Review of Plant Biology*, Vol.54, (June 2003), pp. 519–546, ISSN 15435008

Boland, D. J.; Brooker, M. I. H.; Chippendale, G. M.; Hall, N.; Hyland, B. P. M.; Johnston, R. D.; Kleinig, D. A.; McDonald, M. W. & Turner, J. D. (2006). *Forest Trees of Australia* (5th ed.). CSIRO Publishing, ISBN: 9780643069695, Collingwood, Victoria, Australia

Bowers, W.; Evans, P. & Spence, S. (2000). Trichomes in *Eucalyptus maculata* citriodon possess a single potent insect repellent. *Journal of Herbs, Spices and Medicinal Plants*, Vol.7, No.4, pp. 85-89, ISSN 10496475

BRACELPA. 2009. Brazilian Pulp and Paper Industry Performance, Available from: http //www.bracelpa.org.br/eng/estatisticas/

Brooker, M. I. H.; Slee, A. V.; Connors, J. R. & Duffy, S. M. (2006). *Euclid,Eucalypts of Australia*. 3rd Edition. CSIRO Publishing, Canberra, Australia, ISBN 9780643093355

Brunherotto, R. & Vendramim, J. (2001). Bioactividade de extratos aquosos de *Melia azedarach* L. sobre o desenvolvimento de *Tuta absoluta* (Meyrick) (Lepidoptera: Gelechiidae) em tomateiro. *Neotropical Entomology*, Vol.30, No.3, (September 2001), pp. 455-459, ISSN 16788052

Burns, A.; Gleadow, R. & Woodrow, I. (2002). Light alters the allocation of nitrogen to cyanogenic glycosides in *Eucalyptus cladocalyx*. *Oecologia*, Vol.133, No.3, (November 2002), pp 288-294, ISSN 00298549

Chen, Z.; Chang, S.; Ho, C.; Chen, Y.; Tsai, J. & Chiang, V. (2001). Plant production of transgenic *Eucalyptus camaldulensis* carrying the *Populus tremuloïdes* cinnammate 4-hydroxylase gene. *Taiwan Journal of Forest Science*, Vol.16, No.2, (June 2001), pp. 249–258, ISSN 10264469

Chennoufi, R.; Morizur, J.; Richard, H. & Sandret, F. (1980). Study of *Eucalyptus globulus* essential oils from Morocco (young and adult leaves). *Rivista Italiana EPPOS*, Vol.62, pp. 353-357, ISSN 03920445

Chludil, H. D.; Vilariño, M. P.; Franco, M. L.; Leicach, S. R. (2009). Changes in *Lupinus albus* and *Lupinus angustifolius* alkaloids profiles as response to mechanical damage. *Journal of Agricultural and Food Chemistry*, Vol.57, No.14, (July 2009), pp. 6107–6113, ISSN 00218561

Chou, J.T.; Rossignol, P. A. & Ayres, J.W. (1997). Evaluation of commercial insect repellents on human skin against *Aedes aegypti* (Diptera: Culicidae). *Journal of Medical Entomology*, Vol.34, No.6 (November 1997), pp. 624–630, ISSN 00222585

Choi, W.; Lee, L.; Park, H. & Ahn, Y. (2004). Toxicity of plant essential oil to *Tetranychus urticae* (Acari: Tetranychidae) and *Phytoseiulus persimilis* (Acari: Phytoseiidae). *Journal of Economic Entomology*, Vol.97, No.2, (April 2004), pp. 553-558, ISSN 00220493

Cimanga, K.; Kambu, L.; Apers, T. S.; De Bruyne, T.; Hermans, N.; Totte, J.; Pieters, L. & Vlietinck, A. J. (2002). Correlation between chemical composition and antibacterial activity of essential oils of some aromatic medicinal plants growing in the Democratic Republic of Congo. *Journal of Ethnopharmacology*, Vol.79, No.2, (February 2002), pp. 213-220, ISSN 03788741

Close, D. C.; McArthur, C.; Paterson, S.; Fitzgerald, H.; Walsh, A. & Kincade, T. (2003). Photoinhibition: a link between effects of the environment on eucalypt leaf chemistry and herbivory. *Ecology*, Vol.84, No.2, (November 2003), pp. 2952–2966, ISSN 00129658

Cunningham, A. B. (1996). People, park and plant use. Recommendations for multiple-use zones and development alternatives around Bwindi Impenetrable National Park, Uganda. People and Plants Working Paper 4. UNESCO, Paris.

Domingues, R. M. A.; Sousa, G. D. A.; Silva, C. M.; Freire, C. S. R.; Silvestre, A. J. D. & Pascoal Neto, C. (2011). High value triterpenic compounds from the outer barks of several *Eucalyptus* species cultivated in Brazil and in Portugal. *Industrial Crops and Products*, Vol.33, No.1, (January 2011), pp. 158–164, ISSN 09266690

Doran, J. & Bell, R. (1994). Influence of non-genetic factors on yield of monoterpenes in leaf oils of *Eucalyptus camaldulensis*. *New Forests*, Vol.8, No.4, (October 1994), pp. 363-379, ISSN 01694286

Dovie, D. B. K. (2003). Rural economy and livelihoods from the non-timber forest products trade compromising sustainability in southern Africa? *International Journal of Sustainable Development and World Ecology*, Vol.10, No.3, pp. 247–262, ISSN 13504509

Duke, J. A. (2004). Dr. Duke's Phytochemical and Ethnobotanical databases. Available from: http://www.ars-grin.gov/duke/

Edwards, P. B.; Wanjura, W. J. & Brown, W.V. (1993). Selective herbivory by Christmas beetles in response to intraspecific variation in *Eucalyptus* terpenoids. *Oecologia*, Vol 95, No.4, (October 1993) pp. 551-557, ISSN 00298549

Einhellig, F. A. (1989). Interactive effects of allelochemicals and environmental stress. In: *Phytochemical Ecology: Alelochemicals, Mycotoxins, and Insect Pheromones and Allomones*. Chou, C. H. & Waller, G. R. (Ed.), pp. 101-116, Academia Sinica Monograph Series 9, Taipei

Einhellig, F.A. (1995). Mechanism of action of allelochemicals in allelopathy. In: *Allelopathy: Organisms, Processes and Applications*. Inderjit, Dakshini, K. M. M., Einhellig, F. A. (Ed.), pp. (96-116), A.C.S. Symposium Series 582. American Chemical Society, ISBN 0841230617, Washington, USA

Eldridge, K.; Davidson, J.; Harwood, C. & Van Wyk, G. (1994). *Eucalypt Domestication and Breeding*. Oxford University Press, ISBN 9780198548669, USA

Eschler, B. M. & Foley, W. J. (1999). A new sideroxylonal from *Eucalyptus melliodora*. *Australian Journal of Chemistry*, Vol.52, No.2, (February 1999), pp. 157–158, ISSN 00049425

Eschler, B. M.; Pass, D. M.; Willis, I. R. & Foley, W. J. (2000). Distribution of foliar formylated phloroglucinol derivatives amongst *Eucalyptus* species. *Biochemical Systematics and Ecology*, Vol.28, No.9, (November 2000), pp. 813–824, ISSN 03051978

FAO. 1995. *Eucalyptus* oil. Chapter 5, In: *Flavour and Fragrances of Plant Origin*, Food and Agriculture Organization of the United Nations, ISBN 9251036489, Rome, Italy

FAO. 2011a. Non-Wood News 22. An information bulletin on non-wood forest products. April 2011, Food and Agriculture Organization of the United Nations, ISSN 10203435, Rome, Italy

FAO. 2011b. *The State of the World's Forests* 2011. Food and Agriculture Organization of the United Nations, ISBN 9879251067505, Rome, Italy

Fernandes, P. & Cabral, J. M. S. (2007). Phytosterols: applications and recovery methods. *Bioresource Technology*, Vol.98, No.12, (September 2007), pp. 2335–2350, ISSN 09608524

Flechtmann, C. A. H.; Ottati, A. L. T. & Berisford, C. W. (2000). Comparison of four trap types for ambrosia beetles (Coleoptera, Scolytidae) in Brazilian *Eucalyptus* stands. *Journal of Economic Entomology*, Vol.93, No. 6, (December 2000), pp. 1701-1707, ISSN 00220493

Florence, R. G. (1996). *Ecology and silviculture of eucalypt forests,* CSIRO Publishing, ISBN 0643057994, Collingwood, Victoria, Australia

Florence, R. G. (2004). Ecology and Silviculture of Eucalypt Forest. In *Eucalypt Ecology. Individuals to Ecosystems*. Williams, J. & Woinarski, J. (Ed.) . CSIRO Publishing. ISBN 052149740x, Cambridge University Press

Foelkel, C. (2009). Papermaker properties of *Eucalytus* trees, woods and pulp fiber. In *Eucalyptus online book*, Available from:
www.eucalyptus.com.br/eucaliptos/ENG14.pdf

Freeman, J.; O'Reilly-Wapstra, J.; Vaillancourt, R.; Wiggins, N. & Potts, B. (2008). Quantitative trait loci for key defensive compounds affecting herbivory of eucalypts in Australia. *New Phytologist*, Vol.178, No.4, (June 2008), pp. 846-851, ISSN 0028646x

Gentry, H.; Mittleman, N. & McCrohan, P. (1992). Introduction of chia and gum tragacanth, new crops for the United States. *Diversity and Distributions*, Vol.8, No.1, (January 1992), pp. 28-29, ISSN 13669516

Gershenson, J. & Croteau, R. (1990). Regulation of Monoterpene Biosynthesis in Higher Plants. In: *Biochemistry of the Mevalonic Acid Pathway to Terpenoids*, Towers, G. H. N. & Stafford, H. A. (Ed.), pp. 99-159, Plenum, N.Y.

Gershenzon, J. & Dudareva, N. (2007). The function of terpene natural products in the natural world. *Nature Chemical Biology*, Vol. 3, No.7, (July 207), pp. 408-414, ISSN 15524450

Gilles, M.; Zhao, J.; An, Min. & Agboola. S. (2010). Chemical composition and antimicrobial properties of essential oils of three Australian *Eucalyptus* species. *Food Chemistry*, Vol.119, No.2, (March 2010), pp. 731-737, ISSN 03088146

Girijashankar, V. (2011). Genetic trasnformation of eucalypts. *Physiology and Molecular Biology of Plants*, Vol.17, No.1, (March 2011), pp. 9–23, ISSN 09715894

Gleadow, R. M. & Woodrow, I. E. (2000). Temporal and spatial variation in cyanogenic glycosides. *Tree Physiology*, Vol.20, No.9, (May 2000), pp. 591–598, ISSN 0829318x

Gleadow, R. M.; Haburjak, J.; Dunn, J. E.; Conn, M. E. & Conn, E. E. (2008). Frequency and distribution of cyanogenic glycosides in Eucalyptus L'Hérit. *Phytochemistry*, Vol.69, No.9, (June 2008), pp. 1870–1874, ISSN 00319422

Gleadow, R. M. & Woodrow, I. E. (2002). Defense chemistry of cyanogenic *Eucalyptus cladocalyx* seedlings is affected by water supply. *Tree Physiology*, Vol.22, No.13, (September 2002), pp. 939-945, ISSN 0829318x

Gleadow, R. M.; Foley, W. J. & Woodrow, I. E. (1998). Enhanced CO_2 alters the relationship between photosynthesis and defense in cyanogenic *Eucalyptus cladocalyx* F. J. Muell. *Plant, Cell and Environment*, Vol.21, No.1, (January 1998), pp. 12–22, ISSN 01407791

Hall, F. R. & Menn, J. J. (1999). *Methods in biotechnology, biopesticides: use and delivery*. Humana Press Inc., ISBN 0896035158, New Jersey, USA

Hammerschmidt, R. & Schultz, J. C. (1996). Multiple defenses and signals in plant defense against pathogens and herbivores. In: *Phytochemical Diversity and Redundancy in Ecological Interactions. Series Recent Advances in Phytochemistry*, Vol.30, Romeo, J.; Saundres, J. & Barbosam P. (Ed.), pp. 121-154, ISBN 9780306455001, Plenum Press, N.Y.

Harborne, J. B. (1998). *Phytochemical Methods. A Guide to Modern Techniques of Plant Analysis*. Chapman & Hall (Ed.), ISBN 0412572605, London

Harcourt, R. L.; Kyozuka, J.; Floyd, R. B.; Bateman, K. S.; Tanaka, H.; Decroocq, V.; Llewellyn, D. J.; Zhu, X.; Peacock, W. J. & Dennis, E. S. (2000). Insect- and herbicide-resistant transgenic *Eucalyptus*. *Molecular Breeding*, Vol.6, No.3, (June 2000), pp. 307–315, ISSN 13803743

Henery, M.; Moran, G.; Wallis, I. & Foley, W. (2007). Identification of quantitative trait loci influencing foliar concentrations of terpenes and formylated phloroglucinol compounds in *Eucalyptus nitens*. *New Phytologist*, Vol.176, No.1, (October 2007), pp. 82–95, ISSN 0028646x

Herms, D. A. & Mattson, W. J. (1992). The Dilemma of plants: To grow or defend. *Quarterly Review of Biology*, Vol.67, No.3, (September 1992), pp. 283–335, ISSN 00335770

Hoadley, R. B. (2000). *Understanding Wood: A Craftman's Guide to Wood Technology*, The Taunton Press, ISBN 1561583588, Newtown, USA

Homans, A. L. & Fuchs, A. (1970). Direct bioautography on thin-layer chromatograms as a method for detecting fungitoxic substances. *Journal of Chromatography A*, Vol. 51, No.2 (September 1970), pp. 327-329, ISSN 00219673

Ibrahim, S. K.; Traboulsi, A. F. & El-Haj, S. (2006). Effect of essential oils and plant extracts on hatching, migration and mortality of *Meloidogyne incognita. Phytopathologia Mediterranea*, Vol.45, No.3, pp. 238–246, ISSN 00319465

Iglesias-Trabado, G. I.; Carballeira-Tenreiro, R. & Folgueira-Lozano, T. (2009). *Eucalyptus universalis. Global Cultivated Eucalypt Forest Map. Version 1.2.* BRACELPA, 2009. Brazilian Pulp and Paper Industry Performance, Available from: http://www.bracelpa.org.br/eng/estatisticas/

Isman, M. B. (1997). Neem and other botanical insecticides: barriers to commercialization. *Phytoparasitica*, Vol.25, No.4, (December 1997), pp. 339-344, ISSN 03342123

Isman, M. B. (2006). Botanical insecticides, deterrents, and repellents in modern agriculture and an increasingly regulated world. *Annual Review of Entomology*, Vol.51, (January 2006), pp. 45-66, ISSN 00664170

Jones, E. T. & Linch, K. A. (2007). Nontimber forest products and biodiversity management in the Pacific Northwest. *Forest Ecology and Management*, Vol.246, No.1, (July 2007), pp. 29-37, ISSN 03781127

Juan, L. W.; Lucia, A.; Zerba, E. N.; Harrand, L.; Marco, M. & Masuh, H. M. (2011). Chemical composition and fumigant toxicity of the essential oils from 16 species of *Eucalyptus* against *Haematobia irritans* (Diptera: Muscidae) adults. *Journal of Economic Entomology*, Vol.104, No.3, (June 2011), pp.1087-1092, ISSN 00220493

Kawaoka, A.; Nanto, K.; Ishii, K. & Ebinuma, H. (2006). Reduction of lignin content by suppression of expression of the LIM domain transcription factor in *Eucalyptus camaldulensis. Silvae Genetica*, Vol.55, No. 6, pp. 269–277, ISSN 00375349

Kliejunas, J. T.; Tkacz, B. M.; Burdsall, H. H.; DeNitto, G. A.; Eglitis, A.; Haugen, D. A. & Wallner, W. E. (2001). Pest risk assessment the importation into the United States of unprocessed *Eucalyptus* logs and chips from South America. General Technical Report FPL-GTR-124. Forest Products Laboratory, Forest Service, US Department of Agriculture, Madison, WI, USA

Krasutsky, F. A. (2006). Birch bark research and development. *Natural Products Reports*, Vol.23, No.6, (December 2006), pp. 919-42

Külheim, C.; Yeoh, S.; Wallis, I.; Laffan, S.; Moran, G. & Foley, W. (2011). The molecular basis of quantitative variation in foliar secondary metabolites in *Eucalyptus globulus. New Phytologist*, Vol.191, No.4, (September 2011), pp. 1041-1053, ISSN 0028646x

Kumar, B.; Vijayakumar, M.; Govindarajan, R. & Pushpangadan, P. (2007). Ethnopharmacological approaches to wound healing-exploring medicinal plants of India. *Journal of Ethnopharmacology*, Vol.114, No.2 (November 2007), pp. 103–113, ISSN 03788741

Lawler, I. R.; Foley, W. J.; Pass, G. J. & Eschler, B. M. (1998). Administration of a 5HT3 receptor antagonist increases the intake of diets containing *Eucalyptus* secondary metabolites by marsupials. *Journal of Comparative Physiology, B:Biochemical, Systemic and Environmental Physiology*, Vol.168, No.8, (November 1998), pp. 611–618, ISSN 01741578

Lawler, I. R.; Foley, W. J. & Eschler, B. (2000). Foliar concentration of a single toxin creates habitat patchiness for a marsupial folivore. *Ecology*, Vol.81, No.5, (May 2000), pp. 1327–1338, ISSN 00129658

Leicach, S. R.; Garau, A. M.; Guarnaschelli, A. B.; Yaber Grass, M. A.; Sztarker, N. D. & Dato, A. (2010). Changes in *Eucalyptus camaldulensis* essential oil composition as response to drought preconditioning. *Journal of Plant Interactions*, Vol.5, No. 3, pp. 205-210, ISSN 17429145

Leicach, S. R.; Yaber Grass, M. A.; Corbino, G. B.; Pomilio, A. B. & Vitale, A. A. (2003). Nonpolar lipid composition of *Chenopodium album* grown in intensively-cultivated and non-disturbed soils. *Lipids*, Vol.8, No. 5, (May 2003), pp. 567-572, ISSN 00244201

Leicach, S. R.; Sampietro, D. & Narwal, S. (2009a). Allelochemicals, In: *Allelochemicals: Role in Plant- Environment Interactions*, pp. 3-22, Studium Press LLC, ISBN 1933699442, Texas, USA

Leicach, S. R.; Sampietro, D. & Narwal, S. (2009b). Allelopathy, In: *Allelochemicals: Role in Plant- Environment Interactions*, pp. 34-41, Studium Press LLC, ISBN 1933699442, Texas, USA

Leicach, S. R.; Sampietro, D. & Narwal, S. (2009c). Environmental modulation, In: *Allelochemicals: Role in Plant- Environment Interactions*, pp. 90-100, Studium Press LLC, ISBN 1933699442, Texas, USA

Leicach, S. R.; Sampietro, D. & Narwal, S. (2009d). Allelochemicals: Localization and Bioactivity Range, In: *Allelochemicals: Role in Plant- Environment Interactions*, pp. 80-89, Studium Press LLC, ISBN 1933699442, Texas, USA

Liu, X.; Chen, Q.; Wang, Z.; Xie, L. & Xu, Z. (2008). Allelopathic effects of essential oil from *Eucalyptus grandis* x *E. urophylla* on pathogenic fungi and pest insects. *Frontiers of Forestry in China*, Vol.3, No.2, (June 2008), pp. 232–236, ISSN 16733517

Lucia, A.; Audino, P. G.; Seccacini, E.; Licastro, S.; Zerba, E. & Masuh, H. (2007). Larvicidal effect of *Eucalyptus grandis* essential oil and turpentine and their major components on *Aedes aegypti* larvae. *Journal of the American Mosquito Control Association*, Vol.23, No.3, (September 2007), pp. 299–303, ISSN 8756971x

MacMillan, C. P.; Mansfield, S. D.; Stachurski, Z. H.; Evans, R. & Southerton, S. G. (2010). Fasciclin-like arabinogalactan proteins: specialization for stem biomechanics and cell wall architecture in *Arabidopsis* and *Eucalyptus*. *The Plant Journal*, Vol.62, No.4, (May 2010), pp. 689-703, ISSN 09607412

Marsh, K. J.; Foley, W. J.; Cowling, A. & Wallis, I. R. (2003). Differential susceptibility to *Eucalyptus* secondary compounds explains feeding by the common ringtail (*Pseudocheirus peregrinus*) and common brushtal possum (Trichosurus vulpecula), *Journal of Comparative Physiology. B: Biochemical, Systemic and Environonmental Physiology*, Vol.173, No.1, (February 2003), pp. 69-78, ISSN 01741578

Marzoug, H. N. B.; Romdhane, M.; Lebrihi, L.; Mathieu, F.; Coudre, F.; Abderraba, M.; Larbi Khoujam M & Jalloul, B. (2011). *Eucalyptus oleosa* Essential Oils: Chemical composition and antimicrobial and antioxidant activities of the oils from different plant parts (Stems, Leaves, Flowers and Fruits). *Molecules*, Vol.16, No. 2, (February 2011), pp. 1695-1709; ISSN 14203049

Matsuki, M.; Foley, W. J. & Floyd, R. B. (2011). Role of volatile and non-volatile plant secondary metabolites in host tree selection by christmas beetles. *Journal of Chemical Ecology*, Vol.37, No.3, (March 2011), pp. 286–300; ISSN 02757540

Mbuna, J. J. & Mhinzi, G. S. (2003). Evaluation of gum exudates from three selected plant species from Tanzania for food and pharmaceutical applications. *Journal of Science of Food and Agriculture*, Vol.83, No.2, (January 2003), pp. 142–146, ISSN 00225142

Menn, J. & Hall, F. (1999). Biopesticides: present status and future prospects. In: *Methods in Biotechnology. Vol.5: Biopesticides: use and delivery*, Hall, F. & Menn, J. (Ed.), pp. 1-10, Humana Press, ISBN 0896035158, Totowa. New Jersey, USA

Messer, A. (1990). Traditional and chemical techniques for stimulation of *Shorea javanica* (Dipterocarpaceae) resin exudation in Sumatra. *Economic Botany*, Vol.44, No.4, pp. 463-469, ISSN 00130001

Miller, D. R. (2006). Ethanol and (-)-alpha-pinene: attractant kairomones for some large wood-boring beetles in south-eastern USA. *Journal of Chemical Ecology*, Vol.32, No.4, (April 2006), pp. 779-794, ISSN 00980331

Moore, B.; Wallis, I.; Palá-Paul, J.; Brophy, J.; Willis, R. & Foley, W. (2004). Antiherbivore chemistry of *Eucalyptus*- Cues and deterrents for marsupial folivores. *Journal of Chemical Ecology*, Vol.30, No.9, (September 2004), pp. 1743-1769, ISSN 00980331

Moran, G.; Thamarus, K.; Raymond, C.; Qiu, D. U. T, & Southerton, S. (2002). Genomics of *Eucalyptus* wood traits. *Annals of Forest Science*, Vol.59, No.5-6, (July-October 2002), pp. 645-650, ISSN 12864560

Moreira, M. R.; Ponce, A. G.; del Valle, C. E. & Roura, S. I. (2005). Inhibitory parameters of essential oils to reduce a foodborne pathogen. *Lebensmittel-Wissenschaft und Technologie*, Vol.38, No.5, (August 2005), pp. 565–570, ISSN 00236438

Mulyaningsih, S.; Sporer, F.; Zimmermann, S.; Reichling, J. & Wink, M. (2010). Synergistic properties of the terpenoids aromadendrene and 1,8-cineole from the essential oil of *Eucalyptus globulus* against antibiotic-susceptible and antibiotic-resistant pathogens. *Phytomedicine: International Journal of phytotherapy and phytopharmacology*, Vol.17, No.13, (November 2010), pp. 1061-1066, ISSN 09447113

Muneri, A. & Raymond, C. (2000). Genetic parameters and genotype-by-environment interactions for basic density, pilodyn penetration and diameter in *Eucalyptus globulus*. *Forest Genetics*, Vol.7, No.4, pp. 321-332, ISSN 1335048x

Navarro, M.; Celine, A.; Yves, M.; Joan, L.; Walid, E. K.; Christiane, M. & Teulières, C. (2011). Two EguCBF1 genes over expressed in *Eucalyptus* display a different impact on stress tolerance and plant development (p). *Plant Biotechnology Journal*, Vol.9, No.1, (January 2011), pp. 50-63, ISSN 14677644

Neilson, E. H.; Goodger, J. Q. D. & Woodrow, I. E. (2006). Novel aspects of cyanogenesis in *Eucalyptus camphora* subsp. *Humeana*. *Functional Plant Biology*, Vol.33, No.5, pp. 487-496, ISSN 14454408

Nishizawa, M.; Emura, M.; Kan, Y.; Yamada, H.; Ojima, K. & Hamanaka, M. (1992). Macrocarpals: HIV-RTase inhibitors of *Eucalyptus globulus*. *Tetrahedron Letters*, Vol.33, No.21, (May 1992), pp. 2983–2986, ISSN 00404039

Ohara, K.; Matsunaga, E.; Nanto, K.; Yamamoto, K.; Sasaki, K.; Ebinuma, H. & Yazaki, K. (2010). Monoterpene engineering in a woody plant *Eucalyptus camaldulensis* using a limonene synthase cDNA. *Plant Biotechnology Journal*, Vol.8, No.1, (January 2010), pp. 28–37, ISSN 14677644

Olajide, O.; Udo, E. S. & Out, D. O. (2008). Diversity and population of timber tree species producing valuable non-timber products in two tropical rainforests in cross river state, Nigeria. *Journal of Agriculture & Social Sciences*, Vol.4, No.2, (April 2008), pp. 65–68, ISSN 18132235

O'Reilly-Wapstra, J.; Freeman, J.; Davies, N.; Vaillancourt, R.; Fitzgerald, H. & Potts, B. (2011). Quantitative trait loci for foliar terpenes in a global eucalypt species. *Tree Genetics and Genomes*, Vol.7, No.3, (June 2011), pp. 485-498, ISSN16142942

O'Reilly-Wapstra, J. M.; McArthur, C. & Potts, B. M. (2002). Genetic variation in resistance of *Eucalyptus globulus* to marsupial browsers. *Oecologia,* Vol.130, No.2, (January 2002), pp. 289-296, ISSN 00298549

O'Reilly-Wapstra, J. M.; Potts, B. M. & McArthur, C. (2004). Genetic variation and additive inheritance of resistance of *Eucalyptus globulus* to possum browsing, Proceedings of IUFRO Conference "*Eucalyptus* in a Changing World", Aveiro, Portugal, October 11-15, 2004 Available from http://eprints.utas.edu.au/7407/

O´Reilly-Wapstra, J.; Potts, B.; Mc Arthur, C. & Davies, N. (2005). Effects of nutrient variability on the genetic-based resistance of *Eucalyptus globulus* to a mammalian herbivore and on plant defensive chemistry. *Oecología*, Vol.142, No.4, (February 2005), pp. 597-605, ISSN 00298549

Osawa, K.; Yasuda, H.; Morita, H.; Takeya, K. & Itokawa, H. (1996). Sideroxylonals A, B and C, dimeric phloroglucinol compounds, have also been found in leaves and flower buds of several *Eucalyptus* species. *Journal of Natural Products*, Vol.59, No.9, (September 1996), pp. 823-827, ISSN 01633864

Östrand, F.; Wallis, I. R.; Davies, N. W.; Matsuki, M. & Steinbauer, M. J. (2008). Causes and consequences of host expansion by *Mnesampela privata* (Lepidoptera: Geometridae). *Journal of Chemical Ecology*, Vol.34, No.2, (February 2008), pp.153-167, ISSN 00980331

Otten, A.; Elpel, D. & Krmatschenko, N. (2007). Adhesives from renewable raw materials. *Coating International,* Vol.40, No.8, pp. 28-32, ISSN 05908450

Paine, T. D.; Steinbauer, M. J. & Lawson, S. A. (2011). Native and exotic pests of *Eucalyptus*: A Worldwide Perspective. *Annual Review of Entomology*, Vol.56, (January 2011), pp. 181-201, ISSN 00664170

Pandey, R.; Kalra, A.; Tandon, S.; Mehrotra, N.; Singh, H. N. & Kumar, S. (2000). Essential oils as potent source of nematicidal compounds. *Journal of Phytopathology*, Vol.148, No.7-8, (August 2000), pp. 501-502; ISSN 09311785

Papachristos, D. P. & Stamopoulos, D. C. (2003). Selection of *Acanthoscelides obtectus* (Say) for resistance to lavender essential oil vapour. *Journal of Stored Products Research*, Vol.39, No.4, pp. 433-441, ISSN 0022474x

Ramezani, H. (2002). Antifungal activity of the volatile oil of *Eucalyptus citriodora. Fitoterapia,* Vol.73, No.3, (June 2002), pp. 261-2, ISSN 0367326x

Rapley, L. P.; Allen, G. R.; Potts, B. M. & Davies, N. W. (2007). Constitutive or induced defences - how does *Eucalyptus globulus* defend itself from larval feeding? *Chemoecology*, Vol.17, No.4, (September 2007), pp. 235-243, ISSN 09377409

Regnault-Roger, C. 1997. The potential of botanical essential oils for insect pest control. *Integrated Pest Management Reviews*, Vol.2, No.1, (February 1997), pp. 25-34, ISSN 13535226

Saad, E.; Hussien, R.; Saher, F. & Ahmed, Z. (2006). Acaricidal activities of some essential oils and their monoterpenoidal constituents against house dust mite, *Dermatophagoides pteronyssinus* (Acari: Pyroglyphidae). *Journal of Zhejiang University, Science. B.*, Vol.7, No.12, (December 2006), pp. 957-962, ISSN 16731581

Sartorelli, P.; Marquioreto, A. D.; Amaral-Baroli, A.; Lima, M. E. L. & Moreno, P. R. H. (2007). Chemical composition and antimicrobial activity of the essential oils from two species of *Eucalyptus*. *Phytotherapy Research*, Vol.21, No.3, (March 2007), pp. 231-233, ISSN 0951418x

Satoh, H.; Etoh, H.; Watanabe, N.; Kawagishi, H.; Arai, K. & Ina, K. (1992). Structures of sideroxylonals from *Eucalyptus sideroxylon*. *Chemistry Letters*, pp. 1917-1920, ISSN 03667022

Schnitzler, P.; Schon, K. & Reichling, J. (2001). Antiviral activity of Australian tea tree oil and *Eucalyptus* oil against *Herpes simplex* virus in cell culture. *Die Pharmazie*, Vol.56, No 4, pp. 343–347, ISSN 00317144

Seyoum, A.; Killeen, G. F.; Kabiru, E. W.; Knols, B. G. J. & Hassanali, A. (2003). Field efficacy of thermally expelled or live potted repellent plants against African malaria vectors in western Kenya. *Tropical Medicine and International Health*, Vol.8, No.11, (November 2003), pp. 1005–1011, ISSN 13602276

Shani, Z.; Dekel, M.; Cohen, B.; Barimboim, N.; Kolosovski, N.; Safranuvitch, A.; Cohen, O. & Shoseyov, O. (2003). Cell wall modification for the enhancement of commercial *Eucalyptus* species. Proceedings of IUFRO tree biotechnology, Sundberg, B. (Ed.), Umea Plant Science Center, Umea, Sweden, June 7-12, 2003

Shao, Z.; Chen, W.; Luo, H.; Ye, X. & Zhan, J. (2002). Studies on the introduction of cecropin D gene into *Eucalyptus urophylla* to breed the resistant varieties to *Pseudomonas solaniacearum*. *Scientia Silvae Sinicae*, Vol.38, No.2, pp. 92-97, ISSN 10017488

Sidana, J.; Rajesh, K. R.; Nilanjan, R.; Russell, A. B.; Foley, W. J. & Inder Pal. S. (2010). Antibacterial sideroxylonals and loxophlebal A from *Eucalyptus loxophleba* foliage. *Fitoterapia*, Vol.81, No.7, (October 2010), pp. 878-883, ISSN 0367326x

Sidana, J.; Sukhvinder, S.; Sunil, K. A.; Foley, W. J. & Singh, I. P. (2011). Formylated phloroglucinols from *Eucalyptus loxophleba* foliage. *Fitoterapia*, Vol.82, No.7, (October 2011), pp. 1118-1122; ISSN 0367326x

Silvestre, A. D.; Cavaleiro, J. A. S.; Delmond, B.; Filliatre, C. & Bourgeois, G. (1997). Analysis of the variation of the essential oil composition of *Eucalyptus globulus* Labill. from Portugal using multivariate statiscal analysis. *Industrial Crops and Products*, Vol.6, No.1, (February 1997), pp. 27-33, ISSN 09266690

Singh, I. P.; Takahashi, K. & Etoh, H. (1996). Potent attachment-inhibiting and-promoting substances for the blue mussel, *Mytilus edulis galloprovincialis*, from two species of *Eucalyptus*. *Bioscience, Biotechnology and Biochemistry*, Vol.60, No.9, pp. 1522-1533, ISSN 09168451

Singh, I. P.; Batish, D. R.; Kaur, S.; Kohli, R. K. & Arora, K. (2006). Phytotoxicity of the volatile monoterpene citronellal against some weeds. *Zeitschrift fur Naturforschung C: A Journal of Biosciences*, Vol.61c, No.5-6, pp. 334-340, ISSN 09395075

Spokevicius, A. V.; Southerton, S. G.; Mac Millan, C. P.; Qiu, D.; Gan, S.; Tibbits, J. F. G.; Moran, G. F. & Bossinger, G. (2007). β-tubulin affects cellulose microfibril orientation in plant secondary fibre cell walls. *Plant Journal*, Vol.51, No.4, pp. 717-726, ISSN 09607412

Steinbauer, M. J.; Schiestl, F. P. & Davies, N. W. (2004). Monoterpenes and epicuticular waxes help female autumn gum moth differentiate between waxy and glossy *Eucalyptus* and leaves of different ages. *Journal of Chemical Ecology*, Vol.30, No.6, (June 2004), pp. 1117-1142; ISSN 00980331

Steinbauer, M. J. & Wanjura, W. J. (2002). Christmas beetles (*Anoplognathus* spp., Coleoptera: Scarabaeidae) mistake peppercorn trees for eucalypts. *Journal of Natural History*, Vol.36, No.1, pp. 119-125, ISSN 00222933

Su, Y. C.; Ho, C. L.; Wang, I. C. & Chang, S. T. (2006). Antifungal activities and chemical compositions of essential oils from leaves of four *Eucalyptus*. *Taiwan Journal of Forest Science*, Vol.21, No.1, (March 2006), pp. 49-61, ISSN 10264469

Takahashi, T.; Kokubo, R. & Sakaino, M. (2004). Antimicrobial activities of *Eucalyptus* leaf extracts and flavonoids from *Eucalyptus maculate*. *Letters in Applied Microbiology*, Vol. 39, No.1, (July 2004), pp. 60-64, ISSN 02668254

Takasaki, M.; Konoshima, T.; Fujitani, K.; Yoshida, S.; Nishimura, H. & Tokuda, H. (1990). Inhibitors of skin-tumor promotion. VIII: inhibitory effects of euglobals and their related compounds on Epstein-Barr virus activation. *Chemical and Pharmaceutical Bulletin*, Vol.38, No.10, pp. 2737-2739, ISSN 00092363

Teulières, C.; Bossinger, G. ; Moran, G. & Marque, C. (2007). Stress studies in *Eucalyptus*. *Plant Stress*, Vol.1, No.2, pp. 197-215, ISSN 17490359

Thacker, J. R. M. & Train M. R. (2010). Use of Volatiles in Pest Control. In: *The chemistry and biology of volatiles*, Herrmann, A., (Ed.), Wiley, ISBN 0470777788, Switzerland

Thamarus, K.; Groom, K.; Bradley, A.; Raymond, C.; Schinleck, L.; Williams, E. & Moran, G. (2004). Identification of quantitative trait loci for wood and fibre properties in two full-sib pedigrees of *Eucalyptus globulus*. *Theorical and Applied. Genetics*, Vol.109, No.4, (August 2004), pp. 856-864, ISSN 00405752

Thumma, B. R.; Nolan, M. R.; Evans, R. & Moran, G. F. (2005). Polymorphisms in cinnamoyl CoA reductase (CCR) are associated with variation in microfibril angle in *Eucalyptus* sp. *Genetics*. Vol.171, No.3, (November 2005), pp. 1257–1265, ISSN 00166723

Tournier, V.; Grat, S. ; Marque, C. ; El- Kayal, W. & Penchel, R. (2003). An efficient procedure to stably introduce genes into an economically important pulp tree (*Eucalyptus grandis* x *Eucalyptus urophylla*). Transgenic Research, Vol.12, No.4, (August 2003), pp. 403-411, ISSN 09628819

Trigg, J. K. (1996). Evaluation of eucalyptus-based repellent against *Anopheles* spp. in Tanzania. *Journal of the American Mosquito Control Association*, Vol.12, No.2, (June 1996), pp. 243–246, ISSN 8756971x

Tripathi, P.; Dubet, N. & Shukla, A. (2008). Use of some essential oil as post-harvested botanical fungicides in the management of grey mould of grapes caused by *Botrytis cinerea*. *World Journal of Microbiology and Biotechnology*, Vol.24, No.1, (January 2008), pp. 39-46, ISSN 09593993

Tyagi, A. & Malik, A. (2011). Antimicrobial potential and chemical composition of *Eucalyptus globulus* oil in liquid and vapour phase against food spoilage microorganisms. *Food Chemistry*, Vol.126, No.1, (May 2011), pp. 228-235, ISSN 03088146

Tzortzakis, N. G. (2007). Maintaining postharvest quality of fresh produce with volatile compounds. *Innovative Food Science and Emerging Technology*, Vol.8, No.1, (March 2007), pp. 111-116, ISSN 14668564

US EPA. (1993). *Wildlife Exposure Factors Handbook.* Volumes I and II. EPA/600/R 93/187 U.S. Environmental Protection Agency, Office of Research and Development, US Environmental Protection Agency, Washington D.C.

Vernardakis, T. (2009). Pine for your inks trees provide organic alternatives to water-based acrylics. *Flexo,* Vol.34, No.3, (March 2009), pp. 46-48, ISSN 10517324

Vilariño, M. P.; Mareggiani, G.; Yaber Grass, M. A.; Leicach, S. R. & Ravetta, D. (2005). Post-damage alkaloid concentration in sweet and bitter lupin varieties and its effect on subsequent herbivory. *Journal of Applied Entomology*, Vol.129, No.5, (June 2005), pp. 233-238; ISSN 09312048

Wallis, I. R.; Smith, H. J.; Henery, M. L.; Henson, M. & Foley, W. J. (2010). Foliar chemistry of juvenile *Eucalyptus grandis* clones does not predict chemical defence in maturing ramets. *Forest Ecology and Management*, Vol.260. No.5, (July 2010), pp. 763-769, ISSN 03781127

Wei, R. P.; Xu, D. (2003). *Eucalyptus Plantations: Research,Management and Development.* World Scientific Publishing Co. Inc., ISBN 9812385576, London, U.K.

Wiedenhoeft, A. (2010). Structure and function of wood. In: *Wood handbook: wood as ingeneering material,* Ross, R. (Ed.), pp. 1-18, USDA Forest Service, Forest Products Laboratory, Madison, Wisconsin, USA

Wildy, D.; Pate, J. & Bartle, J. (2000). Variations in composition and yield of leaf oils from alley-farmed oil mallees (*Eucalyptus* spp) at a range of contrasting sites in the Western Australian wheatbelt. *Forest Ecology and Management,* Vol.134, No.1-2, (September 2000), pp. 205-217, ISSN 03781127

Willfor, S.; Nisula, L.; Hemming, J.; Reunanen, M. & Holmbom, B. (2004). Bioactive phenolic substances in industrially important tree species. Part 2: Knots and stem wood of fir species. *Holzforschung*, Vol .58, No.6, (October 2004), pp. 650–659, ISSN 00183830

Williams, J. E. & Woinarski, J. C. Z. (1997). *Eucalypt Ecology: Individuals to Ecosystems,* Williams, J.E. (Ed.), Cambridge University Press, ISBN 052149740x, Cambridge, UK

Williams, L. (1962). Laticiferous plants of economic importance. I: balata-chicle, gutta percha and allied guttas. *Economic Botany*, Vol.16, pp. 17-24, ISSN 00130001

Wingfield, M. J.; Slippers, B.; Hurley, B. P.; Coutinho, T. A.; Wingfield, B. D. & Roux, J. (2008). Eucalypt pests and diseases: Growing threats to plantation productivity. *Southern Forests*, Vol.70, No.2, pp. 139-144, ISSN 20702620

Wong, J. L. G.; Thornber, K.; Baker, N. (2001). *Evaluación de los recursos de productos forestales no madereros.* Food and Agriculture Organization of the United Nations, ISBN 9253046147, Rome, Italy

Yaber Grass, M. A.; Leicach S. R. (2011). Changes in *Senecio grisebachii* pyrrolizidine alkaloids abundances and profiles as response to soil quality. *Journal of Plant Interactions,* Available from
http://www.tandfonline.com/doi/abs/10.1080/17429145.2011.591504

Yamada-Watanabe K.; Kawaoka A.; Matsunaga K.; Nanto K.; Sugita K.; Endo S.; Ebinuma H.; Murata N. (2003). Molecular breeding of *Eucalyptus*: analysis of salt stress tolerance in transgenic *Eucalyptus camaldulensis* that over-expressed choline oxidase gene (codA). *Proceedings of IUFRO tree biotechnology*, Sundberg, B. (Ed.), Umea Plant Science Centre, Umea, Sweden, June 7-12, 2003

Zini, C. ; de Asis, T. ; Ledford, E. ; Dariva, C. ; Fachel, J. ; Christensen, E. & Pawliszyn, J. (2003). Correlations between pulp properties of *Eucalyptus* clones and leaf volatile using automated solid-phase microextraction. *Journal of Agricultural and Food Chemistry*, Vol.51, No.27, (December 2003), pp. 7848-7853, ISSN 00218561